Numerical control
users' handbook

Numerical control users' handbook

Editor: W. H. P. Leslie

 McGRAW-HILL

London · New York · Toronto · Sydney · Johannesburg
Mexico · Singapore · Düsseldorf

Published by

McGRAW-HILL Publishing Company Limited

MAIDENHEAD · BERKSHIRE · ENGLAND

07 094216 1

PRINTED AND BOUND IN GREAT BRITAIN

Preface

In preparing this book, we felt that the time was now ripe to record the advances which have been made in the utilization of NC* since *Numerical Control in Manufacturing*† was published in 1963. Since then, the major advances have been concerned less with the design principles of NC equipment and machines than with the production of a much wider range of machines. The rate of increase of installation of NC machines continues to grow each year, both in Europe and in the USA. On the one hand this results in an ever-increasing number of new NC users, and on the other hand it increases the number of existing users who have now sufficient experience and equipment to encourage them to embark on more ambitious plans to use NC effectively.

The reader who seeks details of the electronic and mechanical design of NC equipment and machine tools, or of cutting tools and holders, is, then, referred to *Numerical Control in Manufacturing* whilst in this NC Users' Handbook we concentrate on the users' problems and are thus able to deal with these problems in greater depth.

The 'user' to whom the handbook is addressed may be on the shop floor or in management, in Technical College or in University, just as long as his interest is in the better use of NC equipment and not in the principles on which the design of NC equipment is based.

There are many systematic ways in which this subject might be subdivided. We have chosen to start in this book by inviting Bill Ferris to write on when and how to use NC and how successfully to introduce it into a factory. He has played a prominent part in building up the technical expertise in the use of NC at the Dunlop Company Ltd in Coventry, and has drawn on his extensive experience gained there and broadened by an extended visit to the USA to study the use of NC. His recent position as Production Manager of Stat Systems Ltd qualifies him to write with some authority also on the use of preset tooling.

Next, the reader is introduced to the wide range of classes of NC machine

---

* NC throughout this book is an abbreviation for numerical control.
† Edited by Frank W. Wilson and published by McGraw-Hill Book Co. Inc., New York.

tools available today, each class being illustrated by one or two machines currently in production. Ron Iredale, the author of this chapter, is well-known in his position as Editor of *Metalworking Production*, a position which makes him exceptionally well qualified to review this topic.

The reader is then initiated into the mysteries of buying an NC machine by Alf Tack, Machine Tool and Equipment Manager of Rolls-Royce Ltd, Aero Engine Division, at Derby. In this position, Mr Tack is perhaps more widely involved in the purchase of NC machines than anyone else in Britain, and this experience is complemented by his earlier work in pioneering the sales of NC machine tools for a well-known machine builder. The potential user is introduced here to the problems he should look out for, as well as the advantages he should expect. The same author continues in chapter 4 to present a useful guide on making out the economic case for the purchase of NC machine tools, although his advice can be applied to any other expensive equipment.

At this stage, when planning this book, it was intended to discuss the manual preparation of control tapes for NC machines. This is a complicated subject, because of the wide variation in practice for different machine tools and control systems. Fortunately, it has been the main subject of *Programming for Numerical Control** and, rather than dismiss the whole topic in one chapter here, we commend that volume to the reader whose main interest is manual programming.

In chapter 5, Bernard Wood of ICSL, London, deals with some of the NC computer programs which have been developed since 1956 in Britain, most of which are valuable to users today. They have set the stage for the more recent developments dealt with in chapters 7 to 11. In chapter 6, we report on the current position of national and international Standards for NC machine tools and control tapes. In 1968, the International Standards Organization completed some five years' work on this subject and is currently tidying up a few outstanding points, so that only now can this work be commented on. Users today must consider carefully why they should not follow these Standards. Enough user demand for equipment to the ISO Standard would encourage manufacturers to move from selling ISO features as optional extras, to making them standard features. There is no doubt that this will eventually happen, but users can speed the process by specifying equipment to the relevant ISO, or equivalent national, Standard.

In chapters 7 and 8, Dieter Reckziegel, Manager of the EXAPT Association, Aachen, Germany, introduces the reader to EXAPT 1 and 2; these are computer programs for drilling and turning, two of the new computer programs for which the user employs APT compatible input language when describing the work which has to be done. The EXAPT processors were developed in three Technical Universities in Germany, under the influence of the four professors mentioned at the start of chapter 7, and it is fitting that one of those who was

---

* Roberts and Prentice, McGraw-Hill Book Co. Inc., New York, 1968.

responsible for the development work and who currently manages the Users' Association should describe how these programs are used.

In chapter 15, the story of this particular development is completed with a description of how the EXAPT program deals with the choice of cutting feeds, speeds, and sequences of moves for drilling, tapping, turning, etc. This chapter has been contributed by the team responsible for the work at the Machine Tool Institute in Aachen, of whom Professor Opitz is internationally known to production engineers. His colleagues in this work, Dr Engelskirchen, Dr Hirsch, and Mr Budde, are already establishing a reputation by their work in this field.

In chapter 9, Derek Welch and John McWaters describe the NELNC processors; they deal principally with the first program, 2C,L, which again uses an APT-based part-programming language but is aimed particularly at contour-milling machines and machining centres. Mr Welsh has been responsible, while working for the Ferranti NC Division at Dalkeith, for the internal design and coding of 2C,L, and Mr McWaters has been responsible for specifying the performance of 2C,L and attending to language-standardization problems arising from this: he acted as Secretary of the European APT language standards committee until it was superseded by the UNCL (Unified Numerical Control Language) committee last year, and is now responsible for maintenance and development of 2C,L.

In chapter 10, Jim Baughman of T.R.W. Inc., Cleveland, Ohio, USA writes on APT, the program developed at MIT in America between 1955 and 1960 and maintained and further developed since its release to industry in 1961 by IITRI in Chicago. Mr Baughman is very well known to APT users in the USA, and also to those European APT users who have visited the States to discuss APT. He has had a long experience in the Aerospace Industry from production to management, with both NC and computer graphics, before moving to the Numerical Control Firm, T.R.W.

Dick Sim, who writes on post processors in chapter 11, has concentrated on the practical problems involved in writing post processors which can work with APT, NELNC, and EXAPT, so that manufacturers need only supply one post processor per machine tool instead of separate versions for each NC processor. Mr Sim is well-known in this field in Europe and is currently chairman of an ISO working group which has been meeting three or four times a year to try to reach international agreement on this problem quickly. His contribution in chapter 11 is, then, as well informed on this topic as is possible.

Miles Ellis, who writes in chapter 12 on how to make the best use of the computer for NC, was closely associated with the work of the ICSL computer bureau in Kidsgrove, England, when it was adding APT and EXAPT to the KIPPS drilling program already operating on the KDF9 there. He explains the various ways in which the computer may be used, and presents the advantages and disadvantages of each.

In chapter 13, Jim Baughman writes again, this time on some of the less common types of NC machine in use in the USA, briefly indicating how each is used, while in chapter 14 Harry Ogden deals with unusual applications of NC measuring systems. Harry is Manager of the Measurement Department of Ferranti's NC Division at Dalkeith, and has a long experience of NC. He deals with digital readout fitted to manually operated machines, inspection machines, and a drawing-measuring machine, and also with a frame-bending machine for bending ship frames up to 60 ft long.

It is appropriate that the last chapter, chapter 16, should deal with one of the trend-setting developments currently under way, the Molins System 24. Theo Williamson was for many years leading the development of the Ferranti Control System. His move to the Molins Machine Company in London has resulted in a new concept in NC, with all the workpiece movements under computer control. The reason behind this system is dealt with by Mr Williamson in this chapter. Although further development of the complete System 24 has been suspended, the associated machine tools have been introduced into industry, and the ideas behind the system, although perhaps premature today, point a way to future development.

Finally, we have included a glossary. As this is a developing subject and discussion on standardization of terms is only at an embryo stage, the glossary reflects the idiosyncrasies of the editor, who hopes it will prove useful.

The authors gratefully acknowledge the co-operation and assistance of the organizations with which they work and also of the various manufacturers who have supplied illustrations and information.

It is with a sense of satisfaction, but also of relief, that we see this book reach completion. In order to get informed and up-to-date contributions, it is necessary to approach the busiest, most committed experts and ask them to take time to prepare their chapter. It has been interesting to contrast the varied response times: in some cases the men with apparently the most problems have completed their tasks ahead of schedule whereas others, with almost the same load, did not manage to start for a long time, but once started, could do the necessary writing in much the same interval.

The editor thanks them all for making his task both interesting and yet straightforward, and would also like to acknowledge the advice, readily given by Harold Burton, when Editor of *Metalworking Production*, in one area where a second opinion was helpful.

W. H. P. LESLIE

Contents

1. Preparing to use Numerical Control

W. C. Ferris

W. C. Ferris, CF, AMBIM, is Works Manager, New Products Division, The Dunlop Co. Ltd. He received his technical education at Coventry Technical College and the Lanchester College of Advanced Technology. He is a founder member of the British Numerical Control Society, and Chairman of the Coventry branch of the British Institute of Management. He has held various production engineering and management positions within the Dunlop Engineering Group, and in 1964 was awarded the Sir Alfred Herbert Scholarship, which allowed study of NC in Sweden, France, and Germany. In 1966 he spent six months in the USA as a Churchill Fellow, studying production engineering techniques and management.

Types of work suitable for numerical control

Many companies now accept NC as a time-proven standard manufacturing technique used profitably for metal cutting, metal forming, drafting, and many other applications.

It is not only used on the complex applications associated with aircraft work, which are 'naturals' for numerical control. On the contrary, a large proportion of all the NC machines used today are applied to relatively simple work typical of many machine shops and toolrooms. It is work which might not normally be thought to justify numerical control.

Among its satisfied users are an ever-increasing number of small companies who now realize that this is a technique which they can use to financial advantage. Larger companies are also beginning to realize that there are bonus advantages which arise from operating multiple installations.

Despite the fact there are many successful NC installations of varying sizes and applications, some potential users are very dubious of just what can be achieved by using NC and often fail to recognize profitable applications. These can be summarized as follows.

(a) Where operations or set-ups are numerous or costly.

1

(b) For complex and varied operations.

(c) When machine run time is disproportionately low compared with set-up time, especially when conventional machining is committed to skilled operators.

(d) For small batch quantities especially of complex parts.

(e) When the part is so complex that quantity production involves the possibility of human error.

(f) Where close and repetitive tolerances are required.

(g) Where individual variations are required on a family of parts.

(h) For parts subject to design changes.

(i) For parts that demand close tolerance control on tooling.

(j) Where tooling costs are a significant portion of unit cost.

(k) When lead time does not permit conventional tooling manufacture.

(l) When tool storage is a problem.

(m) Where inspection costs represent a large portion of total costs.

Often the individual justification of NC machines is made on one or two of these considerations. However, the purchase of an NC machine brings with it all of the potential interrelated advantages of the NC concept. The level of savings to be effected is entirely dependent upon the user, and can range from a minimum where NC is being used as an operator aid, to a maximum where NC is being used as a management technique.

Many unsuccessful attempts have been made to draw up general rules and formulae which would facilitate decisions regarding which machine is most appropriate to what type of work and in what batch sizes.

The only satisfactory solution, it seems, is to carry out a detailed study of the overall manufacturing scene, taking in all aspects of the production process—machining, tooling, inspection, handling, etc.—and then to carry out a series of time and cost studies of present method versus new method. Even this technique is impossible to use if the production engineer is not fully aware of what NC machines are available and of their capabilities.

However, there are several general factors which have been clearly established and which can guide potential users of NC when searching for areas of application.

Drilling operations

From the beginning of NC machine tools, one of the commonest and most successful applications has been the drilling operation. Most of the initial developments consisted of an NC co-ordinate table set under the arm of a radial drilling machine (Fig. 1.1). Without doubt most of the early publicity on this type of application centred around the savings to be achieved by the elimination of jigs and fixtures and, although this is basically correct, user experience has shown this to be somewhat misleading in terms of the scale of its effect on savings. Many of the case studies carried out have shown that an operator using

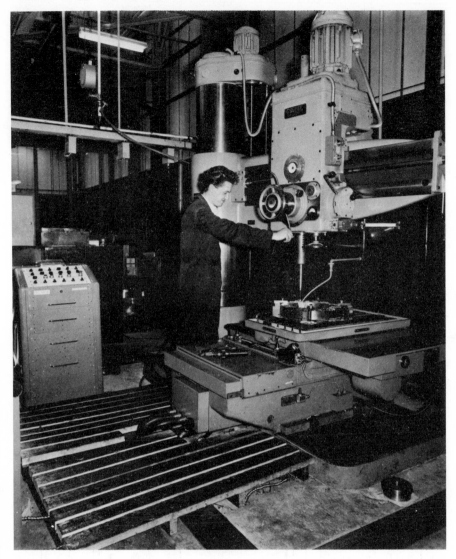

Figure 1.1 Co-ordinate NC table on radial drill.

a radial drill and a jig can often position between holes more quickly than can be done by an NC system (this was particularly true with the earlier relatively slow positioning systems). Even if the NC system could better the time, then the higher machine hour rate of the controlled machine would still make the machining cost greater. This is actually illustrated in Fig. 1.2 where, however, an overall saving for NC is achieved by a subsequent reduction in fitting time.

Workpiece	Material			
Steel door	Mild steel			
Numerically controlled method	Portal frame drilling machine			
Conventional method	Radial drilling machine			
Operations	Drill, tap and counterbore 137 holes, $3/16 - 1\frac{3}{4}$ in. dia.			
Production quantity	60 per batch; 12 batches per year			

Production details	Numerical control		Conventional	
	Time (h)	Cost (£)	Time (h)	Cost (£)
Programming/planning	8	6	–	–
Data preparation				
Jig and tool design and manufacture	20	15	220	250
Pre-production time/cost	**28**	**21**	**220**	**250**
Marking-off and setting-up	3·25	4·90	2·75	3
Machining				
Inspection	similar for both methods			
Handling	–	–	–	–
Fitting and Assembly	24	24	40	40
Processing time/cost	**27·25**	**28·90**	**42·75**	**43**
Total for 720	**19648**	**20829**	**31000**	**31210**

A point of particular interest in this example is that although the machining time and cost are greater under NC, the overall processing cost is reduced because of the reduction in fitting and assembly time. This is achieved mainly because, although all the doors have a number of holes which are common, each door has several holes which are specially placed to suit customers'

requirements. It is a fairly simple matter to add these extra holes to the tape program, and maintain the quality on the finished workpiece. Under conventional methods the extra holes have to be marked out and are therefore not produced to the same degree of accuracy as those produced by jig drilling (or by tape control).

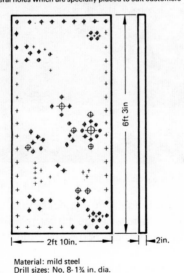

Material: mild steel
Drill sizes: No. 8- 1¾ in. dia.

Steel door

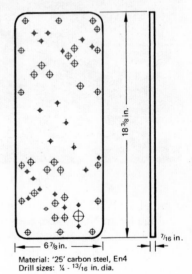

Material: '25' carbon steel, En4
Drill sizes: ¼ - 13/16 in. dia.

Baseplate

Figure 1.2   Reduced fitting and assembly costs with NC.

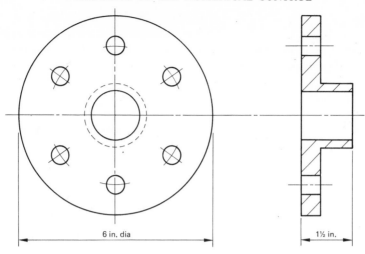

Workpiece	Material			
Flange cover	Brass			
N. C. method	Co-ordinate table and radial drill			
Conventional method	Radial drill			
Operations	Drill 6, $\frac{13}{32}$ in. dia. holes on $4\frac{1}{2}$ in. P.C.D. at 60°			
Production quantity	8 per batch			
	Numerical control		Conventional	
Production details (per batch)	Time (min)	Cost (£)	Time (min)	Cost (£)
Programming/planning	12	0·25	–	–
Data preparation	–	–	–	–
Jig and tool design and manufacture	–	–	–	–
Pre-production time/cost	**12**	**0·25**	**–**	**–**
Marking-off and setting-up	7	0·15	33	0·50
Machining	8·5	0·21	10	0·175
Inspection	–	–	–	–
Handling	–	–	–	–
Fitting and assembly	–	–	–	–
Processing time/cost	**15·5**	**0·36**	**43**	**0·675**
Total for 6 batches	**105**	**2·41**	**258**	**4·05**

Figure 1.3 Reduced marking-off time with NC. NC drilling versus conventional marking-out and drilling.

Almost certainly the best application for the drill and NC co-ordinate table is where there would be no jig anyway, probably because the quantities are small, and the saving can be achieved by the elimination of the marking-off operation and, most probably, during the drilling cycle itself, since it is not as easy to locate centre 'pops' on a manually operated machine as it is to find a locating bush in a jig plate. Figure 1.3 illustrates savings from both these aspects. In addition, subsequent developments in the form of NC indexing turrets have enabled further inroads to be made on floor-to-floor times by improving on manual tool-changing times, although in respect of indexing speeds there is still, in many cases, room for improvement in order to offset the extra cost of the addition of a turret. There is no doubt, however, that the concept is right since, with the inclusion of feed and speed changes on the tape, the machining cycle is then fully automatic and in no way dependent on operator efficiency and diligence.

Boring operations

With boring operations, the machining cycle is usually dominated by the operation itself, unlike drilling where the positioning time and the machine time are of the same order. For instance, on a heavy-duty boring machine it is not uncommon to spend, say, two hours actually boring a few holes and only a matter of perhaps 10 minutes, by manual methods, in the position setting between hole centres. Thus, although the addition of NC might cut the positioning time by 90 per cent (to 1 minute), the total cycle time will only be reduced by about 7 per cent, an amount which will make it difficult to justify the extra £3,000 to £8,000 for the NC system.

It is probably through appreciation of this situation that many of the boring-machine manufacturers are now offering, and indeed selling, a good number of their machines equipped only with digital readout systems as opposed to full NC. When the machine is used strictly for boring operations, particularly on small batches, the borer equipped with digital readout (at probably something under half the cost of full NC) can very often prove a very effective and economic machine.

This by no means precludes the use of full NC on heavy boring machines, but where it is fitted it can be exploited to greatest economic advantage by including in the machining cycle other machining operations—notably drilling—which can take advantage of any speed-up in positioning time and, generally even more important, will eliminate one or more handling operations. The part shown in Fig. 1.4 is a good example of this situation, and it can be seen from the accompanying cost-comparison table that the actual cash savings result mainly because the number of handling and set-up operations in the NC method are very much reduced by comparison with the conventional technique, which involves both a horizontal borer and a radial drill.

Machining centres

The machining centre is capable of performing a number of basic machining operations such as milling, drilling, boring, etc. The idea is, of course, to carry out the maximum amount of machining with the minimum number of set-ups.

Workpiece	Material
Gear case	Cast iron
N.C. method	Horizontal boring machine
Conventional method	Horizontal boring machine and radial drill
Operations	Bore 3 main bores. Face, drill, and tap 32 holes
Production quantity	1 per batch; 46 batches

Production details	Numerical control		Conventional	
	Time (h)	Cost (£)	Time (h)	Cost (£)
Programming/planning	22	16·50	–	–
Data preparation	1·5	0·50	–	–
Jig and tool design and Manufacture	–	–	120	140
Pre-production time/cost	**23·5**	**17**	**120**	**140**
Marking-off and setting-up	–	–	6	6
Machining	27	61·50	42	67
Inspection	0·5	0·5	1·5	1·50
Handling (crane)	2	2	9	9
Rectification	–	–	4	4
Processing time/cost	**29·5**	**64**	**62·5**	**87·5**
Total for 46	**1380·5**	**2961**	**2995**	**4165**

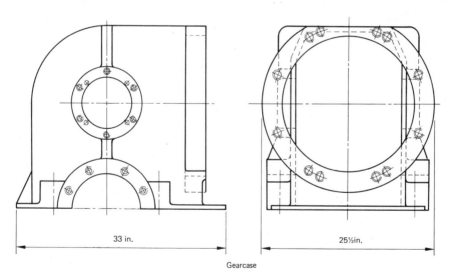

| 33 in. | 25½in. |

Gearcase

This example illustrates most clearly the economic possibilities of NC boring. The machining costs do not differ very much because some of the conventional machining was done on a drilling machine at a much lower machine hour rate. However, it was found more economical, when using NC to leave the gearcase on the boring maching and drill the flange holes at one setting. This resulted in a saving in handling time of 7 h per component. In addition, a saving in both inspection and rectification time was achieved when using numerical control.

Figure 1.4   Reduced setting and handling time with NC.

The term 'machining centre' is very often associated with automatic tool changing, probably because the first machine to really warrant the title was equipped with an automatic tool changer. Machining centres, however, are available which have only a single spindle and rely on manual tool changing. Their effectiveness and economy, by comparison with automatic tool-changing machines, is almost solely dependent on the lower frequency of tool changes which are necessary for suitable work and the lower cost of the machine. When comparing the single-spindle, manual tool-change machine against the auto-tool changer, the floor-to-floor time of the former is entirely in the hands of the machine operator, who is responsible for changing the tools. In this respect, and in view of the often very high cost of the auto-tool-changer, a very effective compromise can often be made by using a six- or eight-station turret machine. Such machines seem ideally suited for the flat-plate type of component which has several sizes of hole and possibly also requires some boring or straight-line milling carried out at the same time.

Turning operations

The development and application of NC lathes has lagged some way behind other types of NC machines, probably because the turning operation has been automated for longer than most other machining operations and has proved itself more than adequate in the large-scale production field. For medium- and small-scale production, the turret, capstan, and copy lathes have similarly proved their worth as semi-automatic machines. The advent of sequence control (often referred to as plugboard control) has enabled these latter categories of machines to become completely automatic, and has therefore made a very effective stand against the introduction of NC.

This has been especially true in the past when the difference between the cost of an NC system, and that of a sequence-control unit, has tended to swing the economic case against NC turning.

The 'opening' for NC, however, is there, and it lies again in what was referred to earlier as the effective utilization of the machine. All conventional automatic machines require a significant time to set up—time which could otherwise be productive. Even the sequence-controlled capstan—almost always economically superior to the manual machine—can, unless preset tooling is used, spend quite a sizeable proportion of its time being set up instead of producing. Indeed, the addition of sequence control has added to the set-up time since the setting up of the plugboard, etc., is all additional to the normal manual set-up procedures. If, however, this type of machine were fitted with numerical control, with the set-up reduced to the time required to insert a tape and adjust the tools, all the saved set-up time could be used for producing more parts. Clearly, however, the NC system to do this would not have to cost a great deal more than the equivalent sequence-control system, since the basic metal-cutting time would be equal for both methods.

Many similar considerations might also apply if economic comparisons were being made between a tracer-controlled copy-turning lathe equipped with sequence control and a similar machine operated entirely by numerical instructions encoded on tape. In this case the points in favour of the NC machine build up as the complexity of the work increases and the batch size reduces. As with the capstan lathe, the sequence-controlled tracer lathe utilizes potentially productive time while being set up, and the NC verson is capable of producing for a higher proportion of its operating time.

Effect of numerical control on lead time and stockholding

The importance and value placed upon the effect of reduction of lead time must depend upon individual circumstances. For firms operating in highly competitive markets, savings in lead time alone could be more than sufficient justification for the purchase of NC.

The conventional process of moving parts from one machine to another is time-consuming. In most cases, the manufacturing lead time is so long that

Figure 1.5 Using standard tooling units.

producing to a given inventory level is essential to provide reasonable delivery dates. When manufacturing with NC and especially with machining centres, it is possible to produce parts as needed—manufacturing to 'order' instead of to an 'inventory level'.

The principal reasons for savings associated with NC are as follows.

(a) Tooling times reduced. With NC it is often possible to use standard tooling elements or standard fixtures, thereby completely eliminating any time delay for tooling (e.g., components can be tooled by using a unit tooling system such as Wharton tooling).

Figure 1.6   Further use of standard tooling.

As tooling for NC need only provide work-holding capabilities, relatively complex components can be tooled using such systems; where a special fixture is required it will be less complex than tooling required on conventional machines to do the same job.

The reduction in tooling demands in both time and money should not be under-estimated, especially with a large installation, as these can be considerable.

(b) Machine setting time is reduced, especially when NC is used in conjunction with preset tooling. In the case of automatic tool-changers equipped with

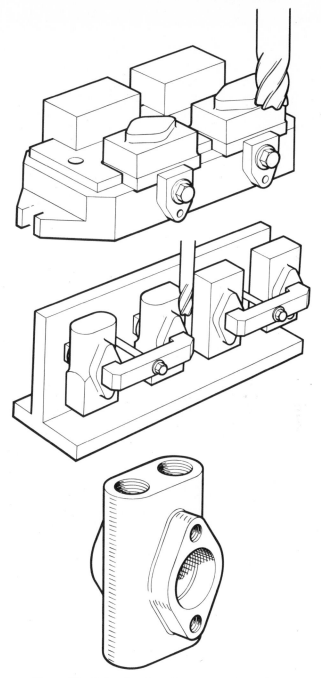

Figure 1.7 Reduced tooling and tape preparation costs.

pallet loading, it is possible completely to eliminate downtime required for machine setting.

(c) Processing time is often reduced by the ability to combine several conventional operations into one NC operation. Figures 1.5 and 1.6 show the Wharton tooling system which enables fixtures to be assembled from standard elements. On first assembly, the fixture can be photographed and a list made of the various elements. The photograph and listing can be stored so that the fixture can be rebuilt at any future date.

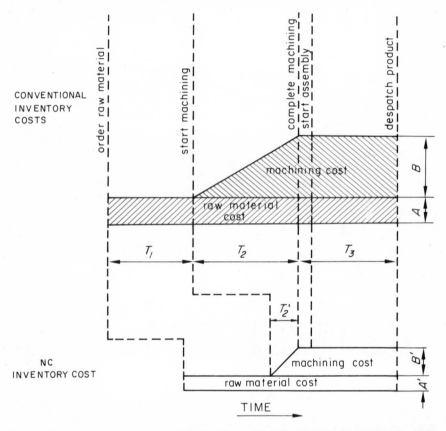

Figure 1.8   Reduced inventory costs with NC. Although it is difficult to quantify all the saving arising from the possible reductions in stockholding and lead time arising through the use of NC, this graph shows the 'work in process' inventory savings possible both in cost and time.

Figure 1.7 shows the machining of aircraft windscreen-wiper bodies. Each type used to have its own tooling, its own fixtures, its own production layout. Each one required, on average, thirteen operations including pantographing the final form (which was unreliable and needed a lot of effort with a fairly small cutter).

Numerical control could do each of these jobs separately and reduce the number of operations to about seven. But twelve of these bodies had the same general configuration, though the position and dimensions of certain machining surfaces were different. A little computation showed that certain combinations of these body designs could be produced with common tapes, other combinations could be produced with common fixtures, and others could be produced with common cutters, fixtures, and tapes.

So, as well as eliminating eight operations, the opportunity was seized to reduce tooling and tape-preparation costs considerably.

Organizing for the introduction and exploitation of numerical control

To achieve the maximum benefit from the use of NC, it is essential to realize that this technique is not 'just another metal-cutting technique'; it is a management technique and discipline, which if applied correctly can only result in improving the overall operation.

The major pitfalls in adopting NC would appear to be:

over-optimism in possible cash savings;
under-pricing of machine time;
inadequate planning and preparation;
lack of management attention at levels that will recognize the peculiarities of this new technique.

When the installation of NC is being considered, it is essential to be prepared to revise accepted conventional ideas of manufacture and rethink the whole factory configuration.

There is always some element of risk involved with new equipment and NC is no exception. The following suggestions will help to minimize this risk.

(a) Prepare the company; assign a high level man to co-ordinate the change to NC, someone who can at least influence, if not make, policy.

(b) Conduct exhaustive cost studies on known parts. Make your own cost estimates. Have test parts programmed and run.

(c) Talk to users of NC but beware of the folklore: generalizations do not always apply.

(d) Ensure personnel are trained in the required disciplines.

No single positive formula can be established for initiating an NC installation, as the type of application and machine tool can vary considerably as can the size of organization. In all cases, action should be taken on the following items.

(a) Acquaint, indoctrinate, and train all departments and individuals involved with NC.

(b) Familiarize trade unions with company plans.

(c) Determine general ground rules and policies relating to operating functions.

(d) Establish services to machines.

(e) Establish reporting system for measuring and comparing performance.

(f) Direct engineering design and tool design towards NC orientation.

(g) Establish 'Part-programming section'.

(h) Consider availability of computer-assist programs and computer service.

(i) Consider installation and maintenance requirements.

(j) Plan a system for quality control.

Numerical control Production Manager

Experience in many companies indicates that there are as many approaches to initiating an NC programme as there are types of factory, types of management organization, and types of management philosophy. Therefore, no single plan can be developed that will satisfy the requirements of all these types.

However, the NC 'Production Manager' approach has been used with considerable success and is one of the notable characteristics of the more efficient NC users. It vests control of the program in a single person. He has a mandate from top management giving him the task and the necessary authority. He plans the program with advice and help of selected personnel. He obtains top management approval. He works through line supervision and selected personnel to attain objectives. Positive response is assured by virtue of his authority. He is normally working through people of his own level or below. Obviously, great care must be taken in the selection of a man for such responsibility. He must have:

machining experience;

knowledge of production engineering, including process planning, tool design, and estimating;

organizing ability;

willingness to rethink accepted techniques and procedures;

ability to 'sell' NC within the organization.

Preferably the NC Production Manager would come from Production Engineering. In this context 'Production Engineering' covers the general functions of process planning, tool specification, fixture design, estimating, and part programming—all of which must give efficient production at the NC machine tool.

The most successful operations in NC have been organized to rotate around this area. The NC programme involves many areas of the factory (Fig. 1.9).

Of course, there is always the possibility of purchasing an NC machine and

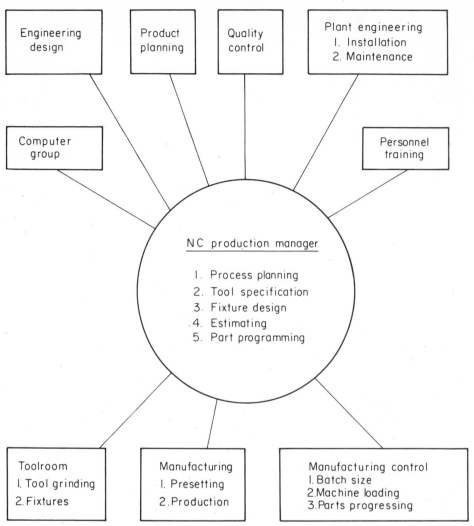

Figure 1.9 Factory departments affected by NC.

simply expecting the shop foreman or superintendent to make it work. This approach could appeal to some shop men, but must be recognized as a very extravagant planning approach and one virtually doomed to failure.

Sources of information

Many sources of information are available.

(a) Sales brochures, literature, films, etc., of both machine manufacturers' and control manufacturers' products.

(b) Presentations by manufacturers.

(c) Excerpts from articles in technical and trade journals.

(d) Visits to NC users.

(e) Manuals, forms, training courses, organization structures, and brochures from other companies.

(f) The British Numerical Control Society. A technical society founded to provide opportunities to contribute to and learn about the application and technology of NC in all industry.

(g) The Ministry of Technology's Numerical Control Advisory and Demonstration Service which is provided by the Production Engineering Research Association, Melton Mowbray, Leicestershire; the Royal Aircraft Establishment, Farnborough, Hampshire, and Plessey N.C. Ltd., High Wycombe, Buckinghamshire.

(h) The National Engineering Laboratory provides assistance to NC users who wish to evaluate computer programming aids.

(i) The APACE unit at Aldermaston provide tuition in the use of some computer programs for NC.

Selection of a specific system can well influence and affect your preparations for NC. As a purchaser, a great deal of information will be given to you by the supplier. This material should be studied and digested very thoroughly in order that potential areas may be brought to light and corrected or circumvented in the planning stage. Almost all of the less desirable features of a system can be worked around and thus minimized in actual practice, if not eliminated outright. Inevitably, systems have undesirable features for certain manufacturing problems; making a system perform satisfactorily is the basic problem of the user.

Installation

Consideration of installation and operation should begin before a machine is bought. Examples of items of concern to the plant engineer normally involve power requirements, major component dimensions, overall dimensions, shipping weight, floor-mounted weight, and facilities required such as water or air, and soil foundation conditions. These are the general conditions when purchasing any machine tool. However, since it is normally expected that superior performance and accuracy will result from an NC installation, a number of particular considerations are in order, but these need not differ substantially from any other machine tool. If extreme accuracy is required, then adverse environments such as dirt, temperature changes, and humidity will affect output of an NC machine tool at least as much as a conventional machine.

Once the machine is installed and a work load provided, utilization can be checked by means of time recorders. These provide valuable information about causes of inefficient machine utilization, which could be due to poor part programming, poor tooling and fixturing, machine breakdowns, lack of work, and so on. It is usually found that, as an NC installation settles down into its stride, machine utilization remains fairly constant and the time recorders simply confirm established operating patterns.

Effect of numerical control on design of workpiece

Design determines the shape, tolerances, material, and specification of a product. In doing so, the designer should be trying to conceive economically producible shapes.

Product configuration

Usually, the simpler the shape, the more economically can a component be produced. This is not necessarily true with NC. Even the most complicated contours can be produced with NC as easily and with no more skill than it requires to produce straight lines; this gives the designer greater freedom in shape specification. Numerical Control has another important effect on product shape. Often when a design has been established and tooling manufactured and set up, it is not economically feasible to change the design. Consequently, improvements may have to wait a long time before being incorporated into a product. Because NC machines are so flexible, changes can be made with a minimum of delay and expense. Parts previously requiring fabrication because of part complexity can now be manufactured as one piece.

Tolerances

It is also the responsibility of design to determine tolerances so that the part will perform its function and can be manufactured at the least possible cost. As tolerances become tighter, conventional manufacturing costs increase. Therefore designers have tried to keep tolerances as loose as possible within a part's operating specification. In the past, designers have put considerable effort into designing parts that would not require tight tolerances and, for that reason, could be made cheaply.

Numerical Control has changed this: with an NC machine tool, tolerance is governed by the accuracy of the machine tool. With NC, it costs no more to specify a tight tolerance than to specify a loose one. This is a double-edged sword, however. If an NC machine is guaranteed to work to a certain tolerance, no operator skill can improve on those tolerances. You may get worse tolerances with NC through negligence, but you will not get better ones.

Dimensioning

Companies using manual-programming techniques could well adopt the method of co-ordinate dimensioning, preferably numbering holes and tabulating co-ordinate values against hole numbers in tabular form. When this is done, part programmers and inspectors work to the same figures.

The same technique can also be used to advantage when using a computer as an aid, but it is then preferable to quote the number of holes, their pitch circle diameter, and the starting point of one of the holes.

Material

Until recently, the choice of material has been independent of NC capabilities. Numerically controlled machine tools cut the same materials as conventional equipment—no better and no worse. However, the Molins System 24 concept is based upon the idea of the designer rethinking material specifications; this subject is dealt with in detail in chapter 16.

Shape

It is essential that designers are made aware of the capabilities of NC well before installation. Often design features which were acceptable or perhaps even desirable with conventional manufacture require to be altered.

Thus, Fig. 1.10 shows that, by keeping the corner radius of the component as large as possible and the fillet radius as small as possible, a large cutter can be used, thereby reducing the number of cuts required to machine the face of the pocket. It should also be noted that a larger corner radius enables a higher feedrate to be used when machining the form of the pocket.

Figure 1.11 shows that, by using common fillet radii, it is possible to reduce

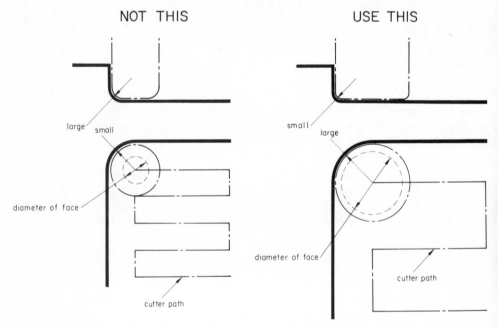

Figure 1.10 Use large corner radius on part.

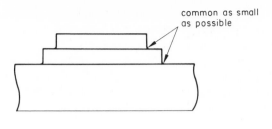

common as small
as possible

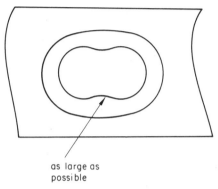

as large as
possible

Figure 1.11 Use one fillet radius on part.

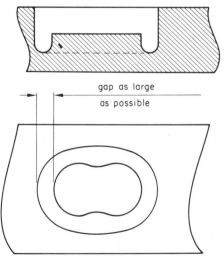

gap as large
as possible

Figure 1.12 Design large clearance between islands and pockets.

the number of cutters used. Fillet radii should also be kept as small as possible to reduce the need for additional face-clearing cuts. The concave radii on the projections should again be as large as possible.

For Fig. 1.12, any gaps should be made as large as possible and cavities should be kept as shallow as possible. This is necessary because standard cutters of small diameter have very small flute lengths. Special small cutters are not only expensive but deflect more under load.

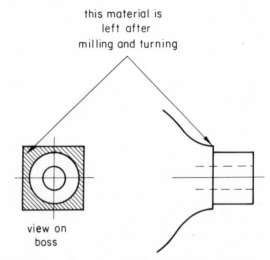

Figure 1.13 Avoid awkward blends or fillets where possible.

Figure 1.13 shows material left after milling and turning. The original drawing specified a blend of these two forms, which required a difficult bench operation. However, the designer was prepared to accept the machined form as shown.

General
General points for the guidance of designers whose work is likely to be produced by NC are:

co-ordinate dimensioning is desirable;
avoid variation in hole diameter;
avoid variation in fillet radii;
avoid tolerance build-up;
consider producing more complex parts, thereby eliminating assemblies;
parts previously cast, forged or fabricated can often be produced by NC with relative ease;
acquire knowledge of cutting-tool characteristics, so avoiding costly special cutters;
acquire knowledge of standard fixtures and machining procedures used.

Computer-programming aids

Basically, NC falls into two categories: positioning and contouring. Numerous calculations are required to instruct continuous-path control equipment. The volume of work required makes computer-programming aids virtually mandatory. Conversely, the need for computer programs for positioning or point-to-point machine tools is not as critical: most users still manually prepare NC tapes for positioning equipment. The trend, however, is to increased computer utilization. Certain computer programs have been written particularly for one combination of machine tool and control system. As the potential number of these combinations is very large, this approach requires extensive programming effort.

To offset this problem, general computer programs have been developed for particular classes of work such as drilling and milling. Some subsequent program is then required to tailor the output of the general program to the specific machine-tool system. The general program is usually referred to as a 'program' or 'processor'. The tailoring program is then referred to as a 'post-processor'. The processor, written only once, typically translates the source language, and calculates geometry and tool-path information. It produces an intermediate output, read by the post processor which tailors the information to suit one specific combination of machine tool and control system. In general, there will be as many individual post processors for a processor as there are unique combinations of control system and machine tool.

Manufacturers of machine tools are now realizing the severe limitations and cost impracticability of maintaining computer-programming aids for specific machine tools. In general, it would appear that machine-tool builders are now refusing to supply specific programs, or do so reluctantly. This attitude reflects the impact and effect that general programs such as APT, ADAPT, and AUTOSPOT have had.

This trend to general programs is worth noting as, a short while ago, a specialized or proprietary program was a selling feature for an NC machine. The prime requisite now appears to be for the provision of post-processors for general programs such as APT, ADAPT, and the National Engineering Laboratory family of programs starting with 2C,L.

General programs can be obtained from the computer manufacturers with the exception of APT and EXAPT, which are available only to subscribers. Post-processors are the responsibility of the machine-tool builder or control builder.

There also appears to be a steady increase in the use of APT-oriented programs based on the widespread existence of FORTRAN compilers on computers of various makes. There are positive indications that in the future there will be less reliance on proprietary programs.

For continuous-path machining, manual programming should not be con-

sidered because of the sheer volume of work involved in calculating change
points and writing out lengthy part programs. Be wary of salesmen who try
to convince you otherwise.

The decision regarding the method of programming for positioning machines
is not as clear, but depends on:

  volume and complexity of work load;
  availability of a suitable computer, either in house or at a convenient com-
    puter bureau;
  availability of general processor for that computer;
  availability of suitable post-processors for the range of machine tools used.

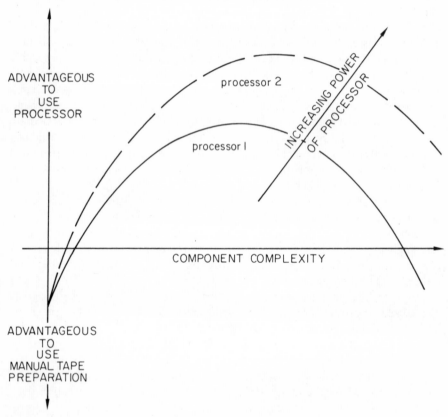

Figure 1.14 Qualitative display of value of using an NC processor. Relationship between the 'worth-whileness' of NC and that of conventional manufacturing and component complexity.

Figure 1.14 seeks to display qualitatively that, for any given processor, the
relationship between the advantages of NC and the complexity of the part have
the general form shown.

Other factors such as batch size are also involved and it is not possible to quantify 'complexity' and 'advantage'. For example, reduced lead time may be very important, but it is often difficult to assign a monetary value to it.

The subjects such as processors, post-processors, use of bureaux, etc., are dealt with in greater detail later in this book.

Staff

The importance of training

Too great an emphasis cannot be placed on the importance and the need for adequate training of staff so as to enable the maximum benefit to be derived from an NC installation.

Unfortunately, many potential users of NC regard the minimum amount of training as sufficient, although they are receptive to the need for adequate preparation and training of computer staff, going to great lengths to develop competent computer specialists. An NC installation requires similar consideration, as it embraces varying degrees of knowledge of the following subjects: control, machine, hydraulics, tooling and fixturing, processing, part programming, production control, and metal cutting. For many companies, the investment in NC equipment far exceeds that for computer installation. When purchasing an NC machine, the training courses and advisory service offered by the supplier should be a major consideration in the evaluation of equipment.

It should be stressed that NC is in its infancy and growing at a very rapid pace. This should be borne in mind when selecting personnel for NC activities. It is far better and easier to initially select candidates of a high calibre and to reduce standards at a later date, if need be, rather than do the reverse. Young people should be selected who are willing to put forth the necessary effort and to accept the challenge of this fast-changing technology.

Selection of personnel for numerical control training

It is most difficult to establish positive guide lines on how to select personnel for the various disciplines associated with NC, especially as training in NC techniques is generally concerned with *re*-training persons who have had basic training in other allied disciplines. So far, there are few training schemes devoted to teaching NC as a basic discipline.

However, there are general considerations which apply to all the associated disciplines, namely:

complexity of machine and control;
application of equipment;
job specification;
size of installation.

Considering these points coupled with the experience of a considerable number of users both in Europe and the USA, it is possible to suggest some desirable tests and means of personnel.

Selecting operators

When selecting an operator one must consider the type of machine he is to control, the application of equipment, and the job specification.

Type of machine Obviously, operators selected to control multi-axis or machining centres should be much more knowledgeable in many aspects of machine shop practice than operators for less complicated NC machines. The value of the machine should also be considered: an unskilled or semi-skilled operator cannot be expected to take charge of a machine costing tens of thousands of pounds.

Application of equipment Identical equipment can be used to effective purpose on completely differing applications, and the manner in which the equipment is controlled can also vary. For example, a machine used on repetitive batch production may be controlled very satisfactorily by a semi-skilled operator who is little more than a machine minder. Yet the same machine used on tool and die work could well require the service of a skilled operator.

Job specification One must also consider what the operator is allowed to do, what you would like him to do, and what services and information are provided for him. The adequate provision of the essential back-up services to the machine all tend to lessen dependency upon the operators.

Answers to these questions should provide a good starting point for selecting candidates, noting that the following qualifications are highly desirable:

knowledge of good machining practice;
knowledge of tooling and tool holding;
knowledge of work holding;
correct use of work- and tool-holding equipment;
knowledge of inspection procedures and methods;
correct use of inspection equipment;
acceptance of new thinking and techniques;
sense of responsibility;
willingness to co-operate. (Note: the success of an NC installation is highly dependent upon team effort.)

Although it is possible to operate certain types of NC machine with an unskilled or semi-skilled operator, invariably results are far superior when a higher grade operator is used. This point can be very significant when considering the overall efficiency of a machine, especially in the case of the more sophisticated machine.

Selecting part programmers

It appears that personnel selected for part programming are generally selected from the semi-professional and professional occupations. Process planners, methods engineers, production engineers, jig and tool draughtsmen, and occasionally operators are selected for this work.

In selecting a part programmer, account must be taken of the following.

(a) The type of machine to be programmed. The degrees of skill required to part program various types of machine bear no comparison. It can vary from being purely a clerical function, to being a highly technical task. The type of machine is not, however, the sole consideration.

(b) The application of the machine. Identical machines can be used on vastly differing complexities of work.

(c) The job specification. Is the part programmer merely to write part programs or is he to provide a complete production engineering service for the machine? The former enables the use of lower grade labour but involves more personnel in preparing the work for the NC machine, at the same time setting up a time-delaying chain. The latter means raising of standards, which involves considerable training, but it does give the part programmer total responsibility for work preparation and in general gives the most satisfactory results.

(d) The size of the installation. This is a most important factor to be considered when selecting part programmers. A multiple installation will have a wide range of types of work, and will require several part programmers. Under these circumstances inexperienced part programmers can work effectively using standard procedures related to 'family groups' of work, while continuing their training in a practical manner and building up their confidence and ability. On the other hand, at a one-machine installation, especially if used for complex machining procedures, the part programmer has to be highly skilled from the outset, as otherwise the overall efficiency of the installation is impaired.

It is essential for the part programmer to determine:

the complete method of the manufacture of the parts;
the tooling requirements for the job;
the fixturing, wherever possible using standard elements or fixtures;
the operation set-up procedure.

He must also:

calculate the co-ordinate values of each operation position or, for continuous-
    path work, the change points (for manual programs only);
select the sequence of operation for the most efficient use of the equipment;
list feeds and speeds for each operation;
write the part program;
prove out tapes.

Most important of all, the part programmer must accept complete responsibility for all parts he has programmed and answer all shop queries related to those parts.

A part programmer should have some formal technical qualifications and preferably have shop experience. It is essential that he has first-hand knowledge of tooling, fixturing, and process planning. Reasoning power comes second only to accuracy. Anything less than 100 per cent correct in a part program is not acceptable. A part programmer must possess an innate sense of logic and, unless he is analytically minded, he will be unable to write a part program which details every step in the machining sequence so that not only the operator but also the NC machine tool will be able to follow his instructions.

It is desirable, therefore, that the part programmer has:

had a technical education, preferably to at least City and Guilds, ONC, or
   HNC standard;
aptitude for mathematics, including geometry and trigonometry;
machine shop experience, with a knowledge of machining procedures;
knowledge of tooling, fixture design, and process planning;
an ability to interpret engineering drawings;
an ability to plan and organize his work;
an analytical mind with good reasoning power;
a good attitude towards work and a strong interest in it;
an ability to be very exact in his working.

There is no set procedure for establishing people in part programming; the most important thing is their attitude to the job and willingness to accept a new job. In some instances, candidates for part programming are given aptitude and mechanical comprehension tests: these measure judgement in business and industrial situations and also mental ability, reasoning power, spatial perception, and clerical skill.

Selecting maintenance staff for control equipment

Electronic maintenance staff are responsible for finding faults in and repairing the electronic control equipment. These men should be alert and clever enough to diagnose faults quickly and know what to do about them.

Maintenance electricians, machine wirers, instrument fitters, and others with electrical experience have been selected for control courses. It is claimed that young men with a strong electronics background are best capable of learning to work on control systems.

In the past, many unsuitable persons have been sent on control courses, with disastrous results due to candidates' suffering emotional disturbances. This is unfair both to the individual and also to the employer who hopes to obtain an adequate maintenance man. As a result of this experience, organizations conducting control maintenance courses now usually demand that potential candidates are subjected to a selection test.

These tests are prepared by the company conducting the control course and administered by the candidate's own company.

Alternatively, there are commercial services available. These commercial testing bureaux will administer tests and conduct interviews with the candidate.

It is desirable that the control maintenance man has:

a technical education, at least City and Guilds in electrical engineering or electronics;

practical electrical and electronic experience, preferably gained by working with machine tools or an equivalent application;

a working knowledge of industrial electronics;

a basic understanding of the principles of digital logic;

knowledge of the proper use of test equipment;

a good attitude to his work;

a desire to learn and further his knowledge in the subject.

Selecting mechanical and hydraulic maintenance staff

This is an area which until recently has received little or no attention with respect to training of personnel. With the increasing use of hydraulics in NC equipment, together with higher demands on mechanical performance, it surely warrants attention.

Success as a conventional machine tool repair man will not necessarily make a good NC maintenance man. Components used in NC are generally more sophisticated and precise: this is in order to build precision into the machine tools. Some maintenance men cannot make the change from rough maintenance to work on extremely close tolerances.

For many maintenance men, hydraulics is quite a new subject, but one which they can understand. The new subject requires more careful attention because the components are delicate and precise instruments. Getting maintenance men to understand this does cause difficulties. Solutions to new problems such as dirt, contamination, proper adjustment, filtering, and fault-finding techniques must be found.

As with control maintenance, potential candidates for maintenance of mechanical and hydraulic systems are subjected to selection tests before they attend courses.

It is desirable that they have:

a technical education, such as City and Guilds in Machine Shop Practice;

a minimum of three years on machine tool maintenance;

a good working knowledge of hydraulics, as applied to machine tools or equivalent;

the ability to read engineering drawings and hydraulic schematics;
a good attitude towards work;
a strong desire and interest to learn and apply correctly what is learned.

Effectiveness of training

The effectiveness of the training given to operators and part programmers is
usually reflected in the utilization, productivity, and profitability of an instal-

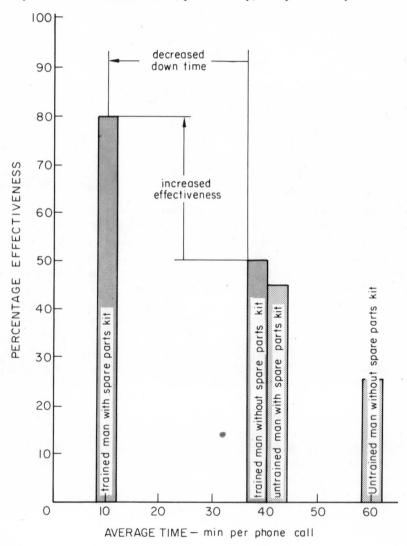

Figure 1.15   NC maintenance efficacy. Effectiveness of training and spare parts kit in solving problems by telephone.

lation. This is usually a domestic concern and the degree of effectiveness should show up in post-audits on the NC equipment.

The effectiveness of maintenance is, however, not solely a domestic issue, as the supplier is often concerned with helping to diagnose faults and the supply of spares. One NC equipment supplier has completed a survey on the effectiveness of their training program, asking a series of questions about their training effort. The survey was taken in relation to the number of telephone calls the Field Service Department received on problems with NC equipment. There were two things that affected the statistics: one was whether the maintenance man was trained or untrained; the other was whether the customer had a spare-parts kit available. With these two variables as a basis, the following questions were asked:

(a) How long does it take to solve a problem over the telephone if the man has not attended training and does not have a spare parts kit? The survey showed that it takes an average of 60 minutes on the telephone and about 25 per cent of the problems were solved.

(b) How long does it take to solve the problem by telephone if the man is untrained but has a spares kit? It takes about 40 minutes on the telephone and about 45 per cent of the problems are solved.

(c) How long does it take if the man has been trained and is without a spare parts kit? This takes, on an average, 40 minutes on the telephone and about 50 per cent of the problems are solved. This is because the trained man understands the terminology and this facilitates effective communication.

(d) How long does it take to solve a problem when a man has been trained and has a spare-parts kit? This takes, on an average, 10 minutes on the telephone and solves 80 per cent of the problems.

Figure 1.15 thus shows that there is a 25 per cent increase in efficiency just by having a spare-parts kit available and a 30 per cent increase due to staff training. From having no kit or trained staff to having both, 55 per cent more of the problems were solved by telephone calls to the supplier. This increase in efficiency is well worth having.

Ministry of Technology training schemes

The Ministry of Technology has established a Numerical Control Advisory and Demonstration Service to accelerate the introduction of NC machine tools in industry and has three functions:

to assist firms by assessing the benefits they can obtain from NC equipment and to advise on the selection of NC machine tools;

to program and machine components submitted by individual firms;

to provide courses and symposia on the advantages and problems involved in using NC machine tools. Subjects include technical and economic aspects, machining techniques, part programming, and computer techniques.

These services are being provided by:

Royal Aircraft Establishment, Farnborough, Hampshire;
Production Engineering Research Association, Melton Mowbray, Leicestershire;
Plessey N.C. Ltd, High Wycombe, Buckinghamshire.

Royal Aircraft Establishment At Farnborough, the RAE workshops have the benefit of almost a decade of experience in operating NC machines for the machining and measuring of a wide range of engineering components and are well equipped with an extensive range of NC equipment which is being added to. Computing facilities are also available to serve the NC equipment.

This enables the RAE to provide technical support in part programming and machining components for individual firms. It is also providing courses with a practical bias for the needs of the production engineer, process planner, and potential part programmers; the primary aim is to help him to assess the application of NC to the products of his own organization.

The first course is an introductory one for personnel who are new to the use of NC machines and their part programming. The second course could be considered as a follow-up stage, being more suitable for personnel with some experience of part-programming techniques. Recruitment for these courses is from potential part programmers, production engineers, and work supervisors up to works manager level.

Course 1 starts with an outline of the fundamental principles of NC and is followed by instruction on positional control with an opportunity to prepare a tape and see it operate a machine tool.

Then follows an introduction to two-axis continuous control as applied to a contour-milling machine, again with a practical demonstration. This technique is further amplified by using different machines and control systems. There are examples of graduated complexity with the staff giving individual tuition. During the course, time is also devoted to talks by engineers from industry on their experience in the use of NC techniques.

Course 2 requires a knowledge up to the standard of Course 1. There is an introductory talk on the selection of components for machining by NC methods and this is followed by periods in which more advanced examples of part programming are discussed. Positional control includes the use of computer-assisted programs, interpolation in a two-axis control program and with three-axis control being introduced for continuous-path machining. Practical demonstrations are also a feature of this course, with visiting lecturers from industry backing up those lectures given by RAE staff.

Firms are invited to follow up these courses by submitting examples of their own workpieces to be part programmed at RAE on a proving basis. Preferably, the firm's representatives who have taken the course will attend to assist.

A manufacturing service is available for the purpose of proof-machining either prototype or batch quantities of parts on numerically controlled machines, in order to typify the advantages of NC.

A free advisory service is offered in which an initial assessment of NC potential is made in a customer's works.

The RAE workshops are also open, by prior arrangement, to display modern methods of machining.

Production Engineering Research Association At Melton Mowbray, PERA is in a unique position to assist firms of all sizes, because it has wide experience of the factors involved in the selection, economic appraisal, and utilization of machine tools.

They have practical experience of management accounting and the application of various investment-appraisal procedures. This is backed up by extensive facilities for the effective dissemination of information and advice through courses, symposia, and mobile units.

The PERA service offers up to three days' free advice on NC machine tools, which includes the preparation of a report making recommendations. Longer studies are undertaken for a fee.

PERA are setting up an NC demonstration workshop with a selection of drilling machines on which customers' parts can be machined to confirm the results of feasibility studies.

Plessey NC At High Wycombe, Plessey have been helped to equip a demonstration workshop with various NC drilling machines on which potential customers for NC machine tools can assess the suitability of particular types of machine for their purposes.

Ministry of Technology programming support

Two Ministry of Technology Establishments have been active in providing support for computer programming aids for NC.

APACE At Aldermaston, AWRE have set up a unit which is unrestricted and to which industry can send staff for short courses in the use of the computer as an aid for NC or for design.

In the NC field, courses are run in the use of APT and of the NEL NC processor. The courses are usually of one week's duration and organized at introductory and advanced stages.

NEL At East Kilbride, the NEL provide a consulting service on the use of their own NC processor and also for APT and EXAPT. Skilled part-programming staff are available to assist the potential user to convert his drawing into a control tape with any of the programming systems, and computer programmers are available to improve programs and consult with computer programmers from users who implement the programs on their own computers.

The NEL staff maintain the NEL NC processor and provide part programming and computer programming up-dates to users.

As there are three different NC computer programs running side by side on one computer, the staff are in a strong position to advise on the relative advantages of each and of the degree of compatability between the language and output of each.

2. Types of NC machine tool

Ronald Iredale

Ronald Iredale, BSc., CEng, FIProdE, MIMechE, is Editor of the journal *Metal-working Production*, which he joined seven years ago as an Associate Editor, after 14 years in industry. He served a production engineering apprenticeship with Electroflo Meter Co., now a division of Elliott Automation, after which, as a Technical Mature State Scholar, he took his degree at King's College, Newcastle, England. He held the position of works manager with Fischer and Porter, manufacturers of process control equipment, and then joined Smiths Industries. His interest in numerical control which, together with other aspects of manufacturing technology, he has fostered during his journalistic career, had its birth while he was working as a Special Projects Engineer with the Production Research and Engineering Unit of Smiths Industries, where his responsibilities included the establishment of a centralized numerical control facility.

With few exceptions, the development of numerically controlled machine tools, their control systems, and the computer programs that go with them, has to date been geared to the requirements of the small-batch producer. This chapter is mainly concerned with the machines that satisfy the requirements of this important sector of the metalworking industries.

It is probably true to say that the designs of NC machine tools for small-batch manufacture are now emerging from their most intensive phase of development particularly regarding configuration, and are likely to be subject in the immediate future to relatively detailed modifications and refinements. The more dramatic developments are likely to emerge from considerations of how NC machine tools can be modified and combined into a variety of machining systems—some with on-line (direct) computer control. Some of these systems will meet the future requirements of the small-batch producer; others will extend the application of the NC machine tool to medium-size batches.

The potential of this market and the demands it will make on NC equipment supplies will be dealt with later in the chapter.

The machining centre

In NC small-batch production this is the age of the 'machining centre'; that is, the single machine which, in a single set-up, will automatically and without transfer of workpieces do a multiplicity of operations on each of a large variety

Figure 2.1   This type of machine, when fitted with automatic workpiece changing, is generally recognized as embodying the full potential of the NC machining centre. (*Courtesy Rolls-Royce Limited*)

of parts, and it is becoming apparent that a good proportion of NC machine tools for small-batch production will be sold under the term 'machining centre'.

Numerical control provides the ideal basis for fulfilling this machining concept, for two reasons. First, because an almost infinite number, and a wide variety, of machining operations can be programmed into a series of control tapes, or even into a single tape. Second, because a new and totally different machining procedure can be provided simply by writing a new tape and making available the requisite cutting tools.

The type of machine which is nowadays claimed to embody the full potential of the NC machining centre is that which automatically changes the cutting tools from a magazine. On a milling and drilling machine the tool is mounted in a workspindle which, if horizontal, has access to all four sides of the workpiece. The latter is mounted on an indexing (or rotating) worktable and, if necessary, can be automatically transferred to and from the worktable on pallets.

This type of machine (Fig. 2.1) is the product of an intensive period of development by American machine-tool makers between 1958 and 1963, which left the European machine tool industry trailing. Since then, Continental manufacturers and, more recently, British manufacturers have closed the gap.

At this stage in the development of the machining centre for milling and drilling, it is to be expected that there is a wide variation of design both in the configuration of the machines and in the mechanisms, particularly of the automatic tool changing.

It is important that the characteristics of these automatic tool-changing machines are analysed because from these characteristics will crystallize the future generations of this type of machine.

As regards configuration, the ancestry of most of these machines can be traced to horizontal borers, milling machines, or drilling machines, and this provides a rough and ready classification by which to subdivide them.

Borers become machining centres

One of the more spectacular of the Continental machining centres with horizontal borer ancestry is the Hüller NCMC 100 travelling-column machine which has a four-tier magazine attached to the rear of the column accommodating 100 tools each up to 75 lb in weight (Fig. 2.2).

The machining head carries two spindles (a heavy-duty milling spindle and a boring spindle), one at the front and the other at the rear, and indexes through 180° to bring either into the working position (Fig. 2.3). This allows the tool to be changed in one spindle while the tool in the other spindle is machining. Idle time for tool changing is thus reduced to the time required to index the head, which is about 3 s.

To bring a tool from the magazine to the work spindle, two transfer units are used, one at the rear of the column and the other on the side on which the spindle head is mounted.

The transfer unit at the rear is on vertical guides to enable it to move from one tier of the magazine to another at a different level, and also to align itself with the other transfer unit which is synchronized to follow the rise and fall motions of the spindle head. Tools are stored in the magazine in numbered positions to which they are always returned.

One of the spindles in the head is a heavy-duty milling spindle and the other a boring spindle. For most operations, either spindle can be used. Only for heavy milling on the one hand (up to 40 in³/min removal rate on mild steel)

Figure 2.2 The Hüller NCMC is one of the more spectacular European machining centres with horizontal spindle. (*Courtesy Karl Hüller G.m.b.H.*)

and fine boring on the other, is it necessary to ensure that the tool is inserted in the appropriate spindle. Generally the program can be arranged so that consecutive tools do not have to be inserted into the same spindle, and thus the advantage of changing tools during a machining cycle can be retained.

Normally, length compensation for 20 tools and radius compensation for 5 tools is provided by the system, but if required these compensations can be provided for all 100 tools.

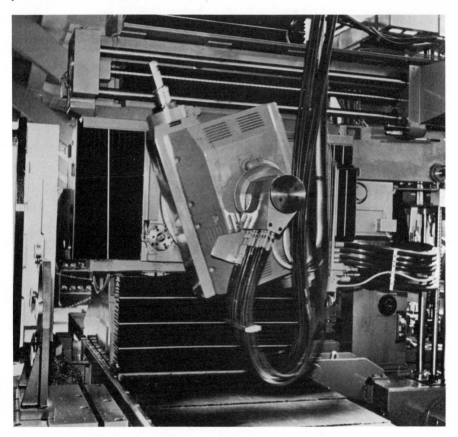

Figure 2.3 The machining head carries two spindles, a heavy duty milling spindle, and a boring spindle. Tools are automatically changed into the idle spindle. (*Courtesy* Metalworking Production)

A good example of the American machining centre with a horizontal spindle is a fairly recent addition to the Kearney & Trecker Milwaukee-Matic range of NC machining centres. Known as the 3B (Fig. 2.4), it has a chain-type tool magazine mounted horizontally on top of the column. Up to 60 preset tools can be loaded at random. On model 2, 3, and 5 Milwaukee-Matics, a drum-type magazine is used.

3

A tool selector coding system on the 3B can keep track of as many as 32,767 tools.

Workpiece capacity on the 3B puts it about midway between the models 2 and 3 Milwaukee-Matics. Maximum tool diameter for automatic changing is 4·5 in. Tools up to 6·5 in. diameter can be used with selective loading and programming, and tools up to 12 in. diameter weighing under 60 lb can be used with certain restrictions (such as loading the chain with an empty socket on each side). Still larger tools can be put in the spindle manually.

The older system on Model 2, 3, and 5 Milwaukee-Matics uses various combinations of ten large- and small-diameter coding rings (in two groups of five)

Figure 2.4   This latest addition to the Milwaukee-Matic range of machining centres has a chain type magazine. (*Courtesy Kearney & Trecker Limited*)

to identify up to 961 different preset tools. The new one simply uses thinner coding rings in three groups of five, plus a new tool selector reading head. Except for the thinner rings, the 3B uses tool holders interchangeable with those of the Model 3.

All Milwaukee-Matics have a double-ended transfer arm which rotates through 180° to transfer a new tool from the magazine to the work spindle while returning the previous tool to the magazine.

A solid-state control system is used both for the machine with 3L capability and for the machine with 2C,L capability. The 3B is available with either 'commercial' ±0·001 in. tolerance, or 'precision' ±0·0005 in. tolerance.

A factory-installed pallet shuttle is available to allow set-up of workpieces on idle pallets and quick interchange between machining cycles. When this shuttle is installed, a second tape reader is added to the NC system to eliminate tape rewind from workpiece interchange time. Duplicate tapes may be used, or different workpieces can be machined alternately.

A two-speed 10 hp hydraulic motor drives the spindle through a 16-speed gearbox, providing 32 speeds from 60 to 2,400 rev/min. In the Milwaukee-Matic machines, a special overload circuit automatically reduces the programmed feedrate until the completion of any operation during which an overload condition develops. Following such an automatic over-ride, programmed feedrate is resumed. If extreme overloads are encountered, a stall protection feature stops all axis movements.

Figure 2.5 A lower-priced 'egg-box' toolstore is a feature of this heavy duty, twin table, Marwin Min-E-Centre. (*Courtesy Marwin Machine Tools Ltd*)

An unusual and inexpensive method of automatic tool-changing is a feature of the British Marwin MEC3 Min-E-Centre (Fig. 2.5).

Fitted to the centre of the table is a two-sided tool store. This comprises removable side-plates with spring-loaded locations in which tools are nested in the X and Y co-ordinates. It is capable of carrying heavy tools and even multi-spindle heads. Capacity is 40 tools, 20 on each of the sides, which can be rotated to face the spindle. Tool deposition and pick-up by the spindle is by the inherent X, Y, and Z motions without additional mechanism. Chip-to-chip time averages 25 s.

Designed for heavy cutting on work up to 2 ft cube, this is a fixed-bed machine with a travelling column carrying a saddle with a horizontal ram in which the spindle is mounted. The principal movements of the machine are thus all in one unit, and the work table is a fixed independent unit that can be arranged to suit requirements.

The machine is equipped with two four-position 20 × 20 in. independent indexing tables, one each side of the centrally mounted tool store, allowing a new workpiece to be set up while the previous component is being machined. Indexing tables with 72 positions and rotary tables are available.

All motions are by hydraulic motors and recirculating ball screws. The spindle is also driven by a hydraulic motor developing 15 hp through a three-speed gearbox giving 230 speeds in a range of 20 to 4,000 rev/min. Rapid traverse is 300 in/min in all directions.

Drillers become machining centres

The other path along which the machining centre has evolved is from the NC co-ordinate drilling machine. When milling capability is provided and automatic tool-changing equipment is added, this type of machine is brought firmly into the machining centre category within a definition that would be accepted by most.

A good example of a Continental machine in this category is the Kolb KBN 80 RM which is indeed described by the makers as a co-ordinate drilling machine. It is a portal-type machine with a drilling capacity of 3·15 in. in steel. It has a stepless range of milling feeds in the X and Y axes, three-axis straight-line NC, a 30-tool magazine with automatic tool change, and a two-position tool carrier to reduce tool-change time. In addition to drilling and milling, such operations as reaming, tapping, boring, and counter-boring can be done on it.

The tool magazine of this machine is in the form of a large-diameter wheel accommodated at the rear of the cross rail. It is actually attached to the spindle head, and traverses with it along the rail. The tools are not transferred directly from the magazine to the spindle. Instead, a tool is placed in a carrier that surrounds the spindle nose. As seen in Fig. 2.6, this carrier has two positions at right-angles to each other. It swivels on a plane at 45° to the spindle axis, so that a tool in the horizontal station is swung into a vertical position beneath the spindle nose by rotation of the carrier through 180°. In this position, the tool is picked up by the spindle as it advances down in rapid traverse. The $6\frac{1}{2}$ in. diameter quill passes through the carrier with the tool in the spindle to perform the operation. On completion of the operation, the tool is returned to the carrier as the spindle moves up to its fully retracted position. By this time a new tool has been inserted into the horizontal station of the carrier ready to be indexed into the position to be picked up by the spindle at its next stroke. Indexing takes 3 s.

The same action brings the used tool into the horizontal station to be exchanged for the tool that will be required next. This exchange is done by a hydraulically operated double-ended transfer arm below the magazine, having an up-and-down motion and a 180° swivelling motion. Grippers at the ends of the arm are on telescopic slides to provide for withdrawal and insertion of tools out of and into the magazine and carrier.

Figure 2.6 The tool magazine on this Kolb 80RM co-ordinate drilling machine is a large diameter wheel fitted to the spindle head at the rear of the cross rail. (*Courtesy* Metalworking Production)

Tools are coded to allow them to be selected in random order. Coding is by binary-decimal combination of electrically conductive and non-conductive rings sensed by contacts.

Simpler tool transfer

A simpler way of transferring tools from the magazine to the spindle is again incorporated in a British machine, the Vero series 2000 machining centre. On this machine, tools are transferred directly from the 20-tool rotary magazine to the spindle without an intermediate gripper arm.

This machine has a 46 × 24 in. table, and has machining capacity for drilling 2 in. in steel and metal removal rate of 6 in^3/min when straight-line milling.

The tool magazine is on the bottom face of the spindle head between the spindle and the column, as seen in Fig. 2.7. It is mounted on a traverse slide and also has an axial movement. During the automatic tool-change sequence, the magazine moves on its slide towards the spindle, engages collars on the tool-holder in the spindle, and then lowers to withdraw the tool. After rotating to bring the next tool into position, the magazine rises again to insert it into the spindle, and then retracts on its slide. The sequence is electro-hydraulically actuated and interlocked with the electro-hydraulic sequence of the machine. Tool change takes 7 s.

Figure 2.7 Mounted on a transverse slide behind the machine's spindle, the magazine of the Vero 2000 machining centre also has an axial movement to facilitate toolchange. (*Courtesy* Metalworking Production)

Tools are loaded into predetermined positions in the magazine, and are always returned to the same position.

An American automatic tool-changer with drilling machine ancestry is the Giddings and Lewis-Fraser 70a-NC-15Y Numericenter (Fig. 2.8).

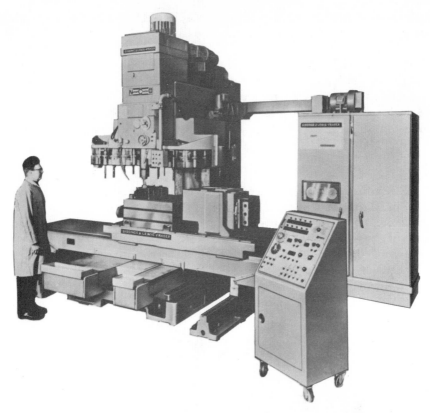

Figure 2.8   Two rotary tool magazines are positioned one on each side of the spindle on this Giddings & Lewis-Fraser NumeriCentre. (*Courtesy Giddings & Lewis-Fraser*)

Two rotary tool magazines are positioned, one on each side of the spindle, so that tool loading and other preparatory operations are easily carried out from the front of this machine. Each of the magazines has a separate hydraulic motor with both rapid traverse and creep feed for positioning the cutting tools under the spindle. Tool changing cycle is 5 s when exchanging tools from alternate magazines, and 8 s when exchanging tools from the same magazine, assuming greatest distance between tools. A resolver determines position and direction of rotation, the shortest path to bring the tool to the exchange position being always selected.

Numerical control is from a Numeripoint SS70 point-to-point and straight-line system, permitting zero offset to any point within the range of slide movements. Override switches provide for correction of programmed speeds and feeds to allow for material variations. Switches are provided to reverse the sense of the co-ordinate data to produce right- or left-hand components.

A sensing control allows rapid approach at 125 in./min until the tool contacts the work, whereupon feed is automatically engaged, and then machining to the programmed depth begins. This eliminates idle time and the need to have predetermined tool lengths.

The machine has a 38 × 90 in. table with traversing motions of 32 and 78 in. The table is mounted on recirculating rollers running on hardened inserts. Main drive is from a two-speed 15 hp motor. Drilling capacity is 3 in. diameter in cast iron.

A third generation

A completely new third generation of machining centres was introduced in 1967 by Molins Machine Co. Ltd, of South East London. In System 24, there are at present two types of twin-spindle machine. The first, an automatic tool-changing machine (Fig. 2.9) is intended for high-speed milling of light alloy. The spindles are driven by oil turbines giving speed ranges between 8,000 and 24,000 rev/min and developing up to 24 bhp.

The second type is intended primarily for hole formation—that is, boring, reaming, drilling, and tapping—with the capability of contour milling, should

Figure 2.9 The construction of this and other high-speed machines in Molins System 24 is based on a new form of combined automatic slide and hydraulic actuator. (*Courtesy Molins Machine Co. Ltd*)

this be desirable. On this machine the spindles are hydrostatically located, and driven by hydraulic motors through a two-speed gear unit. Speeds are changed automatically by programmed instructions on the magnetic control tape. The speed ranges are 0 to $\pm$ 1,800 rev/min, and 2,200 to 5,200 rev/min. The lower end of the lower range is intended mainly for tapping.

The construction of these twin-spindle machines and of the single-spindle six-axis machine is based on a new form of combined hydrostatic slide and hydraulic actuator and this forms the basic method of location and control of all three slides. The machines and their use are more fully described in chapter 16.

Lathes with automatic tool-changing

Turning machines with automatic tool-changing are also strong contenders for the machining centre concept.

Since its introduction, the NC turning machine has developed a variety of forms and capabilities more rapidly than any other NC product.

Vertical types, centre lathes, flat-bed turret machines, vertical and angled bed chuckers with one and two turrets, short-bed machines, and long cross-slide lathes have been closely followed by automatic tool-changing lathes and vertical machines with ancillary side-mounted milling/drilling heads.

A good example of this rapid advance in NC lathe capability is the British built 'turning centre' made by Alfred Herbert (Fig. 2.10).

In configuration it looks like a large twin-bed chucker with a vertically aligned automatic tool-changing magazine fixed at its right-hand end.

Figure 2.10 A good example of the rapid advance in NC lathe capability is this Herbert automatic toolchanging turning centre with milling capability. (*Courtesy Alfred Herbert Ltd*)

3*

But the similarity stops there; for tucked away behind the spindle nose and built snugly into the turning centre are two features which give the machine unique versatility.

First, on the rear slide is an auxiliary power drive of $1\frac{1}{2}$ hp to which rotating tools can be automatically coupled.

The second facility is a precise, tape-controlled spindle indexing system which will position and clamp the spindle at any programmed position in increments of one ten-thousandth of a revolution.

With these facilities, and attachments such as drill spindles, multi-drill heads, and right-angle heads, the machine can combine co-axial and cross drilling, tapping, flat milling, keyway slotting, and precision spline milling to polar co-ordinates within the normal turning and boring cycle of operations without disturbing the set-up.

The spindle control also provides the means of relating slide movements accurately to spindle rotation for thread chasing.

Both sides have tape-controlled (continuous path) movements in transverse and longitudinal axes, but control is shared so that only one slide is employed on cutting at a time. Normally, while one is cutting, the other slide is tool-changing. After tool-change, the inoperative slide parks close to the workpiece so that it is ready to begin the next machining operation as soon as the other completes its cutting cycle.

The tool-change cycle is initiated by the tape, but from there on is a function of the machine cycle and proceeds automatically while the control tape directs movements of the other slide and the spindle.

All tools other than rotating tools can be used on either slide and, in fact, any tool can be used on, say, the lower slide first, then subsequently transferred via the common 13-tool magazine into the vertical slide.

As an alternative to a tool, either slide can carry a work-support centre.

For tool-change, the slide concerned traverses to the rear of its slideway. The magazine rotates past a tool-code reader which then positions it, with the next required tool in line, near the withdrawn tool slide. A tool-transfer unit then simultaneously withdraws the new tool from the magazine and the old tool from the slide, rotates through 180° and exchanges the tool positions. This simultaneous transfer means, incidentally, that the machine can at any time have 15 tools available, two in the tool slides and 13 in the magazine.

Tools can be placed in the magazine in any sequence, since each tool-holder carries its own detachable recognition key. This is withdrawn with the tool-holder and remains with it throughout the time the tool is used on the machine.

Precise location and rigidity of the tool in its holder are ensured by a multi-tooth coupling, and pull-back into position is provided by a dovetail annulus connection. Clamping is assisted by a ball trap with wedging action.

The principle of mounting the tool magazine on the cross slide is followed by Max Müller on their MDW-10-NC lathe. In this case, the magazine is attached at the side of the slide farthest from the chuck. The magazine, seen in Fig. 2.11, is an indexing drum with nine radial compartments of nests for the tool blocks.

The tool blocks slide in and out of these nests on rollers into the tool carriage in a direction parallel to the spindle axis. Tool-changing time averages 6 s.

Figure 2.11   The tool magazine on the Max Müller MDW 10 is an indexing drum mounted on the cross slide (*Courtesy* Metalworking Production)

The effect of the weight of the magazine and the tool-holders on the steeply inclined slide is compensated by a hydraulic counterbalance.

This lathe is just one of many that Continental makers are fitting with automatic tool-changing.

Analysing lathe tools

The greater proportion of the NC turning machines available are short-bed chuckers. This design of machine was based on an analysis carried out by Professor Opitz and his co-workers at Aachen, by PERA, and by MTIRA in Britain into component statistics on normal commercial work. The analysis showed that over two-thirds of turned components are discs with the ratio of length to diameter less than unity.

Many of these chucking lathes have either vertical beds or beds inclined from the horizontal for rapid loading, easy tool change, and good swarf clearance.

But not all NC lathe applications have a requirement for automatic tool-changing. Neither do all makers agree on the need for it.

Many years experience in the tooling of plugboard machines and recent knowledge gained in the grouping of workpieces have shown that, as a rule, a limited range of turning and boring tools with the versatility they acquire within NC is sufficient for a large proportion of workpieces. Heyligenstaedt, for example, equips its Heynumat F chucker (Fig. 2.12) with a three-fold swivelling

Figure 2.12 A three-station turret with external cutting tools and a four-station turret with boring tools are said to meet most turning requirements on this Heynumat F. chucker. (*Courtesy Heyligenstaedt & Co.*)

holder with external tools and a four-fold swivelling holder with boring tools. Only because of wear on the cutter edges are the tips turned or the tools changed on this lathe. The cutting edges are accurately set outside the machine by means of a tool setting fixture; chips or swarf do not hinder the tool clamping and the tools remain in the swivelling holder.

The design of the standard tools to meet these requirements has also been carefully considered, and preferred sets of tools are recommended according to the nature of the work. For example, the Pittler Pinumat (Fig. 2.13) is said to be designed for disc-like components machined in batches of up to 50 from

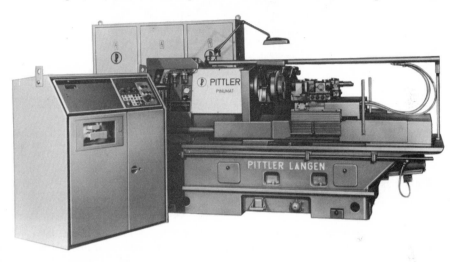

Figure 2.13 The Pittler Pinumat is said to be designed for disc-like components machined in batches of up to 50 off. (*Courtesy Pittler (Great Britain) Ltd*)

unformed blanks. The main feature of the machine is a long cross slide on which up to five tools can be located by means of precision dowels. Some of these tools consist of two disposable tips mounted on a boring bar, one tip for turning and one for facing. The tools can be used for machining of inner and outer surfaces by reversing the spindle rotation so that in effect each tool covers four functions. In this way, only two permanently clamped tools—one for roughing and one for finishing—are claimed to do the work of eight individual tools.

The USA provides the best examples of centre lathes and turret lathes with horizontal bed constructions. Typical is the Jones and Lamson machine (Fig. 2.14).

Lathes designed for shaft work are generally based on the design of copy lathes with vertical or inclined beds and tools mounted over the workpiece on vertical slides. The Sculfort NC lathe for shaft work is a vertical-bed construction with a vertical slide which carries a four-station turret.

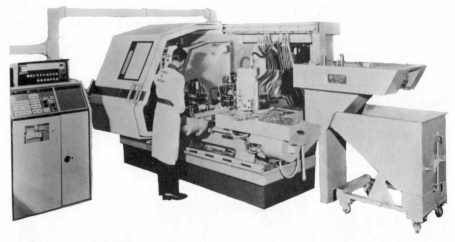

Figure 2.14 Typical of a horizontal bed turret lathe is this American Jones & Lamson NCTL machine. (*Courtesy Buck & Hickman Ltd*)

Some retain copy-lathe templates for contouring. Two such lathes are the Heyligenstaedt Heycomat 2 and the Ernault Somula S Pilote. The Ernault Somula has two slides: one on the front for rough turning under a point-to-point tape control; one at the rear for finish copy turning from the template to fine limits.

The latest trends, however, indicate that these compromises are disappearing and more and more makers are fitting full tape control for contouring work.

Modular construction

A philosophy that is having considerable impact on the development of machine tools is that of designing a family of machine tools which employ in varying forms many of the same sub-assemblies and in certain cases individual parts. This modular construction concept not only provides manufacturing economies but also allows a number of variants to be built to suit the requirements of the user. This design technique is illustrated in Fig. 2.15 which shows an NC lathe of the Churchill-Redman Red Century series. Other machines in the series include a NC bar-turning lathe, a NC shaft lathe, and a plugboard sequence-controlled chucking automatic.

Simpler machines more popular

Very versatile numerically controlled machines with automatic tool-changing are expensive and, as with turning machines, simple or less complex milling, drilling, and boring machines will prove more profitable for many users.

Figure 2.15 Modular construction enables different NC lathes in the Churchill-Redman Red Century range to be constructed from many of the same pieces. (*Courtesy Tube Investments Limited*)

These might be turret drilling and milling machines (using a limited number of fixed tools) or millers, borers, and drillers with single vertical or horizontal spindles using quick-change, preset tooling. These machines are fitted with various degrees of control, ranging from 2P through to 3C and, occasionally, 4C or 5C.

By far the greater proportion of machine tools sold in Europe and America is, and will continue to be, of this simple type. For example, drilling machines, such as shown in Fig. 2.16, hold 40 to 50 per cent by value of the market. In number, they account for around 75 per cent. Next come single-spindle milling machines, including die millers, with 2C or 3C–5C control; then horizontal borers, ram type borer-millers, and jig borers.

What price, then, the machining centre? Strangely enough it still remains odds-on favourite; partly because the makers of these simpler machines are choosing the term 'machining centre' for them.

Whether they are right to do so is a very debatable point. It all depends upon whether the term 'machining centre' is applied to a concept of manufacture, or to a particular type of numerically controlled machine. The tendency is to attempt to limit the term to machines with automatic tool-changing from a magazine. Yet when it is regarded as a concept of manufacture the spread of this concept, at whatever cost in vagueness, is in fact of benefit. It embodies within it the idea of putting NC to work in circumstances in which it is most likely to reduce the manufacturing costs for the small-batch producer.

The concept is being scaled down to suit the requirements of a greater number of users. Even now, a few machine-tool makers are offering not a single all-purpose machining centre but a 'spectrum' of them. What seems to be important is that, within the users' range of production, as much machining as possible is done in one set-up on each workpiece, using the simplest of tooling and with each machine cutting metal for as much as possible of its active time.

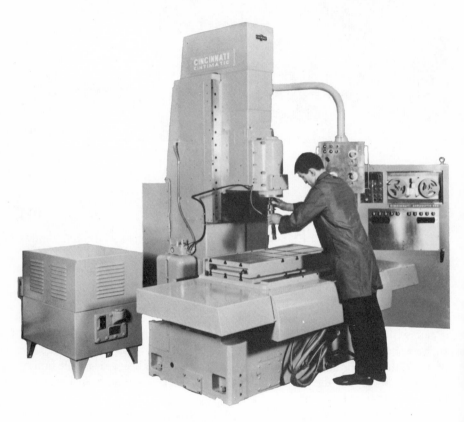

Figure 2.16   The drilling machine holds 40 to 50 per cent of the NC market. In numbers it accounts for around 75 per cent. (*Courtesy Cincinnati Milling Machine Co.*)

Advances in construction

The developments that have taken place in the field of NC machine tools have demanded the positioning of cutting tools and workpieces to a very high degree of precision. This has encouraged increasing use of hydrostatics or recirculating rollers on the slideways, and the use of recirculating ball nuts in the leadscrews. The machine shown in Fig. 2.17 has hydrostatic slideways and recirculating ball nuts on the leadscrews. Also, the structural elements of this machine were subject to model testing to obtain optimum stiffness—a practice that is now being widely adopted by British machine-tool makers.

Figure 2.17 This Cramic Superprofiler has hydrostatic slideways, recirculating ball leadscrews, and a structure designed from the application of model tests. (*Courtesy Cramic Engineering Company Ltd*)

NC linked lines

A growing number of users have proved the value of NC to their satisfaction and have invested in several machines. From these they have often obtained benefits which do not apply when a single machine is used.

These users have shown that, with NC, shift work can be organized more easily, maintenance can be tackled as a collective problem, and operators can be shared between machines. A single manager can be employed, who encourages designers to think in terms of NC, and builds up a team of 'part programmers'.

Some of these users employ a computer for the off-line preparation of control tapes, and to this end special programs and a family of general-purpose programs for international use are being developed (see chapters 7 to 10).

Those companies who have now moved from having a single NC machine on the shop floor to having several might wonder what the next step should be in the manner of using these machines to greatest effect. And probably the answer will be to turn the collection of individual machines into an NC 'machining complex' in which groups of machines, not interconnected, will be under the control of a single control unit (or even computer) which can share its time between the machines.

A number of computer control systems, known as Direct Numerical Control (or DNC) have been developed in the United States, such as General Electric's 'Data Controller', Sundstrand's 'Omnicontrol', and Bunker-Ramo's 'System 70'.

An outstanding example of on-line computer control is the linked NC production line known as System 24.

System 24 is a 'third-generation' series of NC machining centres (such as that shown in Fig. 2.9) with automatic tool-change, each of which is designed to perform a limited range of work at high efficiency. From these individual machines, Molins were building up a linked-line transfer system along which work, throughout a 24-hour day, would automatically progress and be machined under the control of a computer. The design principles of this system are described in greater detail in chapter 16.

Despite the fact that Molins have curtailed development of the full 7-machine computer controlled system, the system points towards future development of manufacturing systems.

Platens, each pre-loaded with a component during an 8-hour manually operated day-shift, will be stored automatically, and then selected by a computer-controlled device which will feed them for accurate location on appropriate machines. A 24-hour work load can be processed on a machine in any order. If one machine does not complete the machining, the platen is temporarily transferred back to storage.

If today is the age of the machining centre, tomorrow may be the age of the 'manufacturing system'. This 'manufacturing system' concept is perhaps the ultimate in versatile automatic NC for small-batch manufacture. It is, however, aimed at raising the efficiency of NC production on a very large variety of parts in small batches, and NC today is not only for the small-batch producer. It is probably projects such as this, however, which caught the interest of the medium-batch production industries and made them consider linking standard NC machines for their purpose.

The first evidence of this sort of thinking among the medium-batch producers came from Borg-Warner at Letchworth. This British branch of a US organization, of its own volition in 1968, linked NC machines by roller conveyor tracks to machine various designs of automatic transmission cases. This was the first stage in the company's projected advance towards an automatic 'variable part' machining concept for its lower-volume products.

This machining method has been developed by Borg-Warner in conjunction with Cincinnati Milling Machines Ltd, of Birmingham, who are developing the concept further in what is termed a 'variable mission' line.

In this projected development, a main conveyor loops round the machine line, with cross-conveyors at each work-station, as shown in Fig. 2.18. The cross-conveyors are long enough to permit a modest amount of queueing of workpieces. Workpieces travel along the conveyor on individual pallets, or fixtures. The base of a pallet is designed to serve the functions of transport and registration (exact location under the tool). Each pallet is given a series of addresses identifying the various work-stations to which its particular work-piece must move. This sort of link-line has the advantage that it can begin as a fairly simple affair, with the machines linked by roller conveyors along which workpieces on platens are pushed manually. Later, power conveyors can be added; the linear induction motor seems to be a very good prime mover for this purpose. Ultimately, there is the possibility of on-line computer control.

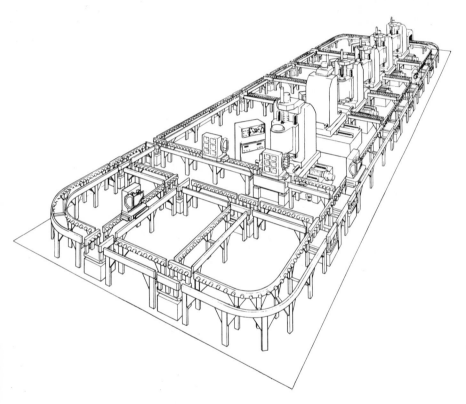

Figure 2.18   This Cincinnati 'Variable Mission' NC machining line is intended to meet the requirements of medium batch producers. Flexibility of NC is combined with higher throughput of multi-station concept. (*Courtesy Cincinnati Milling Machine Co.*)

In such NC machining lines, turret drilling or drilling-and-milling machines, or even machining centres and turning machines, will find new applications. Thus, greater accent will be placed on designing NC machine tools as modular units capable of being assembled to provide a variety of configurations. This thinking is well illustrated in the machines designed to fit into the Stavely Integrated Manufacturing System (SIMS), the first of which was built for Rolls-

Royce Ltd as a complete machining system for parts for the RB.211 airbus engine. This trend is also in evidence as shown in the moves, mentioned earlier, by a number of other machine tool makers to adopt unit construction to provide a range of NC machines with a variety of capabilities.

The author would like to record his appreciation to the McGraw-Hill Publishing Company Ltd for permission to draw freely from material published in the journal *Metalworking Production*.

3. Buying an NC machine tool

Alfred Tack

Alfred Tack, AMBIM, is Machine Tool and Equipment Manager, Aero Engine Division, Rolls-Royce Ltd. He was with the Kearney & Trecker organization for 26 years, for the last 7 years specializing entirely on NC. Soon after bringing the first Milwaukee-Matic Machining Centre to Europe in 1959 he realized the difficulties associated with obtaining industrial acceptance of a costly new manufacturing technology. He developed the discounted cash flow system to enable the production engineer to convert in a simple manner the times and distances of his profession into the percentages of after-tax return-on-investment understood by the accounting profession, and thereby lowered the acceptance barrier. His book—*Factors for the Appraisal of Machine Tool Economies*—was published on limited circulation by Kearney & Trecker. He is a founder member of the British Numerical Control Society, lectures frequently on the many and varied facets of NC, and has taken part in NC discussions on radio and television. Since joining Rolls-Royce Limited in 1966, his machine tool interests have widened to encompass the many manufacturing disciplines encountered in aero engine manufacture, including NC—which Rolls-Royce have developed until they are now the largest users outside the USA.

Buying an NC machine tool does not differ in any major respect from buying any other machine tool. If there are any differences at all, they arise from the higher capital cost of the machine and the lower experience of the user. The higher capital cost factor is well recognized in industry which is being besieged by new technologies. Whereas a few years ago the cost of labour was predominant, today that cost is fading into insignificance. For instance, a man may be employed for 2,000 hours per year at a direct cost to the company of, say, £0·625 (12s 6d) per hour. His cost will vary little whether he is operating equipment of high or low capital cost. Let us compare this situation with two machine tools, one of the last generation, say, a radial drilling machine costing £2,500, and one of the new generation, say, an NC turret drilling machine costing £20,000. If the depreciation policy of the company is that machines should be straight-line depreciated at 10 per cent per annum, the annual cost of the radial drill would be £250 and the NC turret drill £2,000. Based on our 2,000-hour year, the hourly cost of the radial drill would be £0·125 per hour (2s 6d) whereas the turret drill would cost £1 per hour. It is easy to see that the operator cost of the radial drill is highly significant. However, the NC turret drill carries a depreciation charge considerably in excess of operator cost.

The hourly depreciation charge quoted for an NC turret drill is low when compared with that of a machining centre, for instance, which may be £100,000 in capital cost.

This equals £10,000 per annum depreciation, or £5 per hour based on 2,000 hours per annum utilization. The operator cost, under these conditions, has become almost negligible with the result that industry must reverse its previous aim of high labour utilization at the expense of low machine utilization. It must now generate systems of internal control which are based on high machine activity with labour accepting the role of a machine-service organization. Operator activity, in the future, will be based on anticipation of machine demand and the satisfaction of that demand to ensure that utilization is maintained at a maximum. It is also true to say that the whole manufacturing organization must be geared to satisfy the demands of the new technology equipment.

Expect teething trouble

The point was previously made that the main difference between buying conventional compared with NC machines was in the area of high capital cost and lower experience of the user. It is the latter which creates the greatest number of problems. In many cases, the potential user hopes that the difficulties of integrating NC into a conventional operating system will not arise, despite having read and having been told the opposite. Myopia, when purchasing NC, is the short route to failure. Every buyer of NC equipment should realize that he is not simply buying another machine tool; he is committing himself to an internal reorganization without which failure will inevitably result.

How, then, can a buyer ensure that his purchase is a viable proposition? He has probably been told that the purchase of NC will overcome all his labour problems, and that floor-to-floor times and tooling costs on jigs and fixtures will reduce to a negligible figure.

Such comments can only be generalizations and must relate to the specific problems of the purchaser. How many potential purchasers of NC have fully evaluated their own problem areas to assure themselves that NC will eliminate their problems? It is far more likely that they have either decided that, like many other new-fangled gadgets, it will die a death or, alternatively, they have been impressed by the fantastic results claimed and have clutched like drowning men at this straw, without being sure where it will take them.

Reasons for purchase

Emotive decisions are unhealthy in any business and, notwithstanding the many successes achieved by NC users, each buyer must institute problem-analysis procedures which will help him to make the decision 'to buy or not to buy'.

A production problem

Corporate policy has decided that each manufacturing organization will produce saleable goods as a means of creating an excess of income over expenditure. It may have further decided that domestic 'in-house' manufacturing is the procedure which will be implemented as a means of assuring that policy. At that stage, corporate policy will not normally enter more deeply into the manufacturing process. Subject to sales forecasts, it is the function of the manufacturing sector of an organization to generate manufacturing procedures which will ensure that the right goods are produced in the right quantity, of an adequate quality, at the right time, and at the right price. The demand on the machine tools—NC or otherwise—is simply fitness for these purposes. Machine tools should not be endowed with any attributes other than those mentioned.

It must be assumed at this point that a problem exists in the manufacturing process, but the cause of the problem need not necessarily be known. Our first objective therefore must be to define this problem as precisely as possible. A useful tool for this purpose would be a check-list of problem areas in the manufacturing sector which could, with some thought, embrace the basic elements of any manufacturing problem. Such a check-list might have the following form:

Is your problem,
Human?
Labour—quantity available
Labour skills—too little, too much or unsuitable
Personalities
Communications
Organization

Quantity?
Per unit time—high or low output
Per batch

Cost?
Planning
Tooling
Labour—direct or indirect
Supervision
Inventory

Quality control
Scrap and/or salvage
Production control
Consumables
Services

Quality?
Accuracy
Surface finish

Technical?
Complexity
Need to produce

The check-list permits the problem to be broken down into its basic elements. By thorough examination of these elements, possible causes can be developed, tested, and verified.

Example

For example, let us assume we have a quality problem: our output is too low, our cost of quality is too high, and our cost of scrap and salvage is too high. On examination, these three elements all have the same cause—the accuracy of our produced component is insufficient. We then ask ourselves, 'Is this a deficiency in the machine tool or in the operator?' Test results demonstrate that both are at fault; the machine tool is worn and out of date and the operator is not sufficiently skilled to compensate for it. Unfortunately, there is a shortage of skilled operators and in any case the component is too complex for even a skilled operator on this machine tool.

Having defined our problem and determined its cause we have three possible courses of action:

Interim action Interim action is usually taken as soon as any suggestion of a problem becomes apparent and probably before the cause has been determined. It can only be regarded as a stop-gap measure, at best; alone it will not solve the problem and could involve wasteful expense. In our hypothetical case, an example of interim action would be the increase of output from the machining area, by the transfer of more operators and machines to the job. This would also involve the provision of more inspectors to maintain the outgoing quality required at the increased level of output. No attempt has yet been made to determine the cause of the problem and to solve it.

Adaptive action Adaptive action would be taken when the cause of the problem is known but corrective measures are not available or are considered to be unfeasible. The object would then be to minimize the effect of the problem without actually eliminating it. In our example, one course of adaptive action would be to market the components under a quality concession at a reduced price. Such action may be viable but only as a short-term measure.

Corrective action Corrective action removes the cause of the problem and is usually the most efficient 'solution'. As often happens, corrective action can follow interim action, when the cause has been determined. It will normally involve more far-reaching consequences but need not necessarily involve large capital outlay.

In our case-study, the solution involving corrective action could be to introduce NC machines which would in effect off-load the skill required of the operators by provision of sophistication in the machining capability. NC might not be the only answer but it may be the most efficient.

Schematically we have arrived at our course of action from recognition of a problem as shown in Fig. 3.1.

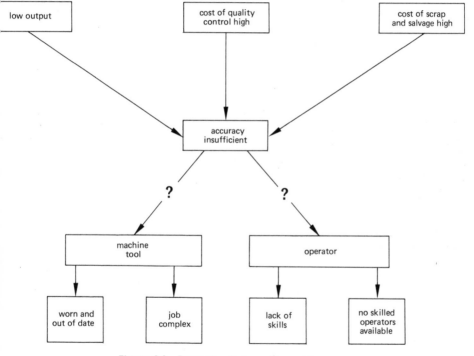

Figure 3.1   Recognition that a problem exists.

Choosing the best solution

Having now decided that corrective action is the way we must go, it is important to establish our objectives as precisely as possible. What are we striving to achieve, both quantitively and qualitively, and what resources do we have available?

Frequently, upon involvement in a particular course of action, objectives may be forgotten in the urgency or complexity of a solution and the end result is far removed from the original concept.

One technique of decision analysis would be performed as follows:

> list the objectives;
> classify the objectives ('musts' and 'wants');
> weight the wants;
> generate the alternatives;
> compare the alternatives according to fulfilment of the 'musts' and fulfilment of the 'wants';
> compute the weighted score (positive) from the 'wants';
> forecast the adverse consequences;
> weight and compare the adverse consequences;
> compute the weighted score (negative) from the adverse consequences;
> make the decision.

Each step of the above procedure will now be examined more closely:

List of objectives   A detailed list must be made of what we are striving to achieve. In other words, we must state the general direction of the course of action we propose to take to alleviate our previously defined problem.

Classify the objectives   Objectives should be divided into two classes: those that we must achieve and those we would like to achieve for an optimum solution. These are termed our 'musts' and our 'wants' respectively. This classification is more easily said than done and no rules can be laid down for it. Consequently, each situation must be tackled on its merits.

Weight the 'wants'   The importance, in terms of achievement of the end result, of each 'want' will not be the same for all our 'wants' and a measure of its importance can be conveniently used as a weighting factor. An arbitrary scale is used, say 0 to 10, against which each 'want' is rated according to its importance. The exact nature of the use of this system will be seen later.

Generate the alternatives   A simple list of all the alternative courses of action should be made with no thought, as yet, as to how each one will measure up to the job. This prevents the possibility of prejudgement which so often clouds the issue and can lead to bad decisions.

Compare the alternatives   Having been impartially listed, the alternatives must now be compared for their potential achievement of our required result. Any one that does not fulfil a 'must' is immediately rejected and no further consideration need be given to it. In the second stage of selection, the choice depends on the relative fitness of each alternative for our 'wants'. This is quantified by the assignment of a score for each alternative according to its comparitive fulfilment of each 'want', again using an arbitrary scale, say 0 to 10 for convenience.

Compute the weighted score (positive) from the 'wants' We now have weighted 'wants', and 'scores' for each alternative against these 'wants', and it is a simple matter to multiply the two together for each alternative to arrive at a series of weighted scores. It would seem most beneficial to choose the alternative with the highest weighted score at this point but this must be tempered by the weight of adverse consequences which would result from this choice.

Forecast the adverse consequences Practically every alternative that can be generated as a course of action to alleviate a problem raises some problem either in its initiation or in its long-term effect. These 'adverse consequences' must be forecasted and listed carefully in order that they may be thoroughly examined.

Weight and compare the adverse consequences An exactly similar technique can be used to assess the performance of each alternative with respect to the adverse consequences as was used in the assessment of achievement of the 'wants'; it is simply a matter of substituting 'degrees of seriousness' and 'probabilities of occurrence' for the fulfilment ratings of the alternatives and the weighting of the 'wants'. It must be noted that the ranges of the arbitrary scales used in this part of the decision analysis must correspond with those used earlier in the assessment of fulfilment of the 'wants'. Otherwise the two factors in the analysis would not be compatible and no direct result from the comparisons could be computed.

Compute the weighted score (negative) from the adverse consequences We now have weighted 'degrees of seriousness' and 'probability ratings' of adverse consequences generated by each alternative and again a weighted score can be derived by the multiplication of each factor for each alternative.

Make the decision In the final analysis, the weighted score of each alternative in its adverse consequences is subtracted from the weighted score of each alternative in its fulfilment of the objectives and the highest score indicates the best choice.

An example of a choice

Let us assume that, to solve our quality problem raised earlier, we are considering the following alternatives:

NC lathes;
plugboard controlled lathes;
conventional centre lathes.

Further, we will assume that only one type of lathe exists for each alternative that is technically capable of machining our complex component.

A simple list of our 'musts' might be:

Musts	NC	Plugboard	Conventional centre lathe
Capability to machine complex component (this would normally list numerous technical features)	yes	yes	yes
No skilled operators required	yes	yes	no
Must be accurate (usually specified)	yes	yes	yes

A simple list of 'wants' might be:

Desired objective	Weight	NC		Plugboard	
		score	weighted score	score	weighted score
Low cost per hour	4	4	16	6	24
Quick tool-change	7	8	56	8	56
Increased output	10	10	100	6	60
Totals			172		140

A simple list of adverse consequences might be:

Adverse consequence	Penalty	NC		Plugboard	
		probability	weighted score	probability	weighted score
More maintenance required	3	7	21	5	15
Support services required, e.g., programming, tape production, and storage	5	8	40	5	25
Strict shop-floor discipline	2	5	10	5	10
Totals			71		50

Total plugboard score = 140 − 50 = 90
Total NC score   = 172 − 71 = 101

From the list of 'must' objectives it can be seen that the conventional centre lathe alternative would be rejected because it would require skilled operators. The weighted scores of plugboard versus NC lathes come out in favour of NC lathes despite the extra weight of adverse consequences attributed to the NC alternative. So, in our case, the course of action should be to purchase NC lathes.

This is by no means the end of our problem and in reality can only be regarded as the starting point for our next and more serious problem. We must ensure the success of our activity by anticipating problems and their causes and taking preventive steps before they even arise. This is particularly important when introducing NC into a manufacturing organization which has no previous experience in this field. It is also more difficult to anticipate the problems without the experience and 'know-how' of established users, but much can be learned by others' mistakes.

Selecting a control system

The selection of a control system from the many now available on the market can be a complex problem. However, there are three prime considerations which should be evaluated. They are, in order of importance:

suitability for purpose;
reliability;
serviceability and availability of service and spares.

Each of the three factors will be separately considered and elaborated as follows.

Suitability for purpose

This is obviously a prime consideration which cannot be subjugated. The control system must, of necessity, be capable of satisfying the technical needs of production.

If the requirement should be for an NC system capable of generating complex 3C contours, it would be pointless to purchase a system limited, in capability, to simple-2P point-to-point positioning. Such a purchasing error is highly improbable; however, machine tool/control combinations ill-matched to the needs of production have often been purchased.

Control systems (like machines) are produced in many varieties. Therefore, carry out a detailed analysis of production requirements and state the 'musts' and 'wants' and consider the adverse consequences before evaluating any control system. It is far better that this is carried out before control system specifications are examined otherwise there will be a tendency to state the 'musts' and 'wants' in a manner which will finally 'prove' that a previously held opinion was correct: do not prejudge the answer.

Besides technical suitability for purpose, cost must also be given consideration. Economics may show that an 'all singing, all dancing system' is justified. The DCF return* may be adequate and the payback period acceptable, but a

* See chapter 4.

lesser system which is still technically suited to the purpose may show a better return. It may be that, by the purchase of an over-sophisticated system, company resources will be over-stretched and result in approval-to-purchase being withheld. Suitability-for-purpose must bear in mind company resources as well as technical requirement.

Chapter 6 deals with the standards which have been developed to help the user to specify the most suitable machine tool and control system for his purpose. It should be remembered that many of the simple features of control systems are optional extras. They must be specifically called for, and they are extra to the basic cost of the machine tool. Thus, a user may be surprised to find the control tape being spewed on the floor because he should have specified that tape-reeling equipment was required.

Control systems are offered as either point-to-point or continuous path. The former implies that it will direct the machine moves to a pre-programmed point but with no control over the path taken between initial and end point. Under the latter, control of two or more axes will be directed and constrained to a previously defined path. Between these two extremes are offered systems which have a limited contouring capability insofar as they will combine the motions of machine axes so as to generate a straight-line path which is not parallel to them. This form of control is normally limited to a motion in two axes. A slightly more sophisticated version will allow the generation of circular arcs in a similar manner.

Within the context of suitability-for-purpose, certain features of controls which are affected by the mystique and misrepresentation which has surrounded NC almost from its inception, are considered below.

Analogue versus digital It can be said that few, if any, control systems are either completely analogue or completely digital. Almost without exception, input of data is in a digital form. Output to machines servos, except on the cheaper and less sophisticated 'bang-bang' systems and the recently applied pulse-motor systems, is analogue. Feedback in closed-loop systems may be of either variety. The arguments which have swayed back and forth around the digital versus analogue controversy are almost completely irrelevant.

Punched tape versus magnetic tape Where point-to-point NC controllers are being considered, there is no case for arguing in favour of magnetic tape. Programming and preparation of punched tapes for such systems is simple and cheap, the volume of data to be stored is quite small, and modifications to programs are simply made on office equipment. In the case of continuous-path controllers, the arguments are less clear-cut. 2C contours of a simple nature such as straight lines, circular arcs, and, possibly, parabolas can be generated within the controller by function generators (interpolating units). The input from punched tape is usually simple and adequate under these conditions.

However, high-order curves in two axes and low-order curves in more than two axes are beyond the capability of normal function generators and, with paper-tape control, it is necessary to resort to linear interpolation where a series of tangents to the curve are generated to produce an approximation to the curve.

The consumption of paper tape under these conditions is often inordinately high, particularly when a large set of short straight lines are needed for accuracy. To supply data at adequately short intervals makes high-speed, expensive photo-electric readers essential. Data rates then become so high that buffer data stores are required in the controller. Magnetic tape controllers normally come in two guises: either the data presentation is similar to that of a paper-tape system but packed with a much higher density or it carries even more detailed path information and eliminates the requirement for an interpolator in the NC controller. In the latter case, the controller is very much simpler (and, therefore, should be cheaper), but this system is not without its disadvantages. Without internal computing facilities, there is no possibility of overriding programmed data or putting in corrections such as cutter diameter compensation or changes of feedrate. Such a controller will normally require off-line computing facilities which are more comprehensive than those found in a normal general purpose computer, and its post-processing costs in the general purpose computer will also be greater. In its favour it must be said that the greater ability of off-line computing and the high data density on magnetic tapes will result in smoother curves without the objectionable appearance of the short straight-line facets which normally result from on-line linear interpolators. The choice between magnetic and punched tape must once more depend upon suitability for purpose.

Linear versus rotary feedback transducers Linear transducers have been produced in many forms but two have stood the test of time and found wide acceptance.

They are the Inductosyn, which is a trade name of the Farrand Corporation, and the Diffraction Grating which was largely developed by Ferranti Limited and is particularly well known in the UK. The accuracy with which they are both produced and their ability to resolve extremely small increments of measurement has made them pre-eminent in the high-accuracy field. Similarly, although many varieties have been produced, only two rotary transducers have stood the test of time: they are the rotary digitiser and the synchro. The latter is rapidly establishing itself as the most widely used system of feedback for NC systems.

The reader may well ask which should be chosen. Rotary devices are normally used for indirect measurement by attachment to the ends of leadscrews. By virtue of their location they are subject to the inaccuracies of screws, screw compression, backlash, and wind-up. Screws, in some instances, suffer from elongation due to elevated temperatures. In each of the foregoing, errors of measurement may occur where the measured distance differs from the actual slide displacement. When linear transducers are used, the errors of transmission are eliminated: the actual slide displacement is measured, so giving greater accuracy of positioning. That is not necessarily the end of the story. Machine tools suffer, in varying degrees, from pitch, roll, and yaw. The clear intention is that the measuring transducer should measure the relative movements of workpiece and tool. In fact, the measurement is indirect and only simulates that relative movement.

If the above conditions occur or the machine suffers from temperature gradients causing differential expansion, the resulting measurement may differ substantially from that indicated at the measuring point. For the highest accuracy, linearity of machine motions and temperature stability are 'musts'.

A common defect of machine tools developed from manually operated machines is a gradual change in the effective Z position of the spindle nose due to a gradual change in temperature as it runs.

Open-loop versus closed-loop systems   The differences between these two systems are very real. The first is similar to shooting on an outdoor range where the marksman has to take into account muzzle velocity, cross winds, and the 'pull' of the rifle. Once the trigger has been pulled the marksman has lost control. The second is analogous to a space rocket which, although accurately sighted, has its position, trajectory, and velocity continuously monitored and corrected as necessary. In NC terms, the open-loop system has no feedback and the closed-loop system utilizes feedback transducers such as the synchro or diffraction grating previously mentioned. The closed-loop system, although more expensive, is highly desirable. The digital hydraulic pulse motor may reduce the need for closed-loop systems in the future and substantially reduce the cost of controllers. Many controllers, although essentially closed-loop, do not rely on a 'live servo' to hold a pre-programmed position. This is particularly so on point-to-point machines where the X and Y are often positioned and then clamped. The position feedback loop then monitors any deviation from programmed position caused by, say, cutting forces, but will not cause a correction to be made.

Reliability

NC machine tools have, in the past, been notoriously unreliable. It has been automatically assumed by the user that this is an electronic fault. This contention is far from the truth, although, undoubtedly some controllers have been unreliable. From the user's viewpoint, a conventional machine which has given superlative service for many years cannot be at fault when it finally loses its handwheels and is married to an NC controller. However, designs of machines made for conventional operation often lack the stiffness which is so essential in NC machine tools. The feed-drive elements such as leadscrews, nuts, and thrust brackets are all high-stiffness springs and too low a stiffness can cause servo-loop instability. The ratio of stiction to friction of slideways which, on manually operated machines appeared to present few problems, exhibits undesirable characteristics of 'stick-slip' which are difficult to control by NC. Inertias of the elements of the servo-loop if not suitably designed may also create control problems. Any or all of these problems are caused by design which is, to a certain extent, tolerable when the machine is under manual control but which is inadequate for NC. Manufacturers of NC controllers have, because of economic pressures, attempted to control machines which exhibit the above undesirable characteristics with, on many occasions, disastrous results. This problem is still with us and the wise buyer will carefully investigate the machine's design or obtain expert independent opinion before placing an order.

Vacuum tubes, relays, transistors, or integrated circuits. Which is the best? Is each reliable? If not, which is? Crimped terminals, soldered terminals, or wire-wrapped terminals. Are they all reliable? If not, which should be chosen? To each of these questions and the many others which spring to mind there is no categoric answer. There are horses for courses and, to further the horse-racing analogy, the would-be user should follow a good, proven stable. The electronic world long ago mastered the art of producing new gimmicks at a very much faster rate than they can be absorbed by the user. Many of these gimmicks have proved to be of great benefit and have advanced the art considerably. However, no manufacturer of repute will risk his reputation by selling unproven gimmicks. Follow the money and back the favourite unless you have competent electronic advisors who can assure you of success when you back an outsider.

Serviceability, and availability of service and spares

No one will suggest that machines and controllers will never fail. The 'perfect' machine has not yet been made. Serviceability is, therefore, of critical importance. Unitized machine design is a desirable feature which should be searched for diligently. The ease of servicing a discrete unit, such as an easily removable quill assembly or gearbox, is worth a great deal in expensive machine downtime. Controllers should, at the very minimum, have ample jack socket test points. Failures should be corrected by the removal of plug-in modules such as printed-circuit boards and replacement from stock. These modules may be serviced by the user or returned to the manufacturer. Certainly, machine downtime should be limited to module fault-finding and replacement. The need for repair or replacement of individual components at the machine should be strictly limited. An American practice, which should be more widely used by European manu-facturers, is the provision of a simple step-by-step fault-finding chart which may turn a useful electrician into a competent electronic serviceman. Full advantage should be taken of the maintenance training courses supplied by controller manufacturers. The user may incur travelling, hotel, and course costs, but they are cheap when compared with the cost penalty of on-the-job machine failure. At least the minimum recommended spares should be stocked. Spares and service which the supplier tells the user will be on a 24-hour availability basis rarely are in practice. The cost in worn and frayed tempers is only exceeded by the cost of lost production from a vital and expensive machine. Remember that machine failures always appear to occur at the most inopportune times and, because of the higher hourly cost of NC machines, downtime is much more expensive.

Preparing for NC

Before even ordering an NC machine tool, it is essential to understand the potential, the implications, the economics, and the problems of NC. It is too late when the machine tool is sitting on the shop floor like a 'white elephant' depreciating at anything between £1 and £10 per hour. The largest part of the learner period should be prior to ordering and during the delivery period.

However skilled the operators, planners, and maintenance staff of a manufacturing organization are for conventional machining, nothing is more certain than that NC will demand re-training of all the associated personnel. This includes tool-design staff, manufacturing and production scheduling staff, shop and other management. The entire approach to machining, tooling, scheduling, and management is very different with NC from the conventional process which it is replacing.

Installation

To examine some of these problems in more detail, we must firstly consider where and how the machine is to be located. It is too easy to say that, because we are replacing four conventional machines with one NC machine, we shall simply put it on the old site. Although it may seem to be too far-sighted to consider that several NC machines may be linked in later life, this may well happen and it would be short-sighted to discount the future completely. With one NC machine in a shop, it is extremely likely that more will follow. Thought should be given to their location from the point of view of work transfer between these machines, as well as to and from them.

The foundations of an NC machine are usually more important than those of a conventional machine. Apart from the additional rigidity which they are often required to provide, they must damp out vibrations and resonances from hydraulic or electric motors. Control cabinets and measuring devices must never be subject to vibration and adequate foundations prevent this. Any power source should strictly be kept as far from the main construction of the machine as possible. This will help to avoid problems associated with temperature stability of the control system, which are particularly prevalent with hydraulic servos.

Similarly, a temperature-controlled atmosphere is advisable particularly when accurately machining light alloys such as magnesium alloys.

The problem here is one of relative expansion causing large measurement errors; magnesium has an exceptionally high coefficient of expansion and with rise in temperature, a magnesium workpiece would expand much more than the steel machine members. The implications of this are obvious and the best solution is to program the component for machining at, say, 20°C and to provide a temperature-controlled environment for the machining area, set at 20°C.

Although this is often impractical from an economic point of view, with careful attention to details, improvements can be effected in this respect. For example, it is a simple and cheap matter to ensure that an expensive and perhaps 'fussy' NC machine need not be subject to a cold blast of air from an open door or window on one side while the other side faces a warm environment.

From the standpoint of the control cabinet, vibration is not good for electronic circuitry and a dust-free atmosphere can reduce electronic breakdowns. As mentioned previously, adequate foundations should prevent the transmission of vibrations from the power source or the machine tool to the control cabinet. A dust-free atmosphere can be more difficult to provide, however, particularly when cast iron is being machined. Consideration must then be given to such requirements as pressurized cabinets, sealed openings, air circulation fans, and de-humidifiers.

Loading the machine

Assuming for one moment that our NC machine has been installed on the shop floor and is now capable of efficient operation, let us consider some of the reasons why it may be idle (which is a sin in itself). It is better, however, that the following points be considered *before* the machine is delivered. They will then not arise as problems, teething or otherwise.

By far the most regular and apparently justifiable reason is that the machine is not cutting because another component is being loaded onto it and set prior to the machining operation. With the high cost per hour and reduced machining times resulting from the use of NC, it is vitally important that careful attention be paid to loading and setting times. Every detail of the loading and setting operation must be examined so as to achieve efficiency and a minimum period of machine downtime. Accurate and quick location of the workpiece is essential; perhaps a small design change would achieve this end. If a datum must be established, this process must also be critically examined again with the view to reducing downtime.

It is possible that, despite streamlining of the loading and setting procedure, its contribution to machine downtime may still be far too lengthy. The answer in that case could be palletized loading. The additional expense incurred at an early stage for palletized loading could be recovered many times over in the saving of downtime and, incidentally, the operator will be occupied actually 'doing something' while the machine is cutting metal.

On the other hand, semi-automatic loading by mechanical-handling equipment may be the solution. This would be true of bulky workpieces that are easily and accurately located, with relatively short machining times in comparison with loading times; NC chucking lathes are obvious candidates for this type of loading equipment.

Work flow

By one means or another, we now have the facility to load and set workpieces on our NC machine in a speedy and efficient manner. But where are these workpieces coming from? If an NC machine has been purchased to increase ouput, as it most certainly will if properly utilized, its input must also be increased accordingly. More raw material is required and roughing capacity (if roughing operations are necessary) must be increased. Without attention to the input of work, it is likely that the machine will quickly devour the existing work and crave for more.

Similarly, because the machine can usually turn out components faster than conventional equipment, an inventory problem could be created with the pile-up of finished or semi-finished parts.

It is obvious that the introduction of NC necessitates careful examination of the whole process of 'throughput' of work from the raw material input stage to the output of the finished goods.

Planning

Having provided the means of efficiently loading the machine, from the standpoints both of handling and of the continued availability of work, the final requirement is for adequate programming support services. Nothing is more frustrating and money-wasting than having the machine 'ready to go' and the workpiece set up, only to find that the necessary control tape has not been prepared. There is no excuse for this inefficiency.

When the decision was taken to introduce NC, a planning organization for NC should have been built up, and the following points considered.

Part-programming facilities   (a) Manual programming: either a body of part programmers should be set up and trained, or the work should be sent out to subcontract. (b) Computer programming: the user should provide his own computer facility if it does not already exist, or he should 'buy time' from a bureau service (see chapter 12).

Post processors   Post-processor programs must be made available to suit the part programming language, the computer facility available (e.g., IBM, S360/50), and the particular NC machine tool. Most machine tool builders offer post processors at various prices, the alternatives being to write the program oneself or to sub-contract it to a programming organization (see chapter 11). It is usually best to buy the post processor with the machine tool.

Provision of tape and tape-punching equipment The type of tape required (magnetic or 'paper') will depend on the selection of the control system for the machine tool. There are many makes and grades of paper tape on the market and the user must decide which tape best fits his needs. If computer programming is used, there is normally either an on-line punch or an off-line card-to-tape converter. If neither facility is available, the computer printer output must be manually transcribed onto tape using a Flexowriter or the like, but this is both inefficient and prone to errors. A hand punch can prove to be useful for the correction or alteration of tapes after the proving run, and a tape-reproducing-and-verifying unit may well be a sound investment.

Likewise, a tape-store or library can prove to be extremely useful when numerous control tape and/or machine tools have to be catered for. It is as well to ensure that the tape-store for master tapes is fireproof as an insurance against loss of the masters which could be expensive and inconvenient to say the least. An effort must be made to use the most up-to-date version of a tape when taking it from store.

The computing of data for magnetic-tape controllers is often offered as a service by the control manufacturers. Ferranti Limited is one such company in Britain.

Acceptance tests

Having performed all the aforementioned preparations satisfactorily, we can now look forward to the installation of our new NC machine tool. It is not, however, simply a matter of having it delivered on a particular date, setting it on its specially prepared foundations and connecting everything up. Before the machine leaves the manufacturers, it must pass a series of tests. These tests can be broken down into three groups as follows:

Static alignment checks The exact nature of such tests must be agreed and stated in writing prior to ordering. The object of static alignment checks is obvious, but the procedure varies from machine to machine. Some guidance on their nature can be gained from what are known as the 'Schlesinger' tests. Series of tests have been devised to suit each basic type of metal-cutting machine (e.g., centre lathe, vertical borer, or knee-type milling machine) which exhaustively and conclusively test the machine for its 'correct' alignment.

Static alignment checks should be performed at the manufacturer's works by, or in the presence of, engineers from the user's company and are carried out strictly in accordance with the previously agreed and stated procedure. Should the machine tool not meet the requirement of any one of the tests, it should not be accepted by the user until such time as the manufacturer has corrected the fault and the machine has been witnessed as having passed the whole series of tests.

Dynamic tests　Again, the nature of any dynamic tests to be performed on the machine tool must be agreed by both the manufacturer and the user. The purpose of such tests is to verify that the performance of the machine/control system meets the criteria quoted by the manufacturer or specified by the user. This would include such points as:

> guaranteed maximum acceleration;
> maximum traverse rates and traverse limits;
> resonant frequencies;
> fail-safe systems;
> repeatability;
> accuracy.

It is advantageous to have one's own electronic engineer perform or witness these tests, preferably the former. But, failing that, an independent party or at least a representative from the users' company should be present.

Here again, the user should be satisfied by every aspect of the tests before he accepts the machine's performance as satisfactory.

Practical cutting tests　As the name implies, these tests are actual cutting tests performed on either special testpieces or sample components supplied by the user. Their purpose is to ensure that the machine will produce parts as it was designed to do, to the required accuracy in the expected time. The passing of cutting tests is not as clearly defined as other types of test and the user must decide whether the machine tool is finally fit for its intended task.

In fairness to the machine tool builder and to his other customers, delays on the builder's test floor should be avoided by the user arranging to have adequate supplies of components available for test by the required date.

When all the above tests have been satisfactorily completed, the NC machine tool is ready for shipment and subsequent installation at the user's works.

Commissioning

Once the machine has been set on its foundations and everything has been connected up, either by the builder's or agent's engineers or under their supervision, the aforementioned alignment, dynamic, and practical cutting test should be repeated. Only when the user is perfectly satisfied that the machine tool is operating in an efficient and satisfactory manner should he accept that he has purchased an NC machine tool.

If he has purchased a prototype machine tool, he should allocate at least year to commissioning.

Maintenance   With the machine tool installed and running, we must not forget, however, that NC machines are just as prone to breakdown as conventional machines, if not more so because of their increased complexity and their use of electronic circuitry. Although the breakdown of a conventional machine may be annoying, the situation is far worse when an NC machine malfunctions. As mentioned earlier, with a running cost of anything between £1 and £10 per hour, it is very expensive to have an NC machine sitting on the shop floor doing no productive work. It is for this reason that maintenance, both electrical and mechanical, and the spares situation are more critical factors when using NC. This is obviously a valuable service particularly when employed in addition to routine maintenance and trouble-shooting procedures.

An important point to remember in connection with NC maintenance is that NC machines only usually have one full year's guarantee, so it is desirable to be self-sufficient as regards maintenance after this year.

Tooling

A consideration that could easily be overlooked when purchasing NC machine tools is that of providing the necessary tools. The procedure and its problems may appear at first glance to be identical to that associated with conventional machine tools but this is not so. Preset tooling is a must with NC, which relies on a program in which all the parameters have been pre-specified. For instance, on a milling, drilling, or boring machine with more than two controlled axes, the part programmer must assume a tool length to allow him to specify Z axis movements and cater for collision conditions. Tool diameter, in these cases, must also be decided at the programming stage and it should not be left to the whim of an operator. There are numerous NC controllers which have provisions for tool length and diameter compensation; however, these facilities should only be used (if at all) with circumspection. 'Cut and try', which tool compensation implies, is a precursor of scrap and relies on operator judgement which, in the past, has proved faulty. Furthermore, the implications of the 'cut and try' technique are that much of the lost production time (due to setting, which we have been trying to avoid on our NC machines) will be returned to the manu-facturing cycle. The NC machine is an expensive device to use for setting tools. It is not economic sense to use a machine costing £20,000 or more for tool setting when this process can be carried out more effectively and cheaply on an 'off-line' tool presetter costing £2,000 or less.

If the user does not have his own electronic engineers on hand, he must enquire of either the machine tool builder or the agent just what services can be offered. When neither of these alternatives is satisfactory, the user must either create his own electronic maintenance department or ensure that some external organization can provide the service he requires. Large users of NC can afford to have their own electronic maintenance organization and, in fact, cannot really afford to be without one. The accent must always be on a quick turn-round time when considering maintenance of NC machines, because the breakdown time must always be kept to a minimum.

Spares It is a wise policy to have as wide a stock of spares for an NC machine tool as possible. In most cases, a set of spares is included in the total price of an NC machine tool, but it is also usually rather inadequate. If, however, a suite of NC machines is installed by one user or he standardizes on one or more particular type of machine, the justification for a large holding of spares is much easier. Remember that stocks of spares for NC machine tools must always include mechanical, electrical, and electronic parts for the machine tool itself and the control system.

Preventive maintenance is receiving much attention these days and rightly so. It is obviously far better to prevent a breakdown, even if it does consume a little of the machine's productive time, than to wait until a major failure results. Large suppliers of NC machine tools are now offering a preventive maintenance service for certain types of machine tools. A typical service would comprise two qualified engineers spending two days checking each machine four times per year and would cost approximately £450 per machine per year.

If the user has no previous experience with preset tooling, he must provide himself with presetting devices and probably a presetting area. In any case this may be necessary for an established user of NC because his presetting load will increase with the purchase of more NC machine tools. It is as well to attempt to standardize on one type of presetter that can perform presetting operations for as wide a range of machine tools as possible. Although presetters are offered by machine-tool manufacturers for one type of machine at what appear to be comparatively low prices, they are rarely adaptable to other makes and types of machine and this policy can prove very expensive in the long run. The aim should be to make the preset tools available to as many machine tools as possible. Incidentally, it may prove advantageous to extend preset tooling to some manually operated machine tools. This is certainly true where a numerical readout is available on such machines (see chapter 14).

Strict disciplines are necessary with tooling for NC machines to prevent the use of a hand-reground tool—for example, on an NC lathe—as this could result in a scrapped component. Operators do swop tools and they do touch-up tools on off-hand grinders; the answer to this problem would seem to be the coding of NC tools for identification. NC tools must be kept separate from tools for conventional machines and should be easily identifiable by the use of a colour code or the like. If the tool stores were only allowed to issue NC tools to NC presetting operators, and NC presetting operators only preset identifiable NC tools for use on NC machines, the problem would not arise.

The value of standardization

The standardization of tools must also receive attention when NC machines are being purchased. No attempt will be made here to propose suitable standards but there are obvious advantages in having standard types of tool for different types of NC machine tools. One standard for a whole range of NC lathes would be beneficial with possible extension to NC vertical turret lathes. This would result in a tool standard for all turning tools, which would allow the interchange of tools between various types and makes of machines. This philosophy can obviously be extended to drilling machines or milling machines and so on.

Hand in hand with tool standardization goes the creation of tool libraries which allows the part programmer to call up any tool by a number or code. Where computer programming of technological data is used, the computer then has access to all the relevant information on that tool in the tool library (see chapters 7, 8, 9, and 15).

Standardization for the purposes of continuity of experience in equipment, maintenance, and minimum spares holdings is well recognized. Standardization for the purpose of work transfer from one NC machine to another is only within the experience of major users of NC equipment. Lack of standardization has been, and will continue to be, the cause of production bottlenecks. A component or operation which has been tooled and programmed for NC is expensive to 'knife and fork' on conventional equipment when the necessity arises due to a production bottleneck or a breakdown of NC.

However, NC equipment, to allow easy transfer, must be compatible to a far greater degree than experience with conventional machines suggests. Not only must the machine functions be similar, but so also must the relationship of component to machine datums. Some controllers have a floating-datum feature or datum-offset facility which reduces the latter problem. Of greater significance is compatibility of data requirements.

There is a variety of tape codes although these have been standardized over the years with the result that the USA, EIA code has become pre-eminent. To ensure a world-wide standard, an ISO code has been produced which is finding wide acceptance and will probably replace the EIA system in the near future. That, unfortunately, is not the end of the problem. Although there is a wide acceptance of the tape coding, the variety of machine function codes shows no signs of decreasing although standards currently exist (see chapter 6). Block formats, such as tab sequential and word address, are strictly non-compatible: tapes produced in either format cannot be interchanged. The recommended standard is the interchangeable tab and address format. For the user who intends using a computer for programming, there is one saving grace. He can post-process his programs to suit the variety of machine/controller combinations he may use. A new user can also insist on conformity to the ISO and BSI Interchangeable standards for code, format, and axes (chapter 6).

It is a pity that, at present, different makes of NC machine tools with the same type of control systems still require different post processors but it may be that sometime in the not-too-distant future one post-processor program might serve for all such machines and that tapes will be interchangeable.

4*

From the foregoing, it is clear that buying an NC machine tool does present a multitude of problems.

Apart from these, the economic justification must also receive detailed attention when an NC machine tool is being purchased (chapter 4).

4. Economic appraisal

Alfred Tack

For career summary see chapter 3.

Many of the pundits have implied, perhaps even stated, that the mere act of purchasing NC equipment is economic. The truth is far less appetizing. In the writer's opinion, the act of NC purchase is only the conversion of highly desirable money into a piece of very expensive hardware, which will depreciate very rapidly in value after leaving the supplier's works. This may seem a very pessimistic viewpoint with which to begin such a chapter. It has only been proffered so as to illustrate clearly that an investment of this type is by no means a licence to print money *ad lib*, but one which will only bear fruit if its potential and needs are fully understood. NC is a dictator: benevolent if its organizational needs are satisfied; a tyrant if they are not.

History will show future generations of technologists that the 'sixties was an age of indecision and of missed chances, when the opportunity to bound into the electronic era was given but not taken. There may still be pessimists who remain to be convinced that NC offers them any more than the traditional machines they have always used. Undoubtedly, one of the major causes of this dilly-dallying has been the lack of an easily understood and valid appraisal technique centred on the relative economics of NC equipment compared with existing manufacturing methods.

Without such an economic appraisal, it is impossible not only to determine whether NC equipment is the right choice or not, but also to gauge whether the peripheral costs and requirements associated with NC have been fully assessed. It is to this end that this chapter is devoted.

It is possible that many new investment-appraisal techniques will be born in this era and that those which do not withstand vigorous scrutiny and the test of time will be rejected. The best of many techniques will eventually be consolidated with the result that capital-investment appraisal will don the cloak of science rather than be the result of an inspired or educated guess. It is importan at this stage, to expand the basic principles that are involved in economi appraisal techniques in general and in particular the technique described i this chapter.

Preliminary comparison

Firstly, we must have some yardstick against which to compare a propose investment, be it NC or otherwise. This yardstick will normally be the existin manufacturing technique, about which the maximum number of facts ai available for the minimum effort.

When using such a yardstick, the fundamental principle should be that absc lute values are replaced by comparative values. This is imperative, since an economic appraisal revolves around the question 'How much better is metho A than method B?', a question which can be answered only by making con parisons.

Thus, any factors which do not differ between the manufacturing technique under consideration can be ignored; e.g., a fixed overhead burden which is no affected by the project should not be taken into consideration. There is n point in calculating costs which remain the same for each possible techniqu representing a saving for neither. If it does not change, it can be ignored.

		NC method		Alternative meth	
		Time (H)	Cost (£)	Time (H)	Cost (
	PRE-PRODUCTION COSTS				
A1	Programming, planning, and rate-fixing	400	400	300	300
A2	Data preparation	100	100	20	20
A3	Jig and tool design and manufacture	250	250	1,000	1,000
A4	Prove out	100	100	200	200
A5	Total pre-production cost	850	£850	1,520	£1,520
	PRODUCTION COSTS				
B1	Marking out (per component)	0·25	0·156	0·30	0·187
B2	Off-machine setting (per component)	0·10	0·063	0·15	0·094
B3	Machine setting (per component)	0·033	0·027	0·083	0·083
B4	Floor-to-floor machining (per component)	0·50	0·688	1·75	1·75
B5	Handling not included in B4 (per component)	0·033	0·167	0·167	0·083
B6	Off-machine inspection (per component)	0·083	0·052	0·112	0·073
B7	Hand or machine finishing (per component)	0·033	0·021	0·05	0·032
B8	Fitting and assembly (per component)	0·033	0·021	0·083	0·052
B9	Total production cost	1·065 h	£1·195	2·695 h	£2·354
C1	Number of components per year	16,000		16,000	
C2	Anticipated life of component	10 years		10 years	
D1	Annual manufacturing cost (B9 × C1)	£19,100		£37,700	
D2	Total cost of production (A5 + B9 × C1 × C2)	£191,850		£378,520	

Figure 4.1 Manufacturing cost comparison sheet

Reduced manufacturing costs

In most instances, NC equipment will be proposed because of possible reductions in manufacturing costs or of the expansion of manufacturing facilities. Even for expansion, cost reduction is a prime consideration. The form shown in Fig. 4.1 has been designed to estimate manufacturing cost reductions; it allows a single component, or group of components, to be costed to establish whether, and to what degree, there is a reduction in manufacturing cost. These calculations will, in most cases, decide whether the proposed NC equipment should be examined for its full economic potentialities or rejected out of hand. It should be noted here that the form is only the first stage in the appraisal; no account has yet been taken of such items as scrap and inventory savings which may accrue through the use of NC. Further, where a group or groups of components have to be considered, extensive preparation of data will be necessary before the manufacturing cost comparison may be completed.

Manufacturing cost comparison sheet

Each item on the form is now considered.

A1: Programming, planning, and rate-fixing   These functions form a variable overhead which should be related directly to the manufacture of the component which incurred the cost. With the advent of NC, more emphasis is being placed on the planning function; the planning department should therefore be established as a 'cost centre' and its cost withdrawn from the general overhead and allocated directly to the component.

A2: Data preparation   This is a function particularly associated with NC in that it concerns the programming involved with such equipment. It may involve simple tape-punching equipment operated by female clerical labour and as such be combined with A1 above, or it may involve the use of computers. In the latter case, there are three distinct possibilities.

(a) An external computer service is used and the cost charged direct by the supplier.

(b) Time will be bought on an external computer, but operated by the user company's labour. The allocated cost will be the time-cost billed by the service bureau plus labour and travelling expenses.

(c) Time will be bought at an agreed cost on an internal computer which may or may not require the addition of a labour charge.

A3: Jig and tool design and manufacture   These functions are entered under a composite heading since many companies place this type of work out to contract toolmakers. If it should be designed and manufactured internally, the tool design office and toolroom should be extracted from the general overhead and set up as cost centres, and the cost of their operations charged direct to the component or assembled product as applicable. The wisdom of this approach becomes more obvious when, by standing orders, competitive tenders are obtained from outside contractors. Efficiency will be improved when orders are based on strictly competitive tendering.

A4: Prove out   This is a procedure for the verification of jigs, fixtures, tooling, programming, planning, and methods, which in most machine shops is ignored as an additional cost. It is a high cost which is charged directly to the component being manufactured. Interpretation of planned methods; unforeseen difficulties with jigs, fixtures, and tooling; lengthy inspection procedures encountered on new jobs and learner curves where production slowly rises to the expected norm: all are part and parcel of the prove-out procedure, and as such must be rigidly timed and costed.

A5: Total pre-production   The total of pre-production time-cost is the total of the above four categories and represents the 'once-off' cost of preparing the components under consideration for production.

B1: Marking out   This is the procedure for verification of a casting or forging and marking datum lines to assist future machining operations. Equally, this may apply to lining out profiles or marking hole centres, etc. for conventional machining.

B2: Off-machine setting   This is a function that applies to the preparation of jigs, fixtures, and/or tooling and should be entered as a time-cost only where labour other than the machine operator is used. If the machine operator undertakes the work, then it must be a procedure which does not hinder the machining cycle. If done within the machining cycle by the machine operator, it can be ignored as an additional cost.

B3: Machine setting   This is the time-cost when the machine is delayed in its manufacturing cycle by the time taken to prepare it for the machining of a new component or batch of components. It will include the delay due to awaiting the inspection of the first-off component and also the resetting time-cost where large batch runs make it necessary. It is the sum of all the setting times involved in the operations being considered. For the calculation of cost, use the machine-hour rate as defined in B4.

B4: Floor-to-floor machining (per component)   This is the in-production time taken for the machining of all the operations under consideration and should be calculated by using the machine-hour rate. This rate may be briefly defined as the sum of the following three factors.

(a) Hourly labour cost. This cost comprises wages, bonuses, premium payments, holiday pay, National Insurance, and other fringe benefits or costs which are directly affected by the employment of labour.

(b) Equipment depreciation per hour. This cost represents the loss in value of the equipment during the time (years) the component is a production commitment, divided by the number of hours spent producing the component in this time.

(c) That part of the overhead burden which is directly affected by the equipment, e.g., insurance, power, maintenance, repairs, etc., but not including equipment depreciation.

B5: Handling not included in B4   This is the inter-operational time-cost of moving the component from machine to inspection, to stores, etc. and will include all handling other than that covered under B4. It will only apply to the operations under consideration.

B6: Off-machine inspection   This is the total time-cost of the first-off inspection and batch inspection involved, for the operations under consideration. This may appear to have been covered under B3, but this is not so: only the idle machine time-cost due to inspection was covered under that heading.

B7: Hand or machine finishing   This is the time-cost related to the amount of hand or machine finishing necessary, which can be related to the operations being considered. It may be that an improved quality of manufacture will eliminate or reduce such operations as de-burring, scraping, etc.

B8: Fitting and assembly   This is the time-cost involved in the fitting of the component under consideration into a sub-assembly or assembly. Many of the modern machining techniques will tend to produce better and more consistent component accuracy, and under these conditions time-cost improvements may be expected during the fitting stage. Some improvements may be expected even from the simple replacement of a worn-out machine by its new counterpart.

B9: Total production   The total of the repetitive time-cost is the total of the previous eight categories and represents the manufacturing cost for the component under consideration.

C1: Number of components per year   The total annual requirement for the component under consideration should be entered.

C2: Anticipated life of component   The anticipated life of the component is the expected production commitment in years for the component.

D1: *Annual manufacturing cost* The annual cost of manufacturing only takes into account the areas dealt within the production cost comparison and is not the total cost. The formula for calculation is B9 × C1.

D2: *Total cost of production* The total cost of production is represented by the calculation A5 + B9 × C1 × C2.

As a general rule, if the anticipated life of the component is large, then errors in the above estimation will be insignificant. The converse is equally true: where the total to be produced is small, the once-off costs can form a considerable proportion of the total manufacturing cost.

Full comparison

On the basis that the comparative manufacturing cost estimates (Fig. 4.1) yield an answer which suggests that the proposed NC method produces components more cheaply, it is now necessary to establish whether it is economical in the broader sense.

Firstly, a comparison of capital and revenue costs must be made. Figure 4.2 is a data sheet for entry of this information, with space provided for differences in cost to be noted. Columns for entry of cost data on a year-by-year basis may also be provided for.

What is capital expenditure?

Before examining the capital expenditure involved, it is worth first defining what is meant by this term. Capital expenditure is the cost of equipment which satisfies the following conditions:

the unit cost of the equipment is in excess of £25;
the expected life of the equipment is in excess of two years;
the equipment is purchased by the company and not hired, leased, or rented.

It is common knowledge that the purchase cost of equipment may be 'capitalized', but it is not so well known as to what is meant by the term 'purchase cost'. In addition to the basic price of the equipment, the following costs may be included in the purchase cost.

(a) The cost of ancillary or auxiliary equipment. This may include the cost of non-consumable tooling and a set of spare parts.

(b) The cost of import duty on foreign equipment.

(c) The financing charge made to the company, due to the forward purchase of foreign currency. An agent for a foreign supplier may insure against currency exchange risk by buying foreign currency and holding at a bank. Interest charges are made by the bank and it is these interest charges that may be capitalized.

		A Conventional equipment	B NC equipment	
Description of equipment		six capstan lathes	two NC lathes	
Capital expenditure		£	£	Additional capital cost B − A
Installed cost of equipment	−1 year −6 months −3 months 0 +3 months +6 months	10,000 10,000 10,000	3,500 35,000 31,500	−6,500 25,000 21,500
	net	£30,000	£70,000	£40,000

Annual expenditure	£	£	Difference in annual expenditure A − B (£/year)
Direct labour costs	11,500	5,100	6,400
Extra direct labour costs	5,400	900	4,500
Indirect labour costs	9,600	12,800	−3,200
Employee benefits	4,000	2,800	1,200
Special to product tooling	1,000	250	750
Consumables	800	350	450
Scrap	5,400	3,800	1,600
Tool maintenance	no change		
Equipment maintenance	100	150	−50
In process inventory	2,800	1,450	1,350
Other expenditure	no change		
Difference in annual expenditure			£13,000/year
Factors			

Once-off expenditure	£	£	Difference in once-off expenditure A−B (£)
Pre-production costs	1,520	850	670
Movements and re-arrangements	400	—	400
Disposals	—	—	
Learner costs	—	3,200	−3,200
Other expenditure	—	—	
Difference in one-off expenditure			−£2,130

Figure 4.2 Expenditure forecast sheet

(d) Carriage, insurance, and freight charges. These are the costs involved in the transportation of equipment from the supplier's works to the company.

(e) The cost of installing the equipment. This may include the cost of providing special services, such as buildings and power supply, but excludes the cost of removal or rearrangement of existing equipment which is to be moved to allow the equipment to be installed.

With all the above considerations in mind, it is now possible to calculate the capital cost of installing the NC equipment. By working from the estimated installation date for the NC equipment, it is a simple exercise to determine when the payment or payments are incurred by the company; the payment date(s) should be quoted in terms of months before or after installation. The number of payments will depend on whether the NC equipment was purchased by a single payment, progress payments, or deferred terms.

An exactly similar exercise may be performed for the capital cost of the alternative equipment that would have been purchased if the NC equipment was not bought.

One alternative may be to continue with the existing equipment until it is worn out, at which time it will be either replaced with similar equipment or reconditioned. It is important in this instance to estimate accurately at what time the relevant payments will occur in relation to the installation date of the NC equipment, but also to be as accurate as possible in estimating the costs involved at that time. For example, the current cost of equipment may vary by several per cent from the cost in a few years' time.

Expenditure forecast sheet

All the various costs having been determined, for both the NC and the alternative equipment and the times at which they occur, these facts can now be transposed onto the data sheet (Fig. 4.2). The costs should be shown against the nearest times with which they agree.

The additional capital expenditure may now be calculated. The net capital expenditure is a summation of all expenditures and will be used later when the payback period is being calculated.

On examining the annual expenditure, it can be briefly stated that this section of the data sheet is a collection of all the revenue expenditures that occur regularly during the operation of equipment. It should be decided at this stage whether escalation of costs with time are going to be considered either separately or altogether. The author believes that, although labour costs and, say, consumable tooling costs will almost certainly rise at different rates at a particular time, the total increase over a period of time will be almost the same. It is therefore recommended that escalation be applied to the total annual expenditure, rather than to individual cases. Further, it has to be decided how far into the future account is to be taken of annual expenditure. For the purposes of this economic appraisal, a period of 10 years is assumed, since this is the normal period over which companies amortize machine tools. If the NC equipment— or its alternative—is for a project of shorter term, it has to be decided whether the equipment can be shifted to the production of different components and still be comparably utilized or whether disposal of the equipment would result. The items under annual expenditure in Fig. 4.2 are as follows.

Direct labour costs   These are those wages paid to the operator which are directly attributable to the operations performed on the component. Where a system of payment by results is used, then these costs may be determined from the hourly rate of an average operator.

Extra direct labour costs   These are the extra costs to the company which are directly related to the operator. It will include such payments made to the operator as overtime, bonus payments, waiting time, etc.

Indirect labour costs   These are the payments made to employees other than the operator of the equipment. It will include such annual costs as setting, component handling, inspection, etc. It may also include costs related to planning, jig and tool design, etc., where these costs recur annually because new designs of components have to be manufactured each year.

Where the equipment is being evaluated for one component or a group of components for the whole life of the project, then some of the costs outlined above will be dealt with later under the heading of 'Once-off costs'.

NB   Depreciation cost data is not included in this stage. It is effectively taken into account in the later stages of the calculation when tax allowances are being determined.

Employee benefits   In addition to wages, the company also provides benefits and makes payments which are directly affected by the employment of labour. Such costs will include holiday pay, National Insurance and Pensions, etc.

Special to product tooling   This is the cost of jigs, fixtures, and tooling for new designs of components which will have to be manufactured each year. Only costs which recur annually should be entered in this section.

Consumables   This cost includes such items as consumable tooling, or other consumables which are used during the manufacturing process.

Scrap   The cost which is required in this section is the annual cost to the company of the components scrapped either by the NC equipment or by the alternative. Very often, scrap costs are indeterminate within a company and are simply viewed as a percentage of good parts produced. It will be necessary to estimate accurately, at the operations stage where they are scrapped, the average cost of the component or components being manufactured. This may be calculated by determining the initial cost to the company of the component, or of its raw material, and then determining the value added to the component up to, and including, the operations being considered. By adding these two values, the cost of a scrap component can be established; once established, the total scrap costs may be estimated.

Tool maintenance   These costs include such recurring expenditures as tool regrinding, tool presetting, and jig and fixture refurbishing.

Equipment maintenance   Equipment maintenance is very often a cost which remains indeterminate until breakdown of equipment necessitates suitable maintenance. It is therefore possible only to estimate the cost of such maintenance and average this cost over a number of years. However, in addition to this type of maintenance, there is often a planned maintenance scheme operated by the company, even if this only entails regular oiling of equipment. Usually it will be necessary to set up a thorough planned maintenance schedule for NC equipment, because of its inherent complexity and the cost to the company of any breakdown.

In-process inventory   In this section, it will be necessary to calculate the 'carrying charge' of in-process inventory. There are three separate inventories which, for this calculation, are considered to be mutually exclusive. The first is the pre-process inventory and includes the storing of raw material, forgings, castings, etc. This is controlled by the company's volume purchase policy and by deliveries from suppliers. A change in manufacturing technique does not normally cause a change in the pre-process inventory. The second is the in-process inventory, which is based on the number of components within the manufacturing process. Manufacturing techniques such as NC may have a profound effect on this aspect of inventory. NC often has the effect of reducing the number of separate component operations and, thereby, reducing manufacturing lead time. A reduction in lead time will effectively reduce in-process inventory. The third is the post-process inventory and is a function of fitting and assembly times and of sales stock.

There are two prime effects resulting from a reduction of in-process inventory. One is the improvement in the liquidity of the company; however, the value of this improvement cannot be simply expressed. The second is the reduction in 'carrying charge', i.e., the cost of financing the in-process inventory.

The following estimates must be determined before a calculation of the in-process inventory carrying charge can begin.

(a) The average cost of the component or components being manufactured at the in-process stage of manufacture. This may be determined by adding the initial cost of raw material and half the value added in the whole manufacturing process. The value added to the component is a summation of the labour costs and overheads, but it excludes the profit made on the component when sold. In order to determine the average value added for the components that are 'in-process', it is simplest to halve the total value added for typical components.

(b) The average number of components being produced per annum for the whole manufacturing process of the component. This may be several times greater than the number of components expected to be in-process for the particular manufacturing operations being considered in this appraisal.

(c) The average length of time the component or components take to complete the whole manufacturing process. This time should be expressed as a fraction of a year.

Multiplying the above three estimates produces a cost which can be regarded as the in-process inventory cost. To determine the carrying charge related to this inventory cost, it has to be decided at what percentage it is chargeable. The carrying charge includes such factors as interest on loans, storage costs, obsolescence, etc. and can be demonstrated to be generally between the limits of 15 and 25 per cent per annum. Once this rate has been established, the carrying charge can be determined by multiplying the in-process inventory cost by this rate.

As has been suggested before, large savings may result from the use of NC equipment where the manufacturing lead time for a component is considerably reduced.

Other expenditure Any other annual cost should be entered here. It may be, for instance, that extra power or a computer service have to be provided for the NC equipment. If this is so, then these costs should be calculated.

All the annual costs having been determined for each of the alternatives, the differences in annual expenditure may be calculated and finally the net difference in annual expenditure (or annual savings) determined. This figure may then be escalated, as outlined under 'Expenditure forecast sheet' above, for the life of the project or for ten years, depending on which term represents the shorter life. Where there are variations in the number of components produced per annum, then either annual costs may be calculated for each year, or more simply a percentage of the annual expenditure may be taken, where the numbers produced fall below that of normal full production. Thus the 4 per cent escalation factor has been applied to the single value of 'Difference in annual expenditure' of Fig. 4.2 to give the series of values of 'annual savings' in Fig. 4.3.

The data sheet is completed by determining the once-off expenditure for the two alternatives. This item also includes certain short-term costs. The following five headings represent the areas in which such expenditure is likely to occur.

Pre-production costs   These costs will include all pre-production costs which have not been accounted for in the previous section on annual expenditure. Where the appraisal is being performed for one particular component which will occupy the equipment for its entire projected life, these costs are particularly relevant.

Where general production work is being considered, pre-production costs for NC equipment would include the cost of re-planning, re-tooling, etc. to produce components already being manufactured on existing equipment.

Movements and re-arrangements   Very often the installation of new equipment involves the movement and rearrangement of existing equipment. Where these costs are significant, they should be indicated.

Disposals   Also associated with the installation of new equipment is the disposal of existing equipment which may result. If applicable, the estimated disposal value of this equipment should be determined.

It is also possible that the new equipment being considered may be disposed of at the end of the project. If this is so, then the estimated disposal value, at the time when disposal will occur, should be entered. The time of disposal should also be indicated.

It is usually sufficient in this section to enter the written down, or depreciated, value of the equipment, calculated for the disposal date.

Learner costs   With the installation of any new equipment, there is always a period of time during which the operator reaches normal efficiency. This is particularly true of NC equipment. In order to estimate these 'learner costs', a method as outlined below may be used; it is an approximate method, but is a sufficiently close approximation for the purpose.

Firstly, determine the time involved from installation of the equipment until normal efficiency is expected. If it is then assumed that efficiency increases uniformly with time, it can then be said that half the production rate was achieved during the learner period. Learner costs can now be estimated as being equal to half the annual savings that would have accrued during the learner period.

Other expenditure   Any other once-off cost should be entered here. Entries in this section are normally expenditures which are related to the purchase of the equipment, but which cannot be regarded as capital expenditure.

The once-off expenditure may now be summed and the difference in once-off expenditure established between the alternatives. This summation completes the information required on the data sheet (Fig. 4.2).

Discounted cash flow

The previous exercises have been designed to establish the difference in cost of capital equipment, annual operating costs, and once-off or short-term cost effects. These facts in themselves are only of small value to management. The questions of 'What rate of return on investment will take place?' and 'When will this project pay for itself?' still remains to be answered. It is to this end that the Discounted Cash Flow (DCF) technique is used. As this chapter is not intended to be a discourse in DCF, suffice it to say that, unlike traditional methods of calculating the rate of return on investment, it looks not only at how great the return will be but when the return will occur. It takes into account the effect of time, and converts all cash flows occurring at future dates into their present-day values.

The data sheet (Fig. 4.3) for the calculation of the project's profitability, and the accompanying interpolation chart (Fig. 4.4) have been designed to reduce the large amount of calculation normally associated with DCF. Previous methods have necessitated the use of trial-and-error techniques to obtain the 'solution rate', and have been the cause of DCF (a very old system) falling into disuse. In principle, capital expenditure on, and net income from, a project are discounted at a series of compound interest rates. The interest rate at which the total of all discounted net earnings equals the total of all discounted capital cash outflows is the 'solution rate'. The project, as well as recovering its original capital cost, will earn compound interest at the solution rate during its anticipated life.

Cash flow chart

The factors shown in Fig. 4.3 are based on the assumption that all income and expenditure occur instantaneously in the middle of the year. Although this is obviously incorrect, the author considers it to be more realistic than the normal procedure, which assumes that all income and expenditure occurs on the last day of the year. It has an averaging effect which reduces the effect of errors in timing.

Against 'Additional capital cost' will be entered from the data sheet (Fig. 4.2) the difference in capital cost of the installed NC equipment over its alternative, under the year or years in which they occur. Similarly, against 'Annual savings' will be entered, in time order, the difference in annual expenditure on NC equipment over its alternative, and against 'Once-off costs' will be entered, again at the appropriate time, the difference in once-off or short-term costs.

To start the calculation, it is first necessary to determine the Capital Investment Grant applicable to the purchase. Having determined the rate, y per cent, at which this grant will be made, you then simply multiply the additional capital cost on the NC equipment by y per cent and enter the resultant Capital Investment Grant under the appropriate time on the data sheet. (It is worth noting

	1 year before	6 months before	3 months before	Start	3 months after	6 months after	During year 1	During year 2
Additional capital cost, A_0			−6,500	25,000	21,500			
Capital Investment Grant 40 per cent								16,000
Annual allowance 20 per cent							4,800	3,840
Annual savings							13,000	13,500
Once-off costs								
Amount subject to tax							2,130	9,660
Corporation Tax 45 per cent							6,070	2,730
Net cash flow, B_0							10,870	26,770

Discounted Cash Flows

	1 year before	6 months before	3 months before	Start	3 months after	6 months after	During year 1	During year 2
DCF factors @ 10 per cent	1·105	1·050	1·024	1·000	0·976	0·952	0·952	0·861
Discounted additional capital cost A_{10} / Discounted Cash Flow B_{10}			−6,650	25,000	20,950		10,350	23,060
DCF factors @ 25 per cent	1·284	1·135	1·062	1·000	0·935	0·885	0·885	0·689
Discounted additional capital cost A_{25} / Discounted Cash Flow B_{25}			−6,900	25,000	20,090		9,610	18,440
DCF factors @ 40 per cent	1·492	1·210	1·095	1·000	0·895	0·810	0·810	0·553
Discounted additional capital cost A_{40} / Discounted Cash Flow B_{40}			−7,110	25,000	19,220		8,800	14,800

Figure 4.3 Cash flow chart

	During year 3	During year 4	During year 5	During year 6	During year 7	During year 8	During year 9	During year 10	During year 11	Total
	3,072	2,458	1,966	1,573	1,258	1,007	805	644		Total A_0 £40,000
	14,050	14,600	15,180	15,780	16,310	17,060	17,740	18,450		
	10,978	12,142	13,214	14,207	15,052	16,053	16,935	17,806		
	4,350	4,950	5,470	5,950	6,400	6,770	7,230	7,620	8,020	
	9,700	9,650	9,710	9,830	9,910	10,290	10,510	10,830	−8,020	Total B_0 £110,050

	During year 3	During year 4	During year 5	During year 6	During year 7	During year 8	During year 9	During year 10	During year 11	Total
	0·779	0·705	0·638	0·577	0·522	0·476	0·428	0·387	0·350	
	7,560	6,800	6,200	5,670	5,170	4,900	4,500	4,190	−2,800	Total A_{10} £39,300 / Total B_{10} £75,600
	0·537	0·418	0·326	0·254	0·197	0·154	0·120	0·093	0·073	
	5,210	4,030	3,160	2,500	1,950	1,580	1,260	1,010	−590	Total A_{25} £38,190 / Total B_{25} £48,160
	0·370	0·248	0·166	0·112	0·075	0·050	0·034	0·023	0·015	
	3,590	2,390	1,610	1,100	740	510	360	250	−120	Total A_{40} £37,160 / Total B_{40} £34,030

Figure 4.3 (continued)

here that the Capital Investment Grant may be claimed at any time in the first year after payment for the equipment has been made, and that actual payment of the grant may be made some months after the claim has been submitted. Where a number of payments are made for the purchase of the equipment, a similar number of grants may be claimed. It is common practice, however, to make one single claim when full payment for the equipment has occurred.)

Proceeding with the calculation, it is a well-known fact that a company is permitted to reduce its liability for taxation by subtracting from its taxable income certain allowances which are directly related to the capital cost of the equipment. This allowance may be claimed annually and is a fixed percentage of the annual decreasing balance. This 'Annual allowance', may be calculated as follows. From the 'Additional capital cost' subtract the 'Capital investment grant'. On this amount, an annual allowance of 'x per cent' may be claimed in the first year. For all subsequent years, subtract the amount claimed for the previous year from the balance for the previous year and multiply this new balance by x per cent. It should be remembered that tax allowances, Investment grants, and Corporation tax are subject to national fiscal policy and are liable to change. Check with your accountant to ascertain the current rates.

Having established the annual allowances that can be claimed, you must now calculate the amount that will be subject to Corporation tax. This may be done by algebraically summing each yearly, or part yearly, column for the following three rows on the data sheet.

(a) Annual savings—this row normally represents an 'inflow' of money and as such should be treated as *a positive amount* subject to tax.

(b) Once-off costs—this row normally represents an 'outflow' of money and as such should be treated as *a negative amount* subject to tax. (Where once-off costs are less for NC equipment than for the alternative, there is a saving and hence a positive amount subject to tax.)

(c) Annual allowance—this row represents money against which tax relief can be claimed and should be treated as *a negative amount* subject to tax.

After this summation, it is then possible to calculate the Corporation tax payable. It is first necessary to determine the rate at which Corporation tax is charged. Once determined, multiplication of the 'Amount subject to tax' by this rate produces the Corporation tax which is payable. Since Corporation tax is normally paid one year in arrears, the tax on year 1 'Amount subject to tax' should be entered one year later under year 2 'Corporation tax' and similarly for subsequent years. It is through this transposition that year 11 is used on the data sheet. The entry for year 1 'Corporation Tax' will, of course, be zero.

The above calculations have all been aimed at determining the net cash flow position for the company, when all tax and allowances, etc. have been taken into consideration.

To determine the 'Net cash flow', it is again necessary to algebraically sum for each yearly (or part yearly) column, for the following rows on the data sheet.

(a) Capital investment grant—this represents an 'inflow' of money and should be treated as *a positive cash flow*.

(b) Annual savings—this row normally represents an 'inflow' of money and as such should be treated as *a positive cash flow*.

(c) Once-off costs—this row normally represents an 'outflow' of money and as such should be treated as *a negative cash flow*.

(d) Corporation tax—this row normally represents an 'outflow' of money and as such should be treated as *a negative cash flow*. (For Corporation tax to be a positive cash flow, the once-off costs and annual allowance for a particular year, when combined, must have outweighed the annual savings for that year. In effect a 'tax rebate' is being claimed.)

The net cash and additional capital cost flows having been established, the next stage is to ascertain what these mean in terms of profitability of the NC investment. That is to say, what rate of return on capital expenditure do the net cash flows represent. The DCF calculation determines this all-important figure as being the compound discount rate which, when applied to the two streams of additional capital expenditure (A_0) and annual cash inflows (B_0), exactly equates the two.

In Fig. 4.3, the annual discount factors that are appropriate to the time period of each column, and to rates of 10, 25, and 40 per cent, are provided in the lower half of the chart. By multiplying each additional capital cost (A_0) and net cash flow (B_0) by the appropriate discount factor for 10 per cent at the time the cash flow occurs, entering the results in the relevant boxes, and repeating for 25 and 40 per cent factors, the discounted additional capital costs and net cash flows can be determined. When this stage is complete, the discounted additional capital costs and net cash flows must be individually totalled for each rate, to arrive at totals A_{10}, B_{10}, A_{25}, etc.

Rate of return interpolation chart

The next stage is to divide totals A_{10} by B_{10}, A_{25} by B_{25}, and so on, and enter the result on the interpolation chart (Fig. 4.4). It is the position where $A = B$ we seek. It would be sheer coincidence if such a point were obtained directly from the previous calculations. By entering, on the interpolation chart, points which are related in the X axis to the calculated resultant A/B, and in the Y axis to the discounting rate, a smooth curve may be drawn through all points. Where the curve intersects the vertical line equivalent to the result A/B = 1, the DCF rate of return percentage may be read directly from the Y axis. A simple means of checking the validity of each point is by means of the curve itself. If any point does not fall on a smooth curve, there is an error in the calculation of that point.

As a guide to an acceptable DCF rate of return for a project, it can be said that it is probable that a 15 per cent return will, in most cases, be required to ensure the efficient use of capital. It should be emphasized here that the DCF rate of return is net of tax.

In addition to the principal consideration of the rate of return which an investment in NC might provide, a company will also frequently want to know for how long its money will be at risk. In other words, it will want to know the payback period. This may be determined quite simply from the data sheet (Fig. 4.3). It is the time that elapses before the sum of annual totals for B_0 equals the 'final' total of A_0. It is usually stated as the time from date of installation of the machine. For the illustrative numbers in Fig. 4.3 this payback period is 2·25 years, and the DCF rate of return is 35 per cent from Fig. 4.4.

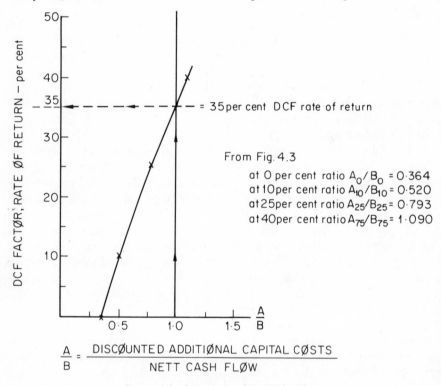

$$\frac{A}{B} = \frac{\text{DISC\O{}UNTED ADDITI\O{}NAL CAPITAL C\O{}STS}}{\text{NETT CASH FL\O{}W}}$$

Figure 4.4   Rate of return interpolation chart.

Presenting the case for purchase

Although some engineers may find difficulty in believing so, industry does not live by the excellence of its products, nor by the magnificence of its designs; neither does it exist solely to give employment. To survive in a highly competitive world, industry must be a viable production unit whose end product is profit: profits without which it would certainly grind to a halt and without which banks, finance houses, and shareholders would lose confidence. Profits are dependent upon having the right product, at the right time, and at the right price and having the outlets and ability to sell that product. The production engineer may not have to decide on the right product, but his prime responsibility is to have it available at the right time and to manufacture it at the right price. It is his job to select the manufacturing techniques which will ensure on-time delivery, and to recommend projects which will result in the lowest cost for a given quality.

Blame cannot be attached to management for failing to see the value of a project if it is poorly presented or if its virtues (so obvious to the production engineer) are neither mentioned nor presented in a correct manner. A case for capital sanction must be sold to management. If we are talking to engineers, we must talk of engineering advantages. If we are talking to essentially financial people, we must stress the financial advantages. It is pointless to generalize. If advantages there are, present them in detail, but always summarize. The accountant will, in all probability, be given the job of checking in detail. Only the summary will land on the boardroom table. Make sure that the summary is yours by using the utmost care in its presentation. If your homework is done properly, this chapter will help you to present your financial case in an acceptable manner. You need no longer be like the tourist who, speaking only his own language, complains that the ignorant foreigner does not understand a word he says. Learn the financial language.

5. Computer production of control tapes

B. J. Wood

B. J. Wood, CEng, MIMechE, BSc(Eng) is Manager, Technical Sales Support with International Computing Services Ltd, London. After graduating from London University in 1947, he spent 8 years with the de Havilland Engine Company, working on the thermodynamics of aero engines, and producing computer programs for this and for NC machines, in particular for the milling of aerodynamic surfaces. He moved to the Ferranti company, engaged in the development of NC programs. Following this, he was associated with the development and sale of computing services for engineers, including NC, during the period which saw the Ferranti/ICT merger and then the ICT/English Electric merger to form ICL, of which his present company is a subsidiary. He is a founder member of the British Numerical Control Society, and its current Chairman.

In all but the simplest applications of NC, a good deal of calculation has to take place between the drawing stage and the cutting of metal. The calculations usually fall into two stages.

(a) Those carried out by the user in order to produce information in the precise form required for punching on to the control tape. This involves, for example, the breaking down of a part drawing into straight and circular segments, or the calculation of co-ordinates of hole centres. It may also be necessary to calculate machining data such as speed, feed, and depth of cut.

(b) Those carried out automatically by the control system during input of the control tape. This stage is particularly significant in the case of contouring systems, and involves principally the breaking down of the straight and circular segments into many small steps in order to provide continuous control of the machine slides.

In the early days of NC, the calculations referred to under (a) were carried out by users as best they could, using slide-rules and desk calculating machines, but, as NC developed, this task became more onerous and threatened in some cases to become a major factor in determining the economics of an installation. It therefore became desirable to use some sort of computing device for these calculations. The calculations under (b) are in any case carried out by a special type of computer, either an 'interpolator' built into the shop-floor equipment or, in the case of magnetic-tape control systems, a separate device. All the user has to do is to supply the information in a rigidly prescribed form. It might therefore be thought that an extension of the special computer would be the best means of dealing with the calculations referred to under (a).

However, the calculations under (b) are of a limited type and do not depend in principle on the component concerned, and it is this limitation which makes it economic to use a *special-purpose* computer, that is, one built to perform only this type of calculation. On the other hand the (a) type of calculations depends entirely on the component and hence varies greatly, not only from user to user, but also within one organization. To cope with this, more flexible computing devices were required, and these were available in the form of *general-purpose* computers. General-purpose computers are capable of carrying out any sort of calculation which can be precisely defined—from, say, problems in nuclear physics and space flight, through engineering design of all kinds, to commercial applications such as payroll or production control. NC calculations are just another job for such computers. Computers are widespread in industry and commerce and computing service bureaux exist which provide facilities for those without their own computers.

The advantages of using a computer for NC calculations may be summed up by saying that it is another step in the automation of the manufacturing process. More specifically, there will be a saving of time and skilled effort in calculations and the chance of errors will be greatly reduced. Moreover, the computer will not only carry out the calculations but will automatically punch the control tape in a form ready for use on the shop floor (or produce the magnetic tape for those systems requiring it). Naturally, the need for computer assistance varies with the type of work.

In point-to-point work, computers can often be useful, especially where large numbers of holes are involved and where feeds and speeds have to be calculated. For contour milling of any complexity, the use of a computer is almost essential, while for three-dimensional work it is indispensable.

A whole new profession has grown up around computers and the NC user may feel somewhat overawed. But it is not necessary for him to be a computer expert, and the purpose of this chapter is to give a general introduction to the subject and examples of the use of computers for several types of NC application.

Computers and programs

Although a detailed knowledge of computers is not required before making use of them, it is necessary to realize where they fit into the system and what information has to be supplied to them. In order to do this, it is helpful first of all to consider how calculations are commonly organized without the use of a computer, by using less-skilled staff for performing the routine part of the calculations. In order for this to be successful, the steps in the calculation have to be prescribed in great detail and set out on a calculation sheet. However, for any particular type of calculation, this has only to be done once; thereafter, whenever this calculation is required, the standard sheet and the current data are submitted to the person doing the calculations. The essential features of such a system are shown in Fig. 5.1. The human being, together with pencil and paper,

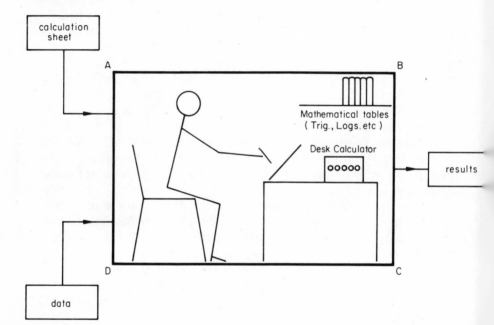

Figure 5.1 Essential features of a calculating system.

slide-rule, desk calculating machine, mathematical tables, etc. shown inside the box ABCD, comprise a basic calculating device. But this cannot in itself produce any result until two further items have been supplied: the calculation sheet describing how to do a particular sort of calculation, and the data. The idea of three distinct things—the basic calculating 'box', the calculation sheet, and the data—is simple, but important, for the situation when using a computer is closely analogous to this. The computer replaces the calculating box of Fig. 5.1 and, before it can produce any results, must be supplied with both the equivalent of a calculation sheet and the data. The former is known as a 'computer program' and has to be expressed in some kind of code to suit the computer, although it provides just the same kind of information as does the calculation sheet.

The technique of writing computer programs requires training and practice, but the NC user will not usually have to become involved in this because suitable programs can generally be obtained ready-made. However, the user will be very much concerned with providing the data for such a program. In NC, the data consists, in its simplest form, of a list of dimensions of parts, but in more advanced applications it also contains descriptive matter (such as CIRCLE, LINE, RADIUS), and can in fact be a fairly elaborate description of a work-piece (or *part*) and of the *program* of operations to be performed. For this reason, the term 'part program' is used instead of 'data'. This must not be confused with 'computer program'.

Before either the computer program or the part program can be put into the computer, they must be transcribed on to either punched cards or punched paper tape. The punching of cards or tape is a routine operation akin to typing, and the NC user will be familiar with the process since it is used when producing without computer assistance.

In the examples which follow, the emphasis is on how the programs are used, rather than on the programs themselves.

Examples of programs

History of NC computer programs

At the present time, a number of programs such as APT, EXAPT, and 2C,L are prominent because they are intended to be of wide application, and these are dealt with in the following chapters. But many programs were produced before the advent of these and some of them are still very much in use. They are intended for a more restricted range of application, e.g., for a particular control system or a particular type of operation, and are important not only because they are in current use, but also because they illustrate the historic development of NC programs. A few of these programs are briefly described below, but it must be emphasized that no attempt is made to describe all the facilities they offer: for this the reader should consult the references given.

Although point-to-point machines are now the most numerous, one of the first applications of NC in this country was to the milling of complex shapes, and in this field the geometric calculations were the major problem. Programs to help with this were developed largely by the aircraft industry, or by control manufacturers with interests in it, or the shipbuilding industry. The Ferranti PROFILEDATA program is in this class.

As point-to-point machines came into use, geometric problems again arose. These were mainly concerned with the calculation of co-ordinates of hole centres arranged in patterns, and the repetition of patterns within a part. For this, programs such as KIPPS and AID were produced by ICSL.

5

With the advent of tool-changing machines, NC began to be used for components requiring miscellaneous sequences of operations such as drilling, tapping, reaming, and straight-line milling. The 'pattern' type of program was not always adequate here because the geometry involved was not necessarily a serious obstacle. The main problem was to determine the sequence in which operations should be performed and to determine the feed and speed to be used for them. A good example of a program for this is MILMAP, developed by ICT (now ICL) between 1963 and 1967, for use with Milwaukee-Matic machines in its own factories. While offering simple geometric facilities, the main features of MILMAP are its tool and material stores, which are tables of speeds and feeds stored in the computer program and used by it to select the cutting conditions for each operation.

NC has only recently been applied to lathes and several programs have appeared for this application. Provided the main problem is one of geometry (as in a contouring lathe), modified versions of some of the milling programs can be useful; but lathes are often used to turn out fairly simple components, in which case the emphasis shifts from geometry to the determination of the cutting technique involved in reducing a blank to a finished part. This is a difficult problem to tackle in a general way since components and preferences for techniques vary so widely. Examples of programs for lathes are AUTOPIT (Pittler), AUTOPOL (IBM Germany), and a program produced by International Computing Services Ltd for a Pinumat lathe to the special requirements of Miehle-Goss-Dexter Ltd.

All NC programs aim to enable the user to prepare his part program with as little effort as possible, by using a notation which is convenient for describing conventional drawings, and there follow some simple descriptions of typical programs, intended to give an indication of the facilities offered to the part programmer. References to sources of detailed information are given.

AID

One of the simpler problems in NC is the calculation of the co-ordinates of the centres of holes which lie in a definite pattern. Typical patterns consist of a pitch-circle or a line of holes (often equi-spaced) or a mesh (matrix) in which the spacing can be different in two directions. In such cases, the essential features of the pattern are described in the part program and the computer uses these to calculate the co-ordinates of every hole.

AID[1] is a typical program of this kind, produced by International Computing Services Ltd. It appeared in 1966 and runs on the Atlas computer. Figure 5.2 shows a hypothetical part designed to illustrate the main features of AID. There are three patterns—a 'matrix', a pitch circle, and a line—and in each case the dimensions given are sufficient to enable the co-ordinates of every hole centre to be calculated. AID does this, and also enables the user to indicate whether more than one operation is required on any of the patterns; for example, it may be required both to drill and to ream some of the holes. This is done by stating the drilling machine turret numbers to be used and, if necessary, stating in the part program the speed and feed to be used with each turret; provided the correct tools are loaded into the turrets the desired result will be achieved. A further feature is that auxiliary functions such as 'coolant on' may be called up. Figure 5.3a shows the main section of the AID part program corresponding to the part shown in Fig. 5.2, and Fig. 5.3b the auxiliary sheet used to specify the speed and feed for each turret called for in Fig. 5.3a.

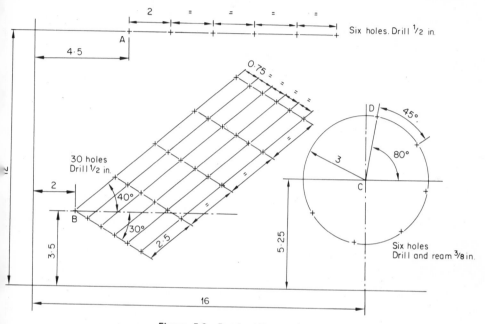

Figure 5.2   Part for AID example.

In this simple example there are five lines in Fig. 5.3a, which may be explained as follows.

Line 1   This is a 'transfer of origin' statement, indicated by the letter T in the column a. In many cases it will not be convenient to make the origin from which the drawing dimensions are given the same as the origin used by the machine control system. The latter may, for example, be at the near left-hand corner of the table, and there might be an obstruction (e.g., another part of the workpiece) which prevents the drawing origin being placed there. In such cases, what is required is to add a constant amount to the drawing X and Y values, a task which would be tedious to do individually. The T statement takes care of this

AID Part program

No.	a	b	c	d	e	f	g	h	i	*	m							n
1	T	6·0	5·0							*								
2	C	16·0	5·25	3·0	80	6	−45			*	1	3						
3	L	4·5	12·0	2·0	0	6				*	2							
4	M	2·0	3·5	2·5	40	5	0·75	−30	6	*	2							
5	P	−6·0	−5·0							*	3							Z
										*								
										*								
										*								
										*								
										*								
										*								
										*								
										*								
										*								
										*								
										*								
										*								
										*								
										*								
										*								

Figure 5.3a Main data sheet

Turret Number	*Speed*	*Feed*
1	805	0·006
2	624	0·008
3	420	0·011
Z		

Figure 5.3b

by causing the computer automatically to add specified amounts to all X and Y values quoted on subsequent lines of the part program. In the example shown, 6 would be added to all X dimensions and 5 to all Y dimensions in lines 2 to 5, which would mean that the drawing origin would be regarded as being at X = 6 and Y = 5 relative to the machine origin. A T statement remains in force until another T statement with different X and Y values appears. Thus, it is possible to change the origin more than once in the course of a part program, a feature which is sometimes useful when the drawing dimensions are not all given from a common origin.

Line 2 This is a 'circle' statement, indicated by C in column a. The values following the C are: col. b, the X value of point C, the centre of the circle; col. c, the Y value of point C; col. d, radius of circle; col. e, angle to X axis at which first hole (D) lies; col. f, number of holes; col. g, angular spacing between holes (in this case a negative value because the progression of holes from D is in a clockwise direction). The m columns contain the numbers 1 and 3 denoting that turrets 1 and 3 have to be used at each of the six holes (turret 1 will hold a $\frac{23}{64}$ in. drill, and turret 2 a $\frac{3}{8}$ in. reamer).

Notice that the dimensions are those from the drawing origin, and that they will be modified by the computer according to the preceding T statement.

Line 3 This is a 'line' statement, indicated by L in column a. The values in the other columns have the following meanings: b, c: X and Y co-ordinates of the first hole A; d: spacing of holes; e: angle of line to X axis (zero in this case); f: number of holes. Turret 2 only is to be used (and will hold a $\frac{1}{2}$ in. drill).

Line 4 This is a 'matrix' statement, indicated by M in column a. The other columns are as follows: b and c; X and Y co-ordinates of the reference hole B; d, spacing of holes along one 'arm' of the matrix; e, angle of that arm to the axis; f, number of holes along that arm, g, h, i, spacing, angle, and number of holes respectively for the other arm. In the m columns, turret 2 only is called for the $\frac{1}{2}$ in. drill).

Line 5 This is a 'point' statement, denoted by P in column a, and in this case is inserted merely to bring the drilling spindle back to a convenient position after finishing its job. The figures −6.0 and −5.0 in columns b and c are the X and Y co-ordinates of the point, and it will be remembered that the T statement in line 1 is still in force, so that the co-ordinates of this position relative to the machine axes are X = 0, Y = 0, i.e., the machine datum. Purely as a convention of the AID program, it is necessary to quote the highest turret number used (3) on this final line. The Z in column n indicates the end of the part program.

Figure 5.3b shows the speed in rev/min and the feed in in./rev which are to

Seq. No.	X co-ord.	Y co-ord.	Speed	Feed	Turret	
V001	X022521	Y013204	S681	F448	T01	
V002	X024457	Y011971				Drill 6 holes on pitch circle 23/64 in.
V003	X024954	Y009729				
V004	X023721	Y007793				
V005	X021479	Y007296				
V006	X019543	Y008529				
V007	X010500	Y017000	S662	F450	T02	Drill 6 holes on line. 1/2 in. (Y values constant)
V008	X012500					
V009	X014500					
V010	X016500					
V011	X018500					
V012	X020500					
V013	X008000	Y008500				
V014	X009915	Y010107				
V015	X011830	Y011714				
V016	X013745	Y013321				
V017	X015660	Y014928				
V018	X016310	Y014553				
V019	X014395	Y012946				
V020	X012480	Y011339				
V021	X010565	Y009732				
V022	X008650	Y008125				
V023	X009299	Y007750				
V024	X011214	Y009357				
V025	X013129	Y010964				
V026	X015044	Y012571				Drill 30 holes on matrix 1/2 in.
V027	X016959	Y014178				
V028	X017609	Y013803				
V029	X015694	Y012196				
V030	X013779	Y010589				
V031	X011864	Y008982				
V032	X009949	Y007375				
V033	X010598	Y007000				
V034	X012513	Y008607				
V035	X014428	Y010214				
V036	X016343	Y011821				
V037	X018259	Y013428				
V038	X018908	Y013053				
V039	X016993	Y011446				
V040	X015078	Y009839				
V041	X013163	Y008232				
V042	X011248	Y006625				
V043	X022521	Y013204	S642	F446	T03	Ream 6 holes on pitch circle 3/8 in.
V044	X024457	Y011971				
V045	X024954	Y009729				
V046	X023721	Y007793				
V047	X021479	Y007296				
V048	X019543	Y008529				
V049	X000000	Y000000	S000	F000	T00	Return to datum.

Figure 5.4 Computer output for AID example

be used with each turret. Not all drilling machines, of course, require the speed and feed on the tape but, for those that do, AID will use the data in Fig. 5.3b to punch them in the correct form (see Fig. 5.4).

Figure 5.4 shows a printout of a typical control tape. In this case the speed (S) and feed (F) have been put into the 'magic three' code. The turret numbers are shown after the letter T. Further notes are given on Fig. 5.4 itself. Since the control tape requires feed in in./min, the F448 code, for example, has been computed as follows: -

$$\text{Feed} = 805 \times 0.006 = 4.83 \text{ rev/min}$$

Thus, magic three code = 448 (see feed code in chapter 6).

AID has been produced in several versions for particular purposes, but the features described have been selected to demonstrate typical capabilities of a fairly simple computer program for NC.

PROFILEDATA

In the UK, one of the earliest applications of NC was to the milling of quite complicated curved contours such as wave-guides and the great variety of shapes occurring in the aircraft industry. These shapes are usually composed of an assembly of straight lines and circular arcs and in addition sometimes include empirical curves, that is, smooth curves through a given sequence of points. One problem in obtaining a control tape for such parts is the calculation of the intersections and tangency points between the various straight and curved segments, since the drawing does not usually give the dimensions of these points directly. Furthermore, some systems require the cutter offset to be calculated before the tape is prepared (as distinct from its being set on the control console) which presents an additional calculating task.

To cope with this type of application, a computer program known as PROFILEDATA was developed between 1962 and 1964 by the Numerical Control Department of Ferranti Ltd. It was one of the earliest in the field, has been continuously enhanced since 1964, and is in widespread use today for machines fitted with Ferranti control systems. Originally it ran on a Pegasus computer, but is now available also on the ICL 1900 and IBM 360 computers. In passing it may be noted that the current Ferranti continuous-path control system uses specially recorded magnetic tape instead of paper tape, and a special-purpose computer known as a curve generator is required after the general-purpose computer in order to produce this. Most users of PROFILE-DATA have their magnetic tapes made by Ferranti from the part programs they submit, but those with access to a suitable computer can purchase a curve generator and thus carry out the whole process themselves.

A full description of PROFILEDATA is given in reference (2). Figure 5.5 shows a very simple part which serves to give an idea of a few of the facilities offered by the program. Apart from the conventional drawing dimensions, it will be seen that the straight lines and circle of which the part is composed have been labelled S2, C5, etc. Similarly the 'change points' where one segment meets another have been labelled P1, P8, etc. The arrows on the lines and circle have a bearing on the geometric definitions described below, but need not concern the reader at the moment. When using PROFILEDATA, the part programmer

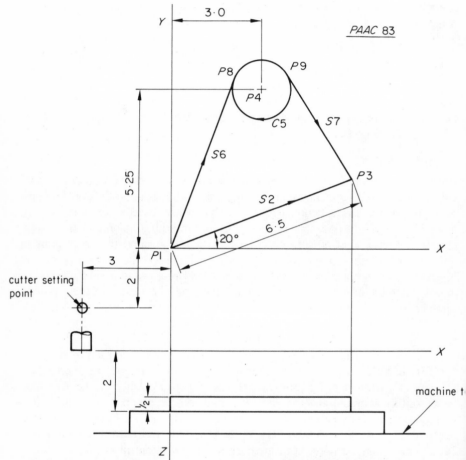

Figure 5.5 Part for PROFILEDATA example.

prepares first of all the geometric definitions in terms of the various labelled segments, and then a cutting sequence which defines machining requirements in terms of the labelled change points (and also gives other necessary information such as feedrates). The computer uses the geometric definitions to calculate the co-ordinates of the change points, inserts these co-ordinates into the cutting sequence and proceeds with the production of the control tape.

Figure 5.6 shows the geometric definitions, and Fig. 5.7 the cutting sequence, for the part shown in Fig. 5.5. These will now be described line by line. A coordinate system must first be established and for convenience X, Y, Z axes are chosen such that the point P1 has X = 0, Y = 0; the position Z = 0 is chosen, again for convenience, to be 2 in. above the mounting block as shown, and in this particular case is the level at which the cutter is set initially. The geometric definitions (Fig. 5.6) then have the following meanings.

IDENTITY CODE	GEOMETRIC DEFS. TAPE No.		TYPE	M.T.D. No.
PAAC	83		GD	222114

DEFINITIONS				
P1	0·0	0·0		
S2	B	20·0	P1	
P3	D	S2	6·5	P1
P4	3·0	5·25		
C5	P4	1·0		
S6	P1	TC5		
S7	TC5	P3		
P8	S6	C5		
P9	S7	C5		
END				

Figure 5.6 PROFILEDATA geometric definitions.

Line 1 States that P1 is a point with X and Y co-ordinates both zero.

Line 2 Means that S2 is a line at an angle (B means 'at an angle') of 20° to the X axis direction and passes through P1. Note that P1 having been defined in line 1 it can be referred to as P1 in any later line without re-definition.

Line 3 Means that P3 is a point at a distance (D means 'at a distance') along S2 of 6·5 in. from P1. Here, two previously defined elements, P1 and S2 have been used.

5*

IDENTITY CODE	PAAC	CUTTING SEQUENCE, TAPE No.	83	TYPE	CP	GEOMETRIC DEFINITIONS TAPE No.	83
M.T.D. No.	222114	STA	DAX	-3·0	DAY	DAZ	0·0	DIA	0·75
					CP -2·0				

LINE No.	DATA
1	TCL RAT 30·0 CXY 1
2	COZ 1·4
3	RAT 3·0 COZ 2·010
4	RAT 6·0 CXY 8
5	CIR XAY CXY 9 PXY 4
6	CXY 3
7	CXY 1
8	RAT 30·0 COZ 0·0
9	TCC COX −3·0 COY −2·0
10	HALT
11	STO.

5.7 PROFILEDATA cutting sequence

Line 4   States that P4 is a point with $X = 3\cdot0$ and $Y = 5\cdot25$, as given on the drawing.

Line 5   Means that C5 is a circle whose centre is at P4 and whose radius is $1\cdot0$ in. Notice that the whole circle is thought of at the moment although only a portion of it is used in the finished part. The subsequent cutting sequence data state how much of the circle is required.

Line 6   Means that S6 is a line which passes through P1 and is tangential to the circle C5 (TC5).

Line 7   Means that S7 is a line which is tangential to C5 and passes through P3.

Line 8   Means that P8 is the point of tangency between C5 and S6.

Line 9   Means that P9 is the point of tangency between C5 and S7.

Line 10   END indicates the end of the geometric definitions.

When the above description is followed through line by line, it will be seen that it is simply a case of writing down in a special way the information which the drawing gives. The computer, taking the information in Fig. 5.6 as data, calculates the co-ordinates of all the change points. Only a few of the ways of defining points, lines, and circles have been used in this example, but PROFILE-DATA provides many more, enabling the part programmer to deal with most situations. All that is necessary is that the conventional drawing shall contain sufficient information to define the part unambiguously.

For machining purposes, the cutter must be set to some position clear of the part initially. The point chosen is $X = -3\cdot0$, $Y = -2\cdot0$, $Z = 0$ as shown in Fig. 5.5 and quoted at the top in Fig. 5.7. The tool diameter, required for calculations of cutter offset is also given; as a left-handed co-ordinate system is used, positive Z is downwards. The cutting sequence may now be briefly explained as follows.

Line 1   TCL means 'tool centre left' and indicates that the cutter is to go round the part in such a direction that its centre lies to the left of the contour when looking in the direction of cutting. Thus, the cutter offset calculation is initiated. RAT 30.0 means a feedrate of 30 in./min which is a rapid rate used only for bringing the cutter in air to the job, CXY 1 means that the cutter is to move from its initial position to the X, Y position corresponding to point P1 (as defined previously).

Line 2   This directs the cutter to move down to a Z value of 1.4, which is just clear of the job.

Line 3   RAT 3.0 changes the feedrate to 3 in./min for boring into the job. COZ 2.010 takes the tool vertically to a Z value of 2.010 which puts its end just below the underside of the plate.

Line 4   RAT 6.0 gives 6 in. min for contouring and CXY8 takes the cutter from P1 to P8 (i.e., along S6).

Line 5   Takes the cutter around C5 to P9. XAY means the circle is in the XY plane and that it is to be traversed in a clockwise direction, and PXY4 denotes that its centre is at P4.

Line 6   Takes the cutter along S7 to P3.

Line 7   Takes the cutter along S2 to P1.

Line 8   Changes the feedrate to 30 in./min and raises the cutter to Z = 0 (i.e., above the job).

Line 9   Takes the cutter to X = $-3 \cdot 0$, Y = $-2 \cdot 0$, that is to the initial setting point. TCC cancels the previous TCL and ensures that the *centre* of the cutter arrives at X = $-3 \cdot 0$, Y = $-2 \cdot 0$.

Line 10   Indicates that the tape deck on the control console is to be stopped on completion of the job.

Line 11   Informs the computer that the end of all the data has been reached.

PMT2

There are some NC jobs for which the calculations required are so complex and lengthy that the use of a computer is essential if the job is to be an economical proposition. Such a problem occurs in the milling of a surface curved in all directions—for example, a turbine blade or a model of a ship's hull. Surfaces such as these are usually defined not by means of a mathematical expression but simply by a mesh of points through which the surface must pass (the co-ordinates of each point being given). The usual 3C way of machining such parts is to take a series of adjacent passes with a ball-ended milling cutter or a radiused end-mill. The chief requirements which dictate the use of a computer may be summarized as follows.

(a) The surface is undefined except at the points given. All that is known is that it must be 'smooth' between the points, and a method of interpolation has to be used which will provide the rather vaguely defined quality of 'smoothness'.

(b) In order to obtain a good finish the passes of the cutter have to be closely spaced. The calculation of these paths therefore will involve the interpolation of perhaps thousands of points between the given ones.

(c) The cutter offset calculation, which in plane milling is a two-dimensional problem, is in this case a three-dimensional one. Furthermore, in order to carry out this calculation, the direction of the 'normal', or perpendicular, to the surface has to be found for a large number of points; this is a tedious calculation.

One of the computer programs for this problem is known as PMT2[3] and has been available since 1962. It is offered by International Computing Services Ltd. A brief history of this program will indicate how development has taken place to meet the demand.

In 1956, Ferranti Ltd had a well established continuous-path control system and a computing service for the preparation of magnetic control tapes for two-dimensional parts including a linear movement in the third axis. At that time, the de Havilland Engine Company Ltd was interested in the possibility of producing, by NC, blade masters for gas turbine engines, and the author, in conjunction with Ferranti, wrote a program for the Pegasus computer which, starting with the fundamental blade design data, produced a tape to cut the surface by means of a series of passes of a ball-ended cutter. In 1960, the program was rewritten by the author at the Computer Department of Ferranti Ltd to take advantage of an improved system of producing magnetic control tapes and to make it of wider application. During the following few years, the program was used, mainly for blade work, and experience with users led to the incorporation of a number of improvements. It continues to be used and now runs on an ICL 1905 computer. A further version, known as AMT2, has been written for the Atlas computer and came into use during 1968. This version can be adapted for use with other manufacturers' control systems.

A typical job for which this program is suitable is shown in Fig. 5.8, in which the axes are shown in isometric form. The surface is defined by a number of cross-sections A to D, each of which is itself defined by a number of points. The cross-sections need not be plane, need not all have the same number of points, and will usually be of varying shape. When assembled together, the cross-sections may be used to define a surface passing through the resultant mesh of points. To give some idea of the number of cutting passes required, a small surface, say two or three inches wide, would require up to 100 passes. Figure 5.8 shows two typical adjacent passes as produced by PMT2. The tool travels in such a manner that the path of the *centre* of the tool is always parallel to the XZ plane. After each stroke, the tool moves across the work a short distance (e.g., 0·030 in.) and then proceeds in the opposite direction along a path which is again parallel to XZ. This process starts at one side of the work and continues until the other side is reached. The resultant surface will show cusps produced by the successive passes of the tool; provided the strokes are sufficiently close together the bench work required for finishing is not too great.

A part program for PMT2 consists largely of lists of the co-ordinates of the defining points, normally X, Y, and Z, although polar co-ordinates may also be given. The user obtains these either by previous calculation, or more often by reading from large-scale drawings. In addition, the following further information has to be given.

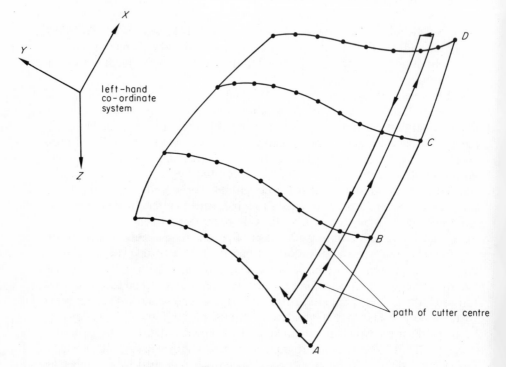

Figure 5.8   PMT2 example.

(a) The co-ordinates of a sequence of points specifying the boundary of the surface. These can be simply the outermost points of the mesh, but a better surface will usually be obtained if the mesh is defined for a short distance outside the boundary.

(b) The size of the cutter to be used. This is required in order to enable the cutter offset calculation to be carried out. In the case of a ball-ended tool, simply the radius is given; the radiused end-mill is defined by two radii.

(c) The spacing required between adjacent strokes. This depends on the finish required and on the cutter size, a large spacing and small tool, for instance, giving large cusps.

(d) The co-ordinates of the point at which the cutter centre will be set before machining starts. Using these values, the program computes and punches onto the control tape the instructions to bring the tool into contact with the work-piece from its starting position. At the end of the job, and intermediately if required by the user, the cutter will be returned to the starting point.

(e) The feedrates desired for cutting metal and for 'cutting air', i.e., for movements to and from the work.

(f) The lengths of the reels of magnetic tape to be used. Ferranti control tapes are available in various reel lengths and usually several reels are required to complete a PMT2 job. PMT2 itself produces its results on punched paper tape which is separately converted to magnetic tape by means of a special purpose computer. (See the beginning of this chapter.) Where a fresh reel of magnetic tape is required, the paper tape contains instructions for the tool to return to its original setting point, so that the operator can change the reel.

In addition to the facilities mentioned, PMT2 provides easy means of rotating and scaling the original data. One example of the need for scaling occurs when forging dies are being machined; the user's data refer to the finished size of the forging but, to allow for contraction, the dies must be made larger by say 2 per cent. Instead of having to multiply all his data, the user can submit it in its original form together with the value of the multiplying factor, e.g., 1·02. The program will then automatically carry out the multiplication on the given dimensions.

The PMT2 program has been used for many surfaces, the majority being turbine and compressor masters and dies, but applications include ship models, wind-tunnel and free-flight aircraft models, special mirrors, and miscellaneous shapes such as a large-scale model of a typewriter key.

MILMAP

MILMAP[4] has already been mentioned as an example of a program which automatically selects feeds and speeds from tables stored in the computer along with the program itself. Although it provides geometric features such as patterns of holes and some contouring capability, this ability to provide feeds and speeds is its main feature.

It was developed by ICT (now ICL) from 1963 onwards for use with the Milwaukee-matic machining centres in their own factory. In 1966, work began on an improved MILMAP program which could utilize all the facilities provided by these machining centres. Typical components for machining are large castings requiring many different operations on various faces of the component all at one set-up: drilling, tapping, spot-facing, counter-boring, and so on; many different diameters are involved and by no means all of the holes lie in patterns. Milling is also involved, and includes face milling, pocket milling, and individual line milling. Contouring requires linear and circular features in XY, YZ or ZX planes, with the working depth varying within the contouring sequence.

MILMAP meets these requirements and also optimizes the part program so as to minimize tool-changing, indexing, and non-machining movements. Facilities for avoiding collisions are also provided. MILMAP runs on ICL 1900 series computers and is also available as a computing service. Postprocessors have been produced for Milwaukee–matic models Ea, 2, 3 and Extended 2, and for the Marwin Max-E-Mill. Others are under consideration.

The necessary machining data are contained in two tables known as the Tool Store and the Material Store, which are kept in the computer's magnetic tape store.

Tool store For each tool required the Tool Store holds a code number, the tool diameter and setting length, the speed and feed when using a 'standard' material (see 'Material store' below), and the code numbers of any preceding tools, e.g., the entry for a tap would hold the code number of the corresponding drill. This 'preceding tool' feature means that the part programmer need only call for, say, a tap and not be concerned with the drill, for such a call will result in the computer producing automatically the drilling operation.

As well as the above information, the Tool Store holds the so-called extension lengths and diameters, i.e., the dimensions of shoulders on the tool and of the tool holder. The purpose of these is to enable collisions between tool and work-piece to be avoided, a job which the program also carries out. (See 'Guard towers' below.)

Material store As mentioned above, the Tool Store gives the normal speed and feed for each tool assuming a standard material, which is medium cast iron in ICL's own version, though users may choose their own. The Material Store gives conversion ratios for feed and speed for other materials.

It is assumed that surface speed for any particular type of tool (e.g., drills) is independent of tool diameter, so that only one conversion ratio is required for each combination of tool type and material. But, in the case of feedrate, the conversion ratio depends also on the diameter, and it is found sufficiently accurate to divide tools into four diameter ranges. Thus, for each combination of material and type of tool, there are four conversion ratios for feedrate and these, together with the speed ratio, form the Material Store. Different users have different ideas about values of speeds and feeds, and this is recognized in MILMAP in two ways: both the Tool and the Material Stores may be drawn up according to users' requirements, and, further, the part programmer may insert correction factors for, say, the drilling of a particular hole or group of holes. Thus, if he feels for some reason that the values derived from the stored tables are not suitable, he can impose his own ideas; variations may be required, for instance, when machining a particularly slender component, or under vibration conditions.

In addition to the above, facilities provided by MILMAP include the automatic generation of instructions for the milling of deep holes, and for milling to a large working depth. The usual change-of-datum facility is also provided. Two further features deserve special mention.

Sequencing   When writing a part program it is usually more convenient to consider each component face in turn and to write down all the statements for the machining of that face, whatever the variety of tools used or operations performed. MILMAP then sorts and optimizes the part program to minimize the tool-changing, indexing, and tool motion. The program ensures that, in general, machining operations are performed in the following order: rough milling, fine milling, contouring, spot-drilling, drilling, boring, reaming, counter-sinking, tapping. For any specific operation or series of operations, this sequencing can be overridden by the part programmer.

Guard towers   It was mentioned under 'Tool store' that the program checks for collisions between tool and workpiece. To enable this to be done, the part programmer must give details of obstructions on the workpiece, i.e., portions which project from the working face toward the tool holder. This is done by giving the dimensions of so-called 'Guard Towers'. A Guard Tower is a rectangular block enclosing the obstruction and is thus an approximation to the true shape of the obstruction; provided it completely encloses the true shape, there will be no danger of collision, for the computer will check every movement and calculate a safe path for the tool.

General features of programs

As well as the programs described above, many others have been produced—such as, for point-to-point and straight-line milling: KIPPS (ICSL), AUTO-SPOT (IBM); for 2C,L type contouring: AUTOPRESS (Pressed Steel Fisher Ltd), CLAM (Hawker Siddeley Aviation), COCOMAT (Rolls-Royce Ltd), ADAPT (IBM, UNIVAC, Honeywell, etc.).

The notation used for part programming—i.e., the language—varies considerably from one program to another. A major distinction is made between 'fixed-format' and 'free-format'; the former denotes that the part program may be written on a standard form with a number of headed columns, implying that the statements are of a fixed length and that the items making up a statement must be in a definite sequence; 'free-format', on the other hand, allows the user to construct statements as required within certain limits. All the programs mentioned in this chapter, except AUTOSPOT and ADAPT, are of the fixed-format type, which is said to be easier to remember; free-format schemes (e.g., ADAPT) may take longer to get used to, but allow more natural terminology to be used and more sophisticated part programs to be written. Examples of these are given in later chapters.

Another distinction between various part programming languages is whether or not they allow items to be defined symbolically and referred to later in the part program. PROFILEDATA, described above, allows this—as shown, for example, by the use of P1 and S2. AUTOPRESS and COCOMAT, however, do not allow this and each item has to be given its numerical value every time it is used.

There are many other features under which programs may be classified and, for a survey of these, the reader is referred to reference (5).

Special-purpose and general-purpose programs

The programs mentioned in this chapter, and several others, were produced for a more or less limited purpose, e.g., for a particular control system or machine tool, to run on a particular computer, or for a special class of component or user. As users began to acquire a variety of NC machines, it became inconvenient to have to use several different part programming languages and perhaps several computers. Because of this, *general-purpose* programs were produced—general in the sense that they used, as far as possible, a common part programming language, they could be run on any modern computer of sufficient size, and they could produce control tape for a wide range of machine tools and control systems. The latter feature is made possible by providing a number of 'post processors' as part of the program; this subject is dealt with in chapter 11, but it should be noted here that even some of the more restricted programs mentioned in this chapter use the post-processor idea to enable them to work with a number of machine tools and control systems.

General-purpose programs are dealt with in chapters 7 to 10, and the reader is also referred to reference (5).

However, it should not be thought that these programs entirely supersede the special-purpose ones, or that no special programs should be produced in future. Quite apart from the fact that users may prefer to continue with a program they know, a special program can have advantages of its own, and it is worth considering in a new application whether a standard program will do, or whether it would be better to obtain a special 'tailor-made' one. Some possible advantages of a special program are as follows.

(a) It may be cheaper to run on the computer, because it will contain nothing extraneous to the application.

(b) It may run on a smaller computer.

(c) Part programming will be easier, and in some cases could amount to little more than the filling in of dimensions on a pre-printed sheet.

(d) It may be able to do much more for the user than could any standard program. Consider, for example, a multi-spindle drill: the number of holes to be drilled with all spindles operating, the number with one drill removed, the number with two removed, and so on, must be determined. This involves considering the pattern of holes and the relative times of drilling and moving from one position to another. No standard program will do this.

Again, the problems of part programming for lathes have already been mentioned in connection with feed, speed, and depth of cut. The automatic determination of these for a given application is difficult for a standard program because it is so dependent on the particular lathe and type of component. In addition, there is the problem of avoiding collisions between slides and workpiece; this is a very difficult problem and probably impossible to solve with a standard program but, for a particular machine and type of component, a special program could greatly simplify the planner's task.

However, a special program will have to be paid for, and usually entirely by one organization. It must also be realized that it will be relatively inflexible and not always easily modified to cope with changed conditions. It is, therefore, vital to consider the specification of such a program at length, in order to ensure that it meets present, and all reasonable future, requirements. If this can be achieved, the cost of the program could be outweighed by the saving in part-programming effort.

Acknowledgements

The author gratefully acknowledges the permission of International Computing Services Ltd to publish material relating to AID and PMT2, of ICL for MIL-MAP, and of Ferranti Ltd for PROFILEDATA. Thanks are due to Ferranti also for help in preparing the section on PROFILEDATA.

References

1. *Point-to-Point work using AID Mk 1*, International Computing Services Limited, Publication CS 436b, 1966.
2. *Profiledata Part Programming Manual*, FSP 313, Ferranti Limited, Numerical Control Division.
3. *Surface Milling using PMT2*, International Computing Services Limited, Publication CS 455, 1966.
4. *MILMAP Numerical Control Program*, International Computers Limited, Technical Publication 4028, 1967.
5. *Programming of Numerically Controlled Machine Tools*, National Engineering Laboratory, NEL Report No. 187, 1965.

6. Standards

W. H. P. Leslie

W. H. P. Leslie, CEng, FIEE, FIMechE, BSc (ElEng) is Head of the Numerical Control Division at the National Engineering Laboratory, East Kilbride, Scotland, and a member of the British Numerical Control Society since its formation. After graduating from Glasgow University in 1940, he started his career at the Royal Aircraft Establishment. He transferred to the Fluids Division of the National Engineering Laboratory in 1951, and became subsequently Head of the Machine Tool and Metrology Division, Technical Adviser of the NEL Automatic Design Sub-Committee, and moved to his present position in 1965. He is currently Chairman of the Institute of Mechanical Engineers Automatic Control Group, and President of Glasgow University Engineering Society. He is a member of several UK committees concerned with NC, and Secretary of the Ministry of Technology Advisory Committee for Machine Tools and Numerical Control; a member of the UNCL (Unified Numerical Control Language) International Committee; and Editor of the Proceedings of the 1st International Conference on NC Programming Languages (PROLOMAT) organized by IFIP/IFAC. He was leader of the British delegation to the ISO NC Computer Programming Working Group in 1968 and 1969.

As with other rapidly developing new subjects, each NC system developer made his own choices for each of the variables which affected the user.

Thus, in Britain, we saw punched cards of various sizes; punched paper tape with 5, 6, 7, and 8 rows of holes, and 0·625, 0·875, and 1·0 in. wide; plastic film 35 mm and 4 in. wide; and magnetic tape 0·25, 0·5, and 1 in. wide—all used to carry the numerical information read by the NC system. Various letters, or no letters, were used to indicate the main movements (usually X, Y, and Z). The movement which would occur, however, for X on a given type of machine tool, varied from one control system to another; even when there was a consensus of opinion (e.g., the use of Z for the axial movement of a drill), some systems used $+Z$ to move the drill into the metal, whereas others used $-Z$.

In the case of numerical specification of speeds, feeds, etc., and the order in which information was specified along the tape (the format), there were as many methods as suppliers.

The EIA Standards

In the USA, where NC machines became used on a wide scale most rapidly, the need for standards was realized and the Electronics Industries Association published a series of standards backed by National Aerospace Standards of a similar form.

Much of the equipment sold to Europe from the USA conformed to these specifications and in due course they were also submitted to the International Standards Organization as a basis for ISO recommendations for NC.

ISO recommendations

Early in 1968, the countries who participated in ISO work agreed to a final form of these standards. They have been issued as:

ISO/R840   *Code for numerical control of machines compatible with the ISO 7-bit character set*

ISO/R841   *Axis and motion nomenclature for numerically controlled machines*

ISO/R1057   *Interchangeable punched tape variable block format for position- and straight-cut numerically controlled machines*

ISO/R1058   *Punched tape variable block format for positioning and straight-cut numerically controlled machines*

ISO/R1059   *Punched tape fixed block format for positioning and straight-cut numerically controlled machines*

ISO/R1056   *Punched tape block format for the numerical control of machines —coding of preparatory functions G and miscellaneous functions M*

British Standard Specification

The British Standards Institution have completely rewritten the above ISO recommendations to make them more easily followed, while maintaining the effect of the texts to be the same. The relevant document, which makes it easier to appreciate the difference between the agreed formats, is about to be issued as a BS on '*The numerical control of machines*'.

In this one document is brought together all the material which will enable a customer and a supplier to discuss those properties of NC machine tools which affect interchangeability of control tapes for machines capable of dealing with similar work. As NC machines are purchased internationally, the ISO agreement gives the purchaser an opportunity of buying the machine which best fits his work without having to introduce new tape-punching or writing equipment, or having to retrain staff in new manual programming methods.

NC computer programming languages

Since November 1967, there has been active discussion internationally under the wing of the ISO on standard languages for NC programming and on a standard output format (or CLDATA) from NC programs.

It is too soon (1970) to record a final recommendation, but the discussions are currently based on the use of APT-like input languages (see chapters 7, 8, 9, and 10) and APT 3 CLDATA (see chapter 11).

Punched tape

Paper tape

Because the tape used for input to NC systems was based on similar tape for telecommunication and computer purposes, and because this tape was nearly always only used a few times, paper was the most common material used for the tape.

In the standards documents, the adjective 'paper' has been used to cover all materials employed for tape of the normal size, even when the material is a plastic or a metal film, or a combination of paper and/or plastic and/or metal.

There are two British Standards dealing with 'paper' tape. BS 3880 entitled *Dimensions of punched tape for automatic data processing* deals with the width of the tape and the relative positions at which holes may be punched. Figure 6.1 illustrates these dimensions. BS 3967, *Representation on one inch punched tape of 6 and 7 bit coded character sets for data interchange*, defines only a few terms.

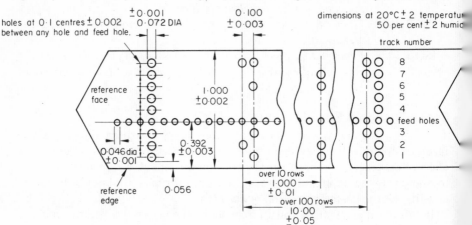

Figure 6.1   Standard tape dimensions for eight-hole tape.

It refers to BS 3880 for dimensions and hole layout and states, in effect, that the reference edge of the tape is the edge nearest the feed holes. An observer looks down on the reference face if it emerges from a tape reader, or punch, towards him with the reference edge to his right. In Fig. 6.1, it is the reference face which is illustrated. BS 3967 also ties up a few loose ends such as indicating that a punched hole corresponds to a binary one (or bit) and that the least significant bit shall be punched in the first track next to the reference edge, with the most significant bit in the seventh track, and the even parity bit in the eighth track (furthest from the reference edge). The originator of the tape should indicate the direction of movement of a tape. The means of doing this is not stated, but it is good practice to buy tape marked by arrows on the reference face to show the direction in which the tape should move through the tape punch or reader. Some tape equipment has an arrow-shaped tape-tearing device which automatically leaves an arrow head or pointed shape at the beginning and a tail shape at the end of each piece of tape. Leaders and trailers, consisting each of about 1 m of Null punched tape (feed-track holes only), should be left at each end of each tape to take the wear when threading or removing tapes.

There are no standards set for the thickness of the tape, its durability and ease of punching, its abrasive effect on punch knives, the amount and type of oil impregnation, its colour, or its opacity (for use with optical readers), but BS 3880 states that there should be at least 925 ft of tape in one continuous length, without a join, in a reel.

The user must inquire into all these and he may uncover the need to compromise. Thus, he may find on the one hand that the punch manufacturer recommends thin, weak, oiled tape to minimize wear on the punch, but on the other hand his NC system manufacturer (who has provided an optical reader) recommends that a thick, durable, unoiled, very opaque tape be used. The oil may smear the optical system in the tape reader but may be needed to lubricate the knives in the punch.

There is a need for working standards for the mechanical and optical properties of tape and work is being slowly done to this end. Unfortunately, tests must be related not only to a range of tape materials, but also to a range of types of tape readers and punches, neither of which are themselves standardized.

At least two types of tape seem to be required. One should be easily punched and lightly oiled, and thus particularly suitable for high speed punching in a computer, yet strong enough for testing out on a machine tool and even for producing a small batch of parts. The other should be tough durable material, completely opaque, free from oil and thus suitable for use for production control tapes after the initial tape testing has proved successful. The tested tape would then be copied in a tape-duplicating system using a punch specially built and maintained for durable paper or metal or plastic tape.

When purchasing a control system, the specification should state that the control tape-reading equipment should be able to deal with both the easily punched (and semi-translucent) tape and the more durable opaque tape.

Tape-punching codeset

Many NC machines were developed, particularly in the USA, using the EIA codeset, details of which are contained in EIA Specification RS 244.

This codeset is illustrated in Fig. 6.2 and was adopted for NC work many years ago. Within the limitations of manual preparation of control tapes, it was found to be very satisfactory.

For at least the past five years, a different codeset has been under development

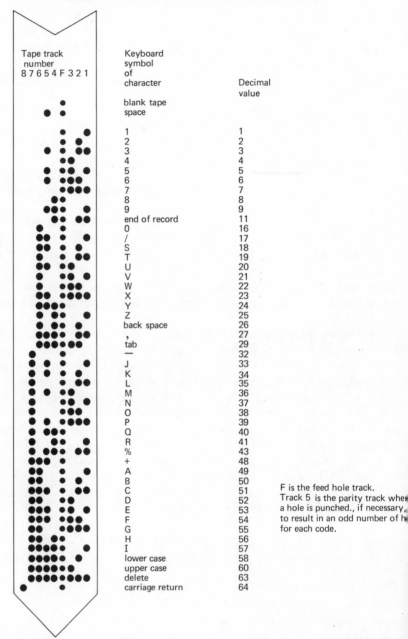

F is the feed hole track.
Track 5 is the parity track where
a hole is punched., if necessary,
to result in an odd number of h
for each code.

Figure 6.2   EIA punched tape codeset.

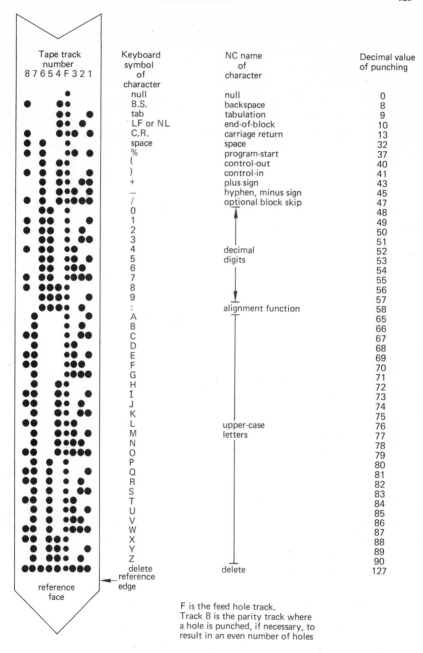

Figure 6.3 ISO NC codeset (seven on eight hole tape).

in the USA, known as ASCII-USASI X3.4–19 (American Standard Code for International Interchange—United States of America Standards Institute).

The ASCII codeset was proposed, as its name implies, for information interchange by punched tape. It was aimed at *all* applications of punched tape, not just NC, so that one type of punched tape equipment would be suitable for computers, typesetting, telecommunications, and NC. The details of the full codeset can be found in the forthcoming British Standard, 'The U.K. 7-bit data code', or in ISO/R646, but for NC applications the full codeset (which contains upper *and* lower case letters and many control codes for telecommunications in addition to numbers) is not required. Figure 6.3 shows the portion of the ISO codeset required for control tapes, as described in ISOR/840, while Fig. 6.4 shows the additional characters '$_*,· = EM$' which are required for NC part-programming tapes if a computer is to be used to help to prepare the control tape.

A new NC user would be well advised to specify the use of this ISO codeset for NC.

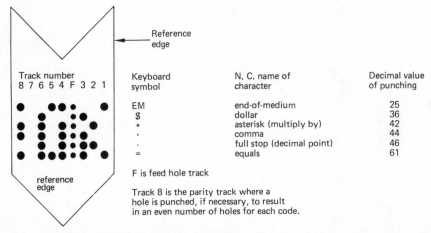

Figure 6.4   Additional ISO codeset characters required for NC computer programs.

The tape-punching typewriter

When purchasing the necessary tape-punching typewriting equipment, the buyer should ensure that it can also cope with the additional part-programming characters. These will have little or no effect on cost when making a purchase, but much more would have to be paid later to add any essential characters missing from the tape-punching typewriter, should it then be desired to use an NC computer-programming system. On some computers, the £ appears instead of the $ but the hole pattern punched on the tape is identical. This is only a minor inconvenience once the typist and the programmer are aware of this different symbol.

There is a temptation when first moving into NC to try to keep the cost of the tape-preparation equipment down, in order to get the maximum number of optional extras on the control system and machine tool. This can be found later to be a false economy.

With a minimum working arrangement of tape equipment it should be possible to:

(a) Punch tape and type a copy of the program by operating the typewriter keyboard.

(b) Read tape and punch a copy of the tape.

(c) Read tape, type a copy of the program, and also punch a tape when desired.

(d) Check that an even number of holes exist on the tape for every character (check for even parity).

(e) Tabulate, when punching and reading the tab code (see Format section in this chapter).

It is possible to purchase equipment which will do much less than this for less money, but such economy will be regretted in the long run. Thus, with equipment which has only a keyboard and a punch, a control tape can be produced. With this it is very difficult to keep a check on where the punching has reached, so that there may be omission or duplication of characters in the final tape. If a control tape has to be corrected, facility (c) above is essential.

Verifying tapes

If the main use of this tape punching equipment is for manually programmed work, and the programs are likely to be many thousand characters long, the tape-punching equipment should also be suitable for verifying tapes.

When verifying, a control tape is produced normally by typing. This tape is then inserted in the reader and the verifying circuit is switched on. A second control tape is then punched by keyboard entry of each character read from the initial handwritten copy of the tape program. As a key is operated, the next code is reached from the completed control tape and compared with the code generated by the key. If they agree, this code is punched on the new control tape. If they disagree, the system locks without punching and displays the codes from tape and key. The operator checks the manuscript from which she is working, decides which source is correct, and then punches the correct code in the control tape and continues with the verification.

Alternatively, two control tapes are independently produced and simultaneously read into a verifier which punches when the codes agree, or stops if they disagree.

It is usually cheaper to verify than to risk damage to machine tools and workpieces when a new manually produced control tape is being tried for the first time.

If the tape is produced by computer, it may possibly be checked, after punching, by being read back into the computer and compared with the stored version. This is a form of automatic verification by the computer.

Axis and motion nomenclature

Standards were developed in the USA aimed at making the names of the various movements of machine tools similar and thus predictable. Two separate problems are involved: deciding whether X, Y, or Z, for example, applies to a particular movement; and then deciding which of the two directions of that movement should be positive and which negative.

The US standard EIA document RS267 was the basis of the international standard ISO/R841 which was adopted early in 1968. The corresponding British document will be the British Standard on '*The numerical control of machines*'.

Principles

The guiding principle is the use of the conventional mathematical right-hand co-ordinate system of Fig. 6.5 within which movements of the cutting tool are

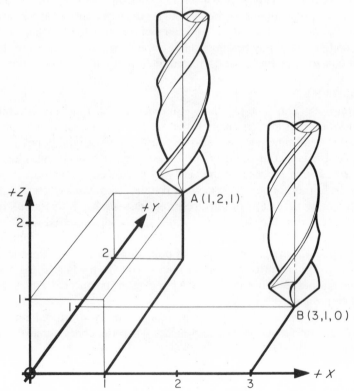

Figure 6.5   Right-hand co-ordinate system and an incremental move of (2, −1, −1) from A to B.

Figure 6.6 Axes on a drilling machine. (*Courtesy A. Herbert Ltd*).

described relative to the workpiece. The part programmer need not know which is positive X direction on the machine tool. He assumes that if he calls for a movement of the cutting tool which involves $+X$, $-Y$, and $-Z$ directions from A to B in the workpiece co-ordinate system as in Fig. 6.5, then movements in the positive X direction and negative Y and Z directions on the machine tool will cause the desired relative movement of tool to workpiece.

If, on the machine tool, an actual movement of the cutter is involved, an un-primed letter is used to describe the movement in a diagram showing machine movements, such as $+Z$ in Fig. 6.6. If the cutter remains stationary but the workpiece moves, a primed letter such as $+X'$ in that figure is used to indicate positive X direction of movement.

Whether a primed or unprimed letter is involved the control system accepts the unprimed letter on the control tape. Thus, if it is desired to move the cutter from A to B in Fig. 6.5 an incremental control tape would specify $X + 2$, $Y - 1, Z - 1$.

On the drilling machine of Fig. 6.6, the table would move to the left in the $+X'$ direction by 2 units, towards the column in the $-Y'$ direction by 1 unit, while the drilling head would fall in the $-Z$ direction by 1 unit. This would re-sult in the desired drill/workpiece move from A to B in Fig. 6.5.

Application to X, Y, and Z

The standards documents indicate by diagrams how most of the common machine tools should have their axes named, and in addition they attempt to frame rules to enable the axes of any machine tool to be named.

The linking principle is that drilling type motions can be performed on many drilling, milling, turning, and boring machines. A drilling movement is thus selected to be in the Z direction on all these machines. A short study of Fig. 6.5 will show that the movement of a drill into a workpiece is in the $-Z$ direction, and away from a workpiece is in the $+Z$ direction. Figures 6.6 to 6.10 show this principle applied to drilling, milling, turning, boring, and machining centres respectively.

Note particularly the lathe in Fig. 6.8. Here the headstock axis is Z and a positive $(+Z)$ movement of the cutting tool is away from the headstock.

In all cases, motion in the $+Z$ direction increases the clearance between the complete workpiece and the tool-holder.

Where there is more than one spindle, one should be selected as the principal spindle. Preferably it should be perpendicular to the work-holding surface; it will then be parallel to the Z axis.

If the principal spindle can be pivoted so that it can be parallel to only one of the rectilinear axes of the machine, then that axis is named the Z axis. If it can pivot parallel to more than one of these axes, then the one perpendicular to the work-holding surface is the Z axis.

The X axis is perpendicular to the Z axis, horizontal wherever possible, and parallel to the work-holding surface. Where this describes two axes, the X axis will be the one which permits the longer movement. If the principal rotating tool is horizontal, then the positive X direction (+X) is to the right when looking from the tool towards the workpiece.

When there is no possibility of drilling by the principal cutting head, as for example in shaping and slotting machines, the Z axis is taken perpendicular to the work-holding surface and the X axis is parallel to the principal direction of cutting and is positive in that direction.

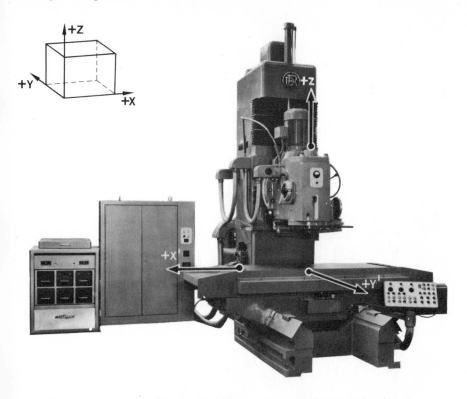

Figure 6.7 Axes on a milling machine. (*Courtesy Droop & Rein Ltd and Ferranti Ltd*).

Where the cutting-power is supplied in the workpiece spindle (as in a lathe), +X motion moves the cutting tool radially outwards from the spindle. If a second such motion exists, it is not labelled +Y because it has the same effect on the workpiece dimensions as +X. The letter +U is used instead, as explained later.

The Y axis, taken with the X and Z axes defined as above, completes a right-hand Cartesian co-ordinate system as shown in Fig. 6.11. (Note: it will save the reader many contortions if he is also prepared to use his left hand with the thumb

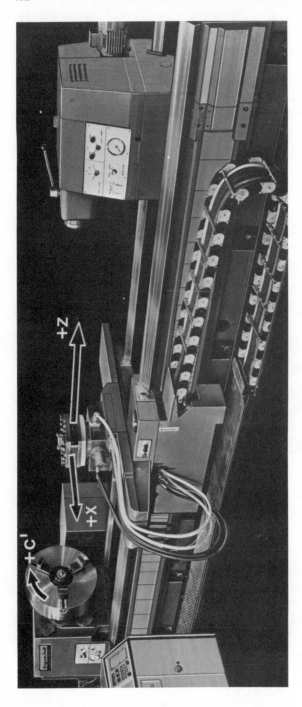

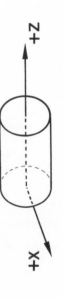

Figure 6.8 Axes on a lathe. *(Courtesy Craven Swift)*

as $+Z$, the forefinger as $+Y$, and the middle finger as $+X$: this has the same effect, but rotations should not be dealt with by the left hand.)

Where there are additional movements parallel to X, Y, and Z, the letters U, V, and W respectively shall be used. A third set of such movements shall use P, Q, and R respectively. Preferably the motions nearest the principal spindle use the primary set X, Y and Z. This is so in the case of the turret lathe, where Z applies to the tool post nearest the spindle and W to the corresponding motion of the turret.

Figure 6.9 Axes on a boring machine. (*Courtesy S.I.P. Ltd*)

Note that R can also be used as the address for a rapid movement of a drilling or milling tool towards a workpiece, associated with preparatory function words G80 to G89 dealt with later in this chapter under 'Control tape format'. The R word then states the distance of the rapid movement. See Fig. 6.14, for example.

Application of rotation

Where the work-table or tool head has to be rotated, A, B, and C are used.

6

Figure 6.10   Axes on a machining centre. (*Courtesy Kearney and Trecker Corp*)

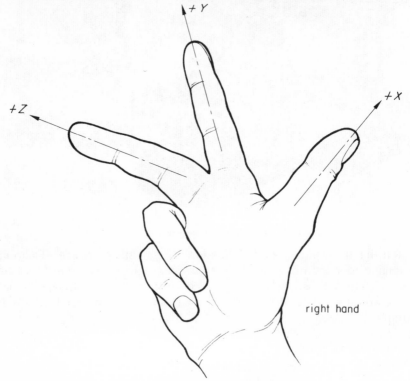

Figure 6.11a   Right-hand co-ordinate system.

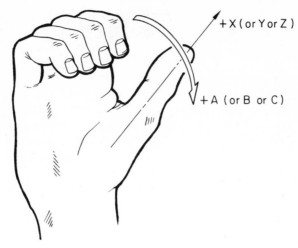

Figure 6.11b   Direction of positive rotation.

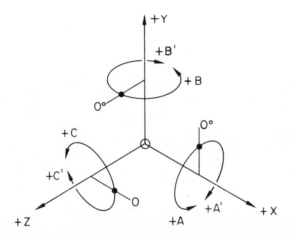

Figure 6.11c   X, Y, Z; A, B, C; A′, B′, C′ in a right-hand co-ordinate system.

They have positive directions which would advance right-handed screws in the positive X, Y, and Z directions. Alternatively, if the right-hand thumb is pointed in the positive X, Y, or Z direction, the fingers will point in the +A, +B, or +C rotational directions respectively as shown in Fig. 6.11b. When absolute angles of rotation are involved (as in the case of an indexed table), rotary angular movements are described from lines parallel to, and in the same direction as, +Y, +Z, and +X, for the rotations A, B, and C respectively as in Fig. 6.11c.

The direction of rotation of all movements except of the cutting spindle are dealt with in this way. The ISO recommendation uses clockwise and anticlockwise to describe the direction of rotation of tool spindles and lathe chuck spindles. Unfortunately, the definition leads to confusion and argument and the illustrated machine tools only show +C direction of rotation as described in the previous paragraph. As long as the terms persist, clockwise and anticlockwise should be regarded as equivalent to negative and positive rotation respectively. This is equally true for primed or unprimed movements. Thus, the +C' rotation in Fig. 6.8 is anticlockwise no matter how it may appear to the reader. It is useful to know that a normal twist drill can be used to cut when a rotation of a tool or workpiece is negative or clockwise, while a left-handed drill is needed for positive or anticlockwise rotation.

The negative rotation for a right-handed tool arises because a movement into the workpiece is in a generally negative direction.

Two further machines are shown in Figs. 6.12 and 6.13. Where there is any doubt about the application of these rules, the reader should consult the standards referred to at the beginning of this section; these are the reference documents and some 25 different types of machine are illustrated.

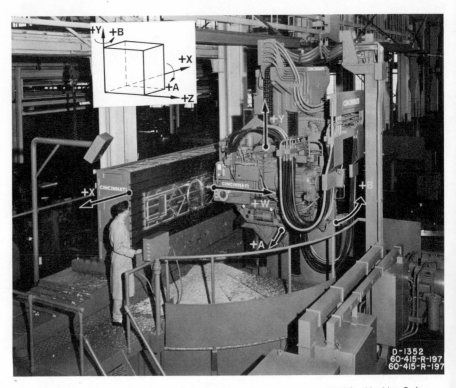

Figure 6.12 Axes on a five-axes contour mill. (*Courtesy Cincinnati Milling Machine Co.*)

Format on control tape

Introduction

Standards were evolved in the USA for the type and order of information on control tapes. The EIA produced specifications for three types of control tape, and with only slight modification these have appeared as the ISO recommendations listed at the beginning of this chapter.

Figure 6.13 Axes on a six-axes contour mill. (*Courtesy Sunstrand Corp.*)

Only the interchangeable variable-block format will be described here, with some indication of the differences permitted in the (non-interchangeable) variable-block format. The fixed-block format is almost tailor-made for each system so that it cannot be described in general. It should also be avoided unless first cost is considered to be the only important factor in an NC system. If the purchase of such systems cannot be avoided, the BS should be carefully studied. The format should be kept as near the interchangeable variable-block format as possible, as a number of fixed-block format machines may otherwise be gradually acquired whose control tapes appear to bear little or no relationship to each other.

Address and tabulation

The interchangeable format is based on the use of two methods of identifying the purpose of each word of information, so that the control tape may be acceptable to systems based on two different ways of deciding on the purpose of each block.

First of all, each word is prefixed by an alphabetic character followed by numerical data (e.g., N071, F435, X017628, T02, etc.). These address letters have the standard meanings shown in Fig. 6.14. It will be seen that they correspond exactly to the axes nomenclature letters. The form of numeric information will be described later.

The second word indicator is provided by a Tab (Tabulation) code between each word. When a word is missing from a block, the Tab codes must still appear until the last useful word in the block is reached. The end of the block is indicated by an 'End-of-block' code.

These Tab and End-of-block codes have dual roles since, in addition to their control system use, they also serve on the tape-preparation typewriter equipment to space out the words horizontally on the typescript (Tabulation) and vertically because the End-of-block code is also the 'New Line' code for the typewriter.

A practical point here is that some typewriters also need a 'Carriage Return' code to get the typehead back to the left-hand side of the typescript. It is, therefore, safe practice to punch a Carriage Return code immediately before a New Line code. The Carriage Return is usually the slowest movement on the typewriter and it can then overlap the New Line movement.

As an example of the use of the dual methods of identifying words, consider the example below, in which the character $\underline{T}$ represents the place where a non-printing Tab code would occur on tape, and $\underline{C}$ $\underline{E}$ where the $\underline{C}$arriage Return and $\underline{E}$nd-of-block codes occur.

      N070X05764Y03628F475S615T05M08$\underline{CE}$

      N071G81Z01472R0500F435$\underline{CE}$

When the Tab codes are inserted, the lines would become

      N070$\underline{TT}$X05764$\underline{T}$Y03628$\underline{TTT}$F475$\underline{T}$S615$\underline{T}$T05$\underline{T}$M08$\underline{CE}$

      N071$\underline{T}$G81$\underline{TTT}$Z01472$\underline{T}$R0500$\underline{T}$F435$\underline{CE}$

Now, when this information is typed on the tape typewriter equipment with the tabulation properly set up, each T code moves the typewriter to the beginning of the next column and presents a more easily read record:

N070		X05764	Y03628				F475	S615	T05	M08
N071	G81			Z01472	R0500		F435			

It will be seen that, purely to simplify checking the typescript, the tabulation character is well worth while, providing that the tape-preparation typewriter can tabulate automatically.

A difference in the non-interchangeable format is that it permits the use of both the address and Tab methods or of either of them.

With address only, the Tab codes are missing and typing is like the first of the three examples above. With Tab only, the information needs to be typed on a tabulating typewriter to be easily read. The example above would then appear as

.	.	.	.	.	.	.	.	.	.	.
070		05764	03628				475	615	05	08
071	81			01472	0500		435			
.	.	.	.	.	.	.	.	.	.	.

Any one Tab character missing would ruin the workpiece, as the control system would read subsequent words in a block into the wrong addresses.

Start of tape and sequence number (N)

The first block of information on a control tape should contain only the program Start code (%) and the End-of-block code as shown in Fig. 6.3. It should be preceded by at least 3 ft of control tape, the first part of which should be all Null codes (feed holes only) so that the identity of the tape can be written on it. Some control systems reject Null codes (because these would be obtained if the control circuit for the tape-punch knives broke down) so that it is best to have about 1 ft of Delete code immediately before the first block.

All other blocks should begin with a sequence number word which normally consists of N followed by three digits. Thus N000 or N065 are sequence number words. The sequence number word is read into the control system and the number is displayed while that block of information is being used. This keeps the operator in touch with the part of the control tape currently being used. (The sequence number word is optional for the non-interchangeable format.)

The letter N can be replaced by the alignment function or rewind stop character (the colon), thus :965. A block marked in this way should have enough information encoded in it (e.g., start spindle clockwise, coolant on, etc.) to enable machining to be resumed from this point after the machine has been shut down. The alignment function code can then be used as a reference rewind stop.

The sequence number word should thus always follow immediately after the End-of-block code of the preceding block. The only permitted exception is the Optional block-skip code (the solidus /). When the operator puts the Optional block-skip switch 'on', the control system will read past any block beginning /N without noting the contents of the block.

The remainder of the information in the block will now be described. The following general conditions apply, however.

(a) In one block there shall not be more than one instruction for one address. (Some relaxation of this is permitted for non-interchangeable format.)

(b) For an application to a particular machine, addresses which do not apply to that machine may be entirely omitted.

(c) Each block shall consist of the sequence number word, followed by the preparatory function word, the dimension words (in the order X Y Z U V W P Q R A B C D and E), and the feedrate word (or words, if D and/or E are used for this purpose), the spindle speed word, the tool number word, and the miscellaneous function word.

(d) A Tab code shall appear between each word but not before the sequence number word nor after the last word in a block.

(e) The End-of-block code (preferably preceded by a Carriage return code) shall follow the last word in each block.

Preparatory function word (G)

The preparatory function word consists of the address character G followed by a two-digit code number. It follows the sequence number word and a Tab code.

There are thus 100 possible preparatory functions, some of which are reserved for contouring. Those most likely to be encountered are:

G00	Point-to-point	A*	Positioning mode of control if several modes are available
G01	Linear	A	Linear interpolation mode of control
G04	Dwell	B*	A predetermined time delay before executing current block instructions
G05	Hold	B	An indefinite delay before executing current block instructions terminated only by operator or interlock switch
G06 to G08	Unassigned		May acquire standard use. Avoid
G13 to G16	Axis selection	C*	Usage described in itemized data for the machine
G22	Coupled motion+	B	Some line motion controls permit automatic coupling of two axes to move in same sense
G23	Coupled motion−	B	As G22 but second motion is in reversed direction
G24	Unassigned		May acquire a standard use. Avoid
G25 to G29	Permanently unassigned		Available for peculiar purposes on any machine

* See note on these letters at the end of this list.

G32	Unassigned		May acquire a standard use. Avoid
G33	Thread	A	Specifies constant-pitch thread cutting
G36 to G39	Permanently Unassigned		Available for peculiar purposes
G40	Cancel cut. com.	D*	Cancel existing cutter compensation instruction. Cancels codes G41 to G52
G41	Cutter left	D	Cutter compensation to be applied to the left when looking in direction of cutter motion
G42	Cutter right	D	Cutter compensation to the right
G43	Compensation positive	B	Add value preset on control panel to relevant co-ordinate dimensions in block
G44	Compensation negative	B	Subtract preset value, etc.
G45	Comp. $+/+$	B	Group of computer compensation codes for use
G46	Comp. $+/-$	B	in line milling. Apply to two axes stated in the
G47	Comp. $-/-$	B	machine-itemized data. The $+$, $-$, and 0 indicate
G48	Comp. $-/+$	B	respectively that values preset on the control panel
G49	Comp. $0/+$	B	have to be added to relevant co-ordinate dimen-
G50	Comp. $0/-$	B	sions, or subtracted from them, or ignored
G51	Comp. $+/0$	B	
G52	Comp. $-/0$	B	
G53	Cancel linear shift	E*	Cancel linear shift instruction. Cancels codes G54 to G59
G54	Linear shift X	E	Datum shift in Line Machining Control Systems
G55	Linear shift Y	E	by amounts preset on control panel assigned to
G56	Linear shift Z	E	the axes stated. Used to shift working area when
G57	Linear shift XY	E	pendulum machining, for tool length compensa-
G58	Linear shift XZ	E	tion when drilling and milling (Z) and tool position
G59	Linear shift YZ	E	compensation (X and Y) when turning
G60	Exact 1	F*	Specifies accuracy desired when (and hence speed of) positioning or line machining
G61	Exact 2	F	Second similar function to G60
G62	Fast	F	Specifies low accuracy allowing rapid positioning
G63	Tap	B	Tapping positioning move—stop spindle rotation on reaching position
G64	Rate change	B	Change to higher or lower feedrate and/or spindle speed specified in this block without pausing

* See note on these letters at the end of this list.

6*

G65			Reserved for future positioning codes
to			
G79			
G80	Cancel cycle	H	Cancels any G81 to G89 code instruction
G81	Cycle drill	H	All cycle instructions call for rapid feed-in of spindle with clockwise spindle rotation ($-$ve) by distance specified by address R. G81 then causes feed-in to Z and rapid to initial R and Z values
G82	Cycle and dwell	H	As G81, with predetermined dwell at bottom of hole
G83	Cycle Peck	H	As G81, with several predetermined rapid moves back to first feedrate position and back to cutting position to clear chips from deep holes
G84	Cycle tap	H	As G81 to bottom. Simultaneous reversal of feed and speed to first feed position and then rapid to start
G85	Cycle bore 1	H	As G84 but no speed reversal
G86	Cycle bore 2	H	As G81 but stop spindle at bottom of hole
G87	Cycle bore 3	H	As G86 but with manual control of feed out of hole
G88	Cycle bore 4	H	As G87 but dwell at bottom before stopping spindle
G89	Cycle bore 5	H	As G85 but with dwell at bottom of hole
G90	Unassigned		May acquire a standard use. Avoid
to			
G99			

The letters in the third column indicate the interaction between the preparatory or G code functions.

The B indicates that the function only applies to the block it appears in.

The A, C, D, E, F, and H indicate that the function continues to apply to all future work until superseded by a further code bearing the same letter in column 3.

A number of G codes have yet to be firmly assigned to continuous-path control purposes. These should be avoided unless their exact function is described in a firm standard. Their probable meanings are listed in the BS. These codes are G01, G02, G03, G08 to G12, G17 to G21, G30, G31, G34, and G35.

Dimension words

The dimension words follow the preparatory function word and a Tab code; a Tab code also appears between each dimension word.

The dimension words must appear in the order: X, Y, Z, U, V, W, P, Q, R,

Address character	Used for
A	Angle about X axis
B	Angle about Y axis
C	Angle about Z axis
D	Additional angle
E	Additional angle
or D	Third feedrate
or E	Second feedrate
F	First feedrate
G	Preparatory function
H	No standard meaning will be assigned
I, J, K	Reserved for contouring control systems
L, O	Not used
M	Miscellaneous function
N	Sequence number of block
X (or U or P)	Main (or subsidiary) motion in X direction
Y (or V or Q)	Main (or subsidiary) motion in Y direction
Z (or W or R)	Main (or subsidiary) motion in Z direction
or R	Distance for rapid Traverse of spindle
S	Spindle speed
T	Tool number

Note: The codes are not used in this order on the control tape.

Figure 6.14 Use of characters for addresses

A, B, C, D, E. Note, however, that D and E can alternatively be used with F as feed function words as described in the next section.

The axes to which the dimension words are related on any particular machine have been described in the axes and motion nomenclature section of this chapter and in Fig. 6.14.

On control tapes for one machine tool, dimension words shall be either all co-ordinate dimensions (absolute position values), or all incremental dimensions (relative to preceding position).

The numbers, as is usual, have the most significant digit first. The decimal point will not appear in the word, but it will have fixed positions in each type of dimension word and the positions are to be defined in the Detailed Format Classification sheet* for the machine.

All incremental control systems and some absolute systems require positive and negative numbers. In such cases the $+$ or $-$ sign must precede the first digit, thus:

$$X + 37521 = +37 \cdot 521, \text{ say}$$
$$Z - 1296 = -1 \cdot 296, \text{ say}$$

Where only positive dimensions are involved, the positive sign must not appear in the dimension word; thus:

$$X37521 = +37 \cdot 521, \text{ say}$$
$$Z1296 = +1 \cdot 296, \text{ say}$$

* See page 153.

The units of the dimensions do not appear anywhere on the control tape but linear dimensions must be expressed in terms of millimetres or inches and angular dimensions must be expressed in terms of degrees or revolutions.

On one machine tool, only one form of linear and one form of angular unit may be employed, as declared on the Detailed Format Classification Sheet.

X37521 may represent 37·521 in. or perhaps 375·21 mm

A98765 may represent 98·765° or perhaps 0·98765 rev

Feed word (F)

A feed function word which applies to only one axis of motion shall, apart from a TAB code, follow immediately after the dimension word for that axis. A feed word, applying to two or more axes of motion in one block, shall follow the last dimension word to which it applies. Thus, if only F codes apply to X, Y, and Z movements and a different feedrate is wanted for Z than for X and Y, two blocks will be needed. One will have X, Y and F and the other Z and F, and the Z move will occur after the X and Y. If different feedrates are required to occur simultaneously, in one block, then the addresses D and E may be used for two further axes, providing that the control system permits this.

Each feed word must consist of the address character F (or E, or D; see Fig. 6.14) followed by an n-digit number. The number n may be 3, 4, or 5.

The n-digit number is known as an arithmetic code (and sometimes referred to in literature in the USA as the 'magic three' code because a 3 is involved in determining the first digit).

Transferring from a feedrate to the code   (a) The first digit of the code is obtained by adding 3 to the number of figures to the left of the decimal point in the feedrate. If there are no figures to the left (feedrate less than 1·0000) the number of zeros immediately following the decimal point is subtracted from 3. Thus

feedrate	first digit of code
1234·567	7
0·1234	3
0·0012	1

(b) The subsequent (n − 1) digits of the code are the actual first (n − 1) digits of the speed rounded to a precision of (n − 1) digits. Thus

	(n − 1) code digits		
feedrate	n = 3	n = 4	n = 5
1234·567	12	123	1235
0·1234	12	123	1234
0·0012	12	120	1200

(c) The complete code is formed by putting the first digit before the (n − 1) digits. Thus

		code	
feedrate	n = 3	n = 4	n = 5
1234·567	F712	F7123	F71235
0·1234	F312	F3123	F31234
0·0012	F112	F1120	F11200

Transferring from code to feedrate   (a) Subtract 3 from first digit of the code. If the result is positive it indicates the number of digits before the decimal point in the feedrate.

If the result is negative the number is the number of zeros after the decimal point before reaching the most significant figure of the feedrate. (b) The remaining digits of the code when the first is removed are the digits of the feedrate. Thus

F543	results in 43
F5432	,, ,, 43·2
F3456	,, ,, 0·456
F016	,, ,, 0·00016
F87654	,, ,, 76540.

Units   The units involved do not appear on the control tape. For a metric machine they are mm/min or mm/rev. For an imperial machine they are in./min or in./rev. However, note that, for thread-cutting, inch screws tend to have whole numbers of threads per inch whereas metric screws tend to have whole numbers of mm per thread. For thread-cutting feedrates, the metric units are thus mm/rev whereas for inch machines they are rev/in.

(The non-interchangeable format recommends the above feedrate code but permits alternative two- and one-digit codes of a look-up table nature.)

Spindle speed word (S)

The spindle speed word follows after all the dimension and feedrate words, and is again immediately preceded by a Tab character.

The spindle speed word consists of the address character S followed by an n-digit number, where n may be 3, 4, or 5. The n-digit number is related to the numerical value of speed in exactly the same way as for feedrate words.

The implied unit for tool or work-table spindle speed is rev/min.

(The one- and two-digit numbers are again permitted for non-interchangeable formats.)

Tool word (T)

The tool function word follows the spindle speed word after a Tab character. It consists of the address character T followed by a code number having the number of digits specified in the itemized data for the machine where the meaning of each code number should be specified.

For a single turret drilling machine, with a six-spindle turret, for example, the tool function words might be T1 to T6.

The tool function word may select a tool either automatically or by an indication to the operator.

A miscellaneous function word, M06, perhaps a few blocks later, may be required to initiate the tool change, when the tool has been located and is available to be changed.

Miscellaneous function word (M)

The miscellaneous function word follows the tool word after a Tab character. It consists of the address character M followed by a two-digit code number.

Miscellaneous functions control machine functions such as spindle rotation, feed, coolant flow, clamping of workpiece, tool change, and workpiece change.

There may once have been a natural allocation of functions to preparatory functions (G) and miscellaneous functions (M), depending on whether the function should be carried out before or after any dimensional instructions in the block. A study of the section on G functions and the following section on M functions will show, however, that this is no longer true. Another possibility is that G functions should be related to cutting movement and M functions to the remainder, but there are again exceptions.

The M function words which have been allocated in the standards are set out below. In the third column, A means after other instructions in the block are completed and B means before the other instructions are obeyed. M functions apply only for the block in which they appear unless there is a letter in the fourth column. If a letter appears in the fourth column the instruction continues in force until countermanded by another instruction bearing the same letter.

A number in the fourth column indicates that related instructions, which also only apply to the block they appear in, will be found with the same number.

An asterisk * in either the third or fourth column signifies that the manner in which the instruction operates should be described in the Itemized Data for the machine.

M00	Program stop	A 1	Stop spindle, feed, and coolant. Requires operator intervention to start
M01	Optional stop	A 1	As M00, but only if the optional stop switch has been set at ON
M02	End of program	A 1 *	As M00, but also reset the control system (e.g., rewind tape to program start character, %, or wind tape loop forward to %, or as stated in itemized data)
M03	Spindle CW	B C	Rotate spindle clockwise, or negative

M04	Spindle CCW	B C	Rotate spindle counterclockwise, or positive
M05	Spindle off	A C	Stop spindle and coolant. Apply brake if available (see also M19)
M06	Tool change	* 2 *	Indicates or causes an automatic change of tools in conjunction with a tool word T. The itemized data should indicate whether the coolant should be turned off or not
M07	Coolant 2 on	B D *	Turn mist coolant on (or other function such as dust collector, or tapping coolant declared in itemized data)
M08	Coolant 1 on	B D *	Turn flood coolant on (or other function as M07)
M09	Coolant off	A D	Turn off coolant which is on because of M07, M08, M13, M14, M50, or M51
M10	Clamp	* E *	Clamp machine slide, workpiece, fixture or spindle as in itemized data
M11	Unclamp	* E *	Unclamp feature clamped by M10
M12	Unassigned		May acquire a standard use. Avoid
M13	CW and coolant	B C D	Turn on coolant and rotate spindle CW, or negative
M14	CCW and coolant	B C D	Turn on coolant and rotate spindle CCW, or positive
M15	Motion+	B 3 *	Feed, or rapid traverse, in positive direction
M16	Motion−	B 3 *	Feed, or rapid traverse, in negative direction
M17	Unassigned		May acquire a standard use. Avoid
M18	Unassigned		As M17
M19	Orient spindle	* C *	Stop spindle at predetermined angular position
M20 to M29	Permanently unassigned	* *	Available for special user requirements
M30	End of tape	A 1 *	Similar to M02 but may in addition be used to switch control to a second tape reader as stated in itemized data

M31	Interlock by-pass	* *	Enables a normally effective interlock to be circumvented
M32 to M35	Constant cutting speed	B C *	Keep constant peripheral cutting speed by adjusting spindle speed inversely with distance of tool from centre of rotating work. Cutting speed for each code stated in itemized data
M36	Feed range 1	B 3 *	First of two feed ranges as detailed in itemized data
M37	Feed range 2	B 3 *	Second feed range similar to M36
M38	Spindle speed 1	B C *	First of two speed ranges as detailed in itemized data
M39	Spindle speed 2	B C *	Second speed range similar to M38
M40 to M45	Gear change or Unassigned	* *	Gear changes specified in itemized data
M46 to M49	Unassigned		May acquire a standard use. Avoid
M50	Coolant 3 on	B D *	Turn on coolant, or other function as for M07
M51	Coolant 4 on	B D *	As for M50
M52 to M54	Unassigned		May acquire a standard use. Avoid
M55	Tool shift 1	B 2 *	Tool zero offset into first of two positions detailed in itemized data. Compensates for positions of alternate tool spindles
M56	Tool shift 2	B 2 *	Similar to M55 but second position
M57 to M59	Unassigned		May acquire a standard use. Avoid
M60	Work change	A 4 *	Change one workpiece for another. Pallet change for example

M61	Work shift 1	B 4 *	Change workpiece position to bring its datum to one of two predetermined positions detailed in itemized data.
M62	Work shift 2	B 4 *	Similar to M61 but second position
M63 to M67	Unassigned		May acquire a standard use. Avoid
M68	Clamp work	* 4 *	Clamp workpiece
M69	Unclamp work	* 4 *	Unclamp workpiece
M70	Unassigned		May acquire a standard use. Avoid
M71	Work angle 1	4 * *	Rotate workpiece into first of two positions detailed in itemized data
M72	Work angle 2	4 * *	As M71, but second position
M73 to M77	Unassigned		May acquire a standard use. Avoid
M78	Clamp slide	A E *	Clamp the machine tool slide detailed in the itemized data
M79	Unclamp slide	A E	Unclamp the slide clamped by M78
M80 to M90	Unassigned		May acquire a standard use. Avoid

Comments to operator

If a control tape is to be stored and reused at infrequent intervals, it is useful for it also to carry comments to the operator. The tape can then be used in a tape typewriter to produce a listing of its contents, including the comments to the operator.

It depends on the control system whether the tape can carry comments.

The control-out '(' and control-in ')' codes (see Fig. 6.3) are used to enclose information not intended for the control system. When the control tape is typed this kind of information appears between brackets. Thus it may appear as:

(Insert 0·875 in. drill in turret position 2 and restart system)

Note that when the tape is rewinding, the ')' or control-in code is reached before the '(' or control-out code. The alignment function or rewind stop ':'

and the program start '%' characters should not appear between the brackets because the control system may not take account that the system is reading backwards. In this case the closing bracket indicates 'control-in' so that the ':' or '%' would be acted on during rewind.

Format summary

Each block starts with an N word, has a Tab character between each word, and has End-of-block character immediately after the last word in the block.

The order of the words in a block are as set out above and, following the N word, only those words which apply to the machine need to accommodated in a block.

For any particular block only those words which give new information need appear.

Words accommodated in a block, but not appearing in a particular block, must be indicated by their corresponding Tab codes up to the last word appearing in that particular block.

The first block on a control tape contains only the program start code (%) and the End-of-block code.

The last block on a control tape contains one suitable Stop code in the following order of preference, M30, or M02, or M00.

There should be about 3 ft of leader and trailer, or if the tape is a loop it should be long enough to be accommodated on the tape reader. If the control system has to read through the joined ends of a loop, the leader/trailer may need to be all delete codes. It may also be useful to have an End-of-medium code (Fig. 6.4) in the trailer so that, when a copy of the control tape is being made, the tape-copying equipment finds where to stop copying. There would, of course, be no point in having the End-of-medium code if the tape-handling equipment were unable to act on it.

The non-interchangeable, and the fixed block, formats have deliberate differences. The BS. or ISO. documents referred to at the beginning of the chapter should be consulted as the authority on the subject both for full details of these formats and of the interchangeable format with which this chapter has dealt.

Comparison of machine tools and control systems

In the USA, a method was developed of estimating quickly the suitability of a particular NC machine for a particular job. This method has been retained in the ISO and BS Specifications for NC. It has not yet been widely used in Europe, but it does form a useful exercise to be carried out by customer and supplier to ensure that a minimum amount of undefined characteristics remain when an an order is placed.

The method, described below, is also very useful to check quickly whether a control tape, prepared for use on one machine, is unsuitable for use on another.

Any difference covered by the classification system is quickly disclosed. It is not, however, possible to guarantee that the second machine is entirely suitable since small differences (not covered by the system) might cause a particular job to be impossible.

Interchangeability

Before examining the three parts of the classification system (Itemized Data which contains Brief and Detailed Format Classification), it is worth considering the main machine features which have to be reviewed in checking for interchangeability, or suitability for making a part. These features are that:

(a) both machines are large enough for the task;

(b) both machines have the appropriate types of control (P, L, C, or M) and the same addresses on corresponding axes;

(c) the codesets on the punched tapes are identical;

(d) function codes and words shared by machines must not have conflicting meanings;

(e) function words not shared by both must appear on the tape for both machines (e.g., one machine may use automatic tool selection, in which case the appropriate codes should appear on a tape which has to operate both machines, but an optional Stop code should also appear so that manual tool loading can occur on the second machine);

(f) the brief format classifications should be similar;

(g) the detailed format classifications should be similar;

(h) the itemized data should not disclose characteristics unsuitable for the task.

In describing (f), (g), and (h), an example will be developed stage by stage for one machine tool. As the brief and detailed classifications are listed on the itemized data sheet for a machine tool, the example for (f) and (g) is also relevant to (h) and used in it.

Brief format classification

This is the name used in the BS document corresponding to 'Format classification shorthand' in the ISO and EIA documents.

This brief format classification is a string of mixed letters and digits formed in the following order:

(a) The letter I for interchangeable variable-block format;

(b) the letter P for positioning

or	L for positioning and line milling
or	C for contouring,
or	D for contouring and positioning;

  (c) the letter M for linear movements expressed in millimetres and decimal
            fractions thereof
       or   I   in inches and decimal fractions thereof;
  (d) the letter R for angular dimensions in decimal fractions of a revolution
       or   D   in degrees and decimal fractions thereof
       or no letter if no angular instructions are used.
[Note that for non-interchangeable systems, the order is:

  (a) as (b) above with the additional alternative of the letter F for fixed-block
        format positioning with or without line milling;
  (b) the letter A for systems with addresses and no tabs
       or   T   for systems with tabs and no addresses
       or   S   for systems with tabs and addresses
       or no letter if neither tab nor address is involved.
  (c) as (c) above;
  (d) as (d) above.]

Then follows for both systems, three digits:

  (e) A digit indicating how many of the axes in (e) have controlled motions,
either by direct numerical position instructions or by indirect function commands (e.g., G82, etc.) which may result in motions to positions preset mechanically by the operator;
  (f) A digit indicating how many axes in (e) have their motion controlled by
digital dimension words;
  (g) A digit indicating how many of the axes in (f) can have their motion
controlled simultaneously.
  The use of this brief format classification can best be illustrated by an example:

I L M D 4 4 2

This indicates:
  by I   that an interchangeable format is used;
      L   that it positions and line mills;
      M that metric units are involved;
      D that angular movements in degrees are involved;
      4   that only 4 of the movements are controlled;
      4   that these 4 movements have their positions numerically specified;
      2   that only 2 can be controlled simultaneously.

Note that the brief format classification does not indicate what kind of
machine tool is involved, nor which axes are controlled, nor how they are controlled.
  This brief format classification would be much more useful if it indicated
which axes were controlled and what type of control was involved. It would be

relatively straightforward to extend the ideas and describe the same machine thus:

$$I, 2, PD, B, LM, X, Y, Z, M, W$$

This would indicate

I	interchangeable format;
2	only 2 axes simultaneously controlled;
PD	positioning control in degree units;
B	on the B axis;
LM	line milling and positioning control in metric units;
X, Y, Z	on X, Y, and Z axes;
M	manual setting;
W	on W axis (W is secondary Z movement).

M, for manual, would indicate that movement of such an axis would be entirely manually, or operator controlled. If the operator set up a series of trip switches or dogs, to determine distances, but the control tape initiated moves to these set up positions, the letter S would be used. Thus, for the above machine, if the Z depths were preset, the proposed code would then be:

$$I, 2, PD, B, LM, X, Y, S, Z, M, W$$

Unfortunately, the existing brief format classification has tradition behind it and the machine must be described, as above,

$$ILMD\ 442$$

and the detail dealt with in the itemized data (see further on) and the detailed format classification.

Detailed format classification

This is the name used in the BS document corresponding to 'Format classification detailed shorthand' in the ISO document and 'Format detail' in the EIA document.

The detailed format classification is a group of letters, digits, and other characters which indicate the addresses used by the control system, the order in which they occur, the number of digits in each word, and the position of the implicit decimal points.

In addition to its use as a quick indication of similarity of controlled systems it has been used, with control systems containing small digital computers, to set the computer up to read in many different formats of control tape, and, for example, to plot graphs of X, Y, and Z motions for a variety of control tapes. For a detailed format classification:

(a) The address character of every address recognized by the control appears in the order in which the addresses are to occur on the control tape. See Fig.

6.14 for a summary of the address characters (but note that for fixed-block format and **for this purpose only** the letter H is used to indicate each word and a comma indicates each space character).

(b) When the control system requires a + or − to appear before a dimensional number in an address word then the address character is followed by a +. If the control system requires neither a + nor a − then the + is omitted after the address character in the detailed format classification.

(c) If the dimension word in a control system uses incremental dimensions, the letter D (presumably for delta) follows the corresponding address character in the detailed format classification.

(d) If a dimension word has no zero suppression the number of digits in the word is indicated by two digits. The first digit indicates the number of digits before the implicit decimal point. The second digit indicates the number of digits after the decimal point.

(e) For the (non-interchangeable) variable-block format only, when leading or trailing zeros are omitted three digits are used to indicate the size of the dimension word. Two digits are as in (d) and the third is a zero before the other two if leading zeros can be omitted, or after them if trailing zeros can be omitted.

The use of these for a dimension word might be:

$$XD + 024$$

indicating that for the X dimension word the dimension is incremental (D), requires a + or − (+), can have leading zeros suppressed (0), has two digits before and four digits after the decimal point (24).

Typical words on the control tape, and their values would then be

```
X − 123456    − 12·3456
X + 12345     + 1·2345
X − 1234      −   ·1234
```

and these would all be incremental moves.

(f) An address character identifying a non-dimensional word will be followed by a single digit which indicates the number of digits in the corresponding control word.

(g) For the purpose of the detailed format classification only, it is necessary to indicate the presence of Tab codes and the End-of-block (Line feed) code which do not have a printed representation. In this context a full stop shows where a Tab code would occur and an asterisk where an End-of-block code would occur.

The complete detailed format classification is indicated by a string of characters with no spaces and is illustrated by the example:

N3.G2.X42.Y42.Z32.R32.B32.F3.S3.T2.M2∗

The meaning of each part is:

Typically

N3	Three-digit sequence number	N071
G2	Two-digit preparatory function	G81
X42	Six-digit X dimension (see above)	X057643
Y42	Six-digit Y dimension (see above)	Y036287
Z32	Five-digit Z dimension (see above)	Z01472
R32	Five-digit R dimension (see above)	R05000
B32	Five-digit B dimension (see above)	B27500
F3	Three-digit feed word	F475
S3	Three-digit speed word	S615
T2	Two-digit tool word	T05
M2	Two-digit miscellaneous function word	M08
.	Tab code positions	non-printing
*	End-of-block code position	

It is again seen that the detailed format classification does not in itself describe the machine, although it is somewhat more descriptive than the brief format classification.

Both classifications are included in the itemized data which completes the description.

Itemized data

The itemized data sheet forms a summary of the machine tool properties to enable a skilled part programmer to manually program parts for the machine and its controls, or to allow supplier and prospective customer to review the suitability of an installation for the customer's work.

The itemized sheet should state the:

(1) Type of machine.

(2) Brief format classification.

(3) Detailed format classification.

(4) Whether axes nomenclature is standard; details of any non-standard features and possible use of D and E codes for feedrate or R for cycle commands (see Fig. 6.14).

(5) Preparatory, miscellaneous, and tool function codes which the system acts on.

(6) Function of any G or M codes which differ from, or are not stated in, the standard list of G and M functions.

(7) Additional information for those G and M codes which are to be amplified on the itemized data sheet.

(8) Speed, feed, and tool codes accepted. What is done with tool codes. How rapid feed is commanded.

(9) Maximum and minimum movements in each axis. Type of control available on each axis. Special features such as mirror image, axis interchange, range of cutter feed and speed compensation, floating zero, zero offset.

(10) The characters from the complete codeset which are recognized and acted on (C.R., L.F., N.L., etc.)

(11) How the control system reacts to the remaining characters—does it ignore them or halt? (e.g., Null and Space codes).

(12) Machine characteristics: Spindle drive power, torque/ speed relationship, maximum safe workpiece weight, swing, and distance between centres for lathes.

Items (6) and (7) above indicate the possible need to specify further information for some M and G codes. These codes are:

G13–16, 22, 23, 25–29, 36–39,
G45–52, 60–62.

Details can be found by studying the table of preparatory, or G, functions on page 140.

M02, 06–08, 10, 11, 15, 16, 19, 20–29,
M30–45, 50, 51, 55, 56,
M60–62, 68, 69, 71, 72, 78.

Details can be found on the table of miscellaneous, or M, functions on page 146.

The timing of the miscellaneous function (before or after carrying out any motion commands in the block) must be stated for

M06, 10, 11, 19, 20–29,
M31, 40–45, 68, 69.

The duration of the miscellaneous function (one block only or until countermanded by which codes) must be stated for

M20–29, 31, 40–45.

Example of application of itemized data

(1) Boring machine, horizontal spindle head sliding on W and Y axes with a quill boring action on Z. Work table slides in X' and rotates in B' direction.

(2) Brief format classification: ILMD442

(3) Detailed format classification:

N3.G2.X42.Y42.Z32.R32.B32.F3.S3.T2.M2∗

(4) Standard axes nomenclature. Similar to Fig. C18 in BS†, R used as auxiliary rapid distance command for cycle codes G81, G82, G87.

† Forthcoming British Standard on ' *The numerical control of machines* '.

(5) The system acts on:

G04, 05, 60, 62, 80, 81, 82, 87
M01, 02, 03, 04, 05, 06, 08, 09, 19.

(6) No non-standard G or M codes.

(7) The M02 will normally cause the tape to rewind. If the tape-rewind switch is set to loop it will cause the tape to wind forward through the trailer and leader. In both cases the tape will stop at the tape Start code '%'.

The M06 code will turn coolant off and stop the machine after completion of the motion called for in the block. The 'Change tool' indicator will light. The operator will read the new tool number (e.g., T73) from the 'Tool number' indicator.

The M19 codes causes the spindle to stop with the locating keyway on top of the spindle after the motion in the block has been completed.

The G60 code causes positioning to 0·01 mm tolerance with final approach to each position in a positive direction. The G62 code causes a quicker positioning to 0·06 mm tolerance with final approach in either direction.

(8) The spindle speed code range is S410 to S710, corresponding to speeds from 1 to 1,000 rev/min.

The feed code range for the X, Y, and Z moves is F325 to F650 corresponding to 0·25–500 mm/min. The rapid, non-cutting, feedrate of approximately 5 m/min is obtained for any feed function code between F710 and F799. As the rotational moves of the table are for positioning only, the rapid feedrate of approximately 3·5 rev/min is obtained without specifying a feedrate.

The tool codes, T00 to T99, are displayed on the control panel until replaced by the next T code read from the tape.

(9) The maximum and minimum moves for each axis are:

X' 4150 and 0·01 mm
Y   1900 and 0·01 mm
Z   990 and 0·01 mm
B' 359·99 and 0·01°
W   1,000 and 100 mm

The W movement of the boring head is controlled by operator manipulation of pushbuttons. Between the W and Z moves the tool spindle reference point (its nose) can reach to table centre and to 1·99 m from the centre. Sign reversal switches, for the control of X, Y and B axes, allow left- and right-hand components to be machined by one control tape.

The control of X, Y and Z is line motion called (L) and the control of B is rotary positioning (P).

The control system acts on '(', ')', '/', and '%' codes, and is not upset by any of the ISO codeset characters. If it encounters G or M codes other than those enumerated above, it lights the 'Unrecognized code' indicator during that block but continues to operate. Null codes can appear anywhere on the tape.

(10) The work-table carries a load up to 7 T and the spindle drive is 17·5 kW.

Conclusion

In purchasing a machine-tool and control system never assume that any required function will happen automatically. An advantage of a standard specification is that it serves as a reminder of the factors involved.

A standard specification for equipment should, by reducing variety in production, cheapen equipment. On the other hand, not every customer wants every optional feature, so that by making provision for all features to be fitted the cost is increased.

Some cheap control systems claim to comply with these standards but may, for example, only arrange to recognize, say, 16 G codes. If these could be any 16 in the range from G00 to G99 this would be very adequate. The cost can be further reduced by wiring the system to recognize 16 specific G codes and no others. The customer specifies, perhaps, that he wants G05–Hold. The supplier states that unfortunately G05 cannot be obtained but that the 'Hold' function can be provided by G25. The customer agrees and is quite content. Suppose that later he buys another similar machine with a different control and he finds that this time 'Hold' only comes on G05; he cannot then exchange control tapes between the machines and has to remake the tape. This is false economy.

For drilling machines the main option offered to the purchaser is the EIA or ISO code option on the tape. There is now no doubt that the EIA code is becoming obsolescent so that a customer new to NC should have no hesitation in demanding ISO codeset. The tape-preparation equipment will remain available throughout the life of the machine tool, and will be suitable for preparing input tapes for computers. Established users of NC will have to make their minds up to change as soon as is convenient if they currently use EIA code.

There are short-term advantages in purchasing to an internationally agreed specification, and in the long term the advantages will become more apparent.

The customer who only purchases control systems which comply strictly with the standards will find it much easier to transfer work around the machine tools in his workshop; much easier to obtain and maintain post processors (see chapter 11); and much easier for programmers and operators to understand their work.

Unfortunately there is less standardization in the case of contouring control systems, except in the US where the contouring tape is very compatible with the positioning and line-milling tape.

7. EXAPT 1

Dieter Reckziegel

Dr.-Ing. Dieter Reckziegel has been Manager of the EXAPT-Association in Aachen, Germany, since it was founded in 1967. The topic of his Doctoral Dissertation, for which he obtained his Dr.-Ing. degree in 1967, was 'The construction of a tool system for NC machine tools'. Before taking up his present appointment, he was assistant to Professor Dr.-Ing. H. Opitz at the Institute for Machine Tools in Aachen.

The EXAPT programming system was developed in Germany through the initiative of Professors H. Opitz, W. Simon, G. Spur, and G. Stute. This chapter deals with EXAPT 1 while chapter 8 describes EXAPT 2 and chapter 15 describes the basis of the processing of the cutting technology statements of EXAPT 1 and 2.

The common features of EXAPT 1 and 2—particularly the language, computer processing, and the construction of part programs—are dealt with once in this chapter, and apply also to 2C,L and APT.

EXAPT 1 is intended to be used as an aid to the preparation of control tapes for drilling machines, and its main features are listed below.

(a) Easily understandable free-format instructions describe the geometry and the machining process.

(b) Programming can be done without special knowledge of the machine tool.

(c) Dimensions can be given directly as they appear on the drawings.

(d) Point patterns, for example lying on pitch circles, can be easily described.

(e) Technological instructions describe the final condition at each machining point. Tools, feeds, rotational speeds, and machining sequences can be automatically determined by the processor.

(f) The automatically determined data can be based on the particular values used by a firm.

(g) The computer is supplied with a combination of part program, and tool, material, and machine cards, and the technological processor. Using a particular

159

firm's cutting values, it can calculate sequences of operations, choose tools, calculate tool paths, avoid tool collisions, and determine optimum cutting values.

The use of EXAPT depends on the user describing his cutting tools and cutting conditions in the particular form required on the tool and material cards. This is dealt with more fully in chapter 15.

Language structure

The information in this and the next two sections of this chapter apply in general to all APT-like languages. This language was evolved in the USA and subsequently adopted as a basis for EXAPT and the NEL NC programming systems.

Permissible characters

The following characters may be used.

The capital letters,

   A B C D E F G H I J K L M N O P Q R S T U V W X Y Z

The figures,

   1 2 3 4 5 6 7 8 9 0

The special characters

   . / + − , = ( ) $ * and space.

Intermediate punctuation—use of special characters

The special characters . / + − * ( ) are used to describe the arithmetical instructions. The special characters also have the following meaning:

Slash /   A slash is used to divide an instruction into a major and a minor part. The major part, known as a major word, is written to the left of a slash. On the right of the slash follows the minor part which can consist of modifiers and/or numbers.

Comma ,   The minor part of a statement is divided into modifiers and/or numbers by commas. No comma or full stop is used at the end of a statement.

Equals sign =   Equals signs are used to separate symbols from the content to be defined, or variables from the word to which they refer.
   Thus

$$B = DRILL/SO, DIAMET, 1, DEPTH, 5$$

Brackets ( )   Nested definitions are enclosed in brackets to indicate that they are complete statements within themselves

Thus

$$GOTO/(POINT/12, 5, 7)$$

or

$$P17 = POINT/12, (3.5 + 3/2), 7$$

Dollar sign $ (a) Single dollar sign $. When punched cards are used for introducing a part program into the computer, a dollar sign is used to indicate
continuation. It is written as the last character of a punched card if a statement
has more than 72 symbols and indicates that the statement is continued on the
next card.

(b) Double dollar sign $$. The double dollar sign is used to separate a state-
ment from any desired comment which can be added without affecting the
computer processor.

Blank   Blanks have no meaning in the language and are ignored. They can be
used at any place in a part program in order to improve clarity.

Elements of the language

Elements of the language are words, numbers, and syntax elements.

Words   Words are built up from letters and figures. The first character must
be a letter. No word may consist of more than six characters. Two word types
are distinguished: vocabulary words and symbols.

(a) Vocabulary words have a distinct significance. They form a vocabulary
that the programmer must learn.

(b) Symbols are freely chosen names. They serve to give a name to any state-
ment to be defined, and enable this to be referred to during the course of the
following program. Symbols must not be identical to vocabulary words. This
can best be achieved by always including a figure in a symbol. Thus POINT 3
or LINE 16 can be used as symbols although POINT and LINE are vocabulary
words.

Numbers   Numbers are formed from figures and in certain cases a mathematical
sign. If decimal fractions appear, they are separated by a decimal point. Leading
zeros may be left out. Numbers without sign are positive.

Syntax elements   Syntax elements are special signs that are used to separate
words and/or numbers and have been described under intermediate punctuation
above.

Types of instruction

Instructions are built up from words, numbers, and syntax elements. The instruction consists of a major part and, if necessary, a minor part that completes the details of the major part. Major and minor parts are separated by a slash; the details in the minor part are separated by commas. A sequence of instructions builds a part program. There are:

> definition statements;
> execution statements;
> additional statements.

Definition statements   Definition statements define a certain arithmetic, geometric, or technological condition. They can be used with or without symbolic reference.

The instructions with symbolic reference have the following construction:

> Symbol = Definition

The symbol is used during the course of the part program to refer to the defined condition by using the symbolic name.

Thus,   P73 = POINT/1, 1, 1

and later   GOTO/P73

In EXAPT, symbols may only appear once to the left-hand side of the equal sign in the course of a part program. Definitions are:

> arithmetic expressions;
> geometric definitions;
> machining definitions.

Execution statements   Execution statements put into effect previously defined technological and geometrical statements.

Additional statements   Additional statements are technological statements giving certain technological data, or program-oriented statements which control processing of the part program in the computer.

Dimensions

Within a part program, no dimensions need be given if the metric system is used. By using the program-orientated instruction UNITS/INCH, it is also possible to work in inch dimensions. The dimensions for numerical values in the part program are shown in Fig. 7.1.

	UNITS/MM	UNITS/INCH
Length	mm	in.
Angle	degree	degree
Feed	mm/rev	in./rev
Rotational speed	rev/min	rev/min
Cutting speed	m/min	ft/min

Figure 7.1　Units assumed in EXAPT.

Nested instructions

In an instruction which may contain a symbol, the symbol may be replaced by a definition enclosed in brackets. The definition enclosed in brackets is called a nested definition. Only a definition instruction suitable for symbolic reference may be nested in this way, but within the nesting the symbol may be omitted. Multiple nesting is also permissible. Nesting takes the following form:

$$\ldots, (\text{Symbol} = \text{Definition}), \ldots$$

or

$$\ldots, (\text{Definition}), \ldots$$

Thus

$$P1 = \text{POINT}/(17\text{-}4/2), 5, 7$$
$$\text{GOTO}/(\text{POINT}/(17\text{-}4/2), 5, 7)$$
$$\text{GOTO}/(P1 = \text{POINT}/(17\text{-}4/2), 5, 7)$$

If a symbol is assigned in a nested definition, the symbol can then be referred to during the part program.

Arrangement of punched cards

If punched cards are used, an instruction is punched in spaces 1 to 72. The remaining eight spaces can be used for card identification. If an expression is so long that it cannot be accommodated on one card, it may be continued on one or more succeeding cards providing the last character on each card having a continuation is a $.

Co-ordinate system

Machine tool co-ordinates and workpiece co-ordinates

Machining positions are defined in a right-handed cartesian co-ordinate system with axes X, Y, Z, as described in chapter 6. The Z axis is parallel to the tool. The zero point of this co-ordinate system is so selected by the programmer that the geometric description of the part is made as simple as possible. Note that the positive Z direction is from the tool point towards the spindle. The part program is independent of the machine tool in which the component will be manufactured. The connection between the machine tool and workpiece co-ordinate systems is given by the following instruction:

$$TRANS/x, y, z$$

where x, y, z are the co-ordinates of the zero point of the workpiece co-ordinate system in the machine tool co-ordinate system. This instruction also indicates how the workpiece must be mounted on the machine table.

Reference co-ordinate system

Apart from the original co-ordinate system, it is also possible to define further reference co-ordinate systems which can be used to avoid the need for dimensional conversions. This may be the only way of doing the work on a machine tool with a rotary table.

Geometric definitions

Co-ordinates

The instruction:

$$ZSURF/z$$

where z = Z co-ordinate defines a plane parallel to the X–Y plane at a distance z from it. The information about a Z plane is used for the definition of points and point patterns. Once a Z plane has been defined, it remains valid for the following point and point pattern definitions until it is replaced by a new ZSURF instruction.

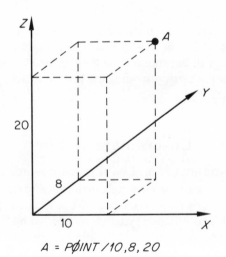

$$A = P\emptyset INT/10, 8, 20$$

Figure 7.2   Point definitions.

Single points

A single point is defined by the following instructions:

$$\text{Symbol} = \text{POINT}/x,\ y,\ z$$

where
$$x = \text{X co-ordinate}$$
$$y = \text{Y co-ordinate}$$
$$z = \text{Z co-ordinate.}$$

If, in the point definition, the Z co-ordinate is omitted, it is replaced by the value laid down by a previous ZSURF instruction. If no Z plane has been defined, the Z value is taken as zero. If the point definition contains a Z co-ordinate, then this information has precedence over the Z value laid down in the ZSURF instruction, when the point is referred to by its symbol.

Straight lines

Straight lines are necessary in EXAPT 1 for the reflection of point patterns. For the definition of a line the following statements are permitted:

$$\text{Symbol} = \text{LINE}/x_1,\ y_1,\ x_2,\ y_2$$

where $x_1,\ y_1$ are co-ordinates of one point
        $x_2,\ y_2$ are co-ordinates of second point.

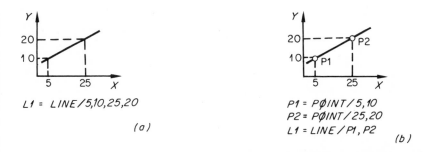

L1 = LINE/5,10,25,20

(a)

P1 = PØINT/5,10
P2 = PØINT/25,20
L1 = LINE/P1,P2

(b)

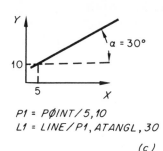

P1 = PØINT/5,10
L1 = LINE/P1,ATANGL,30

(c)

Figure 7.3   Line definitions.

$$\text{Symbol} = \text{LINE}/p_1, p_2$$

where p_1 and p_2 are symbols of points.

$$\text{Symbol} = \text{LINE}/p_1, \text{ATANGL, a}$$

where p_1 is symbol of a point
'a' is angle between line and x axis.

Figure 7.3 shows the use of these statements.

Circles

Circles are necessary for definitions of point patterns which lie on pitch circles. The following statements may be used:

$$\text{Symbol} = \text{CIRCLE}/x, y, r$$

where   x, y are co-ordinates of centre of circle
r is radius of circle.

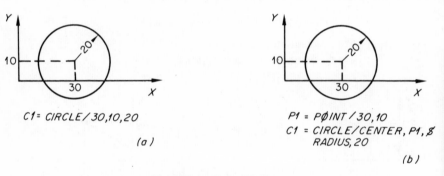

Figure 7.4   Circle definitions.

$$\text{Symbol} = \text{CIRCLE}/\text{CENTER}, p_1, \text{RADIUS}, r$$

where p_1 is symbol of the point forming the centre of the circle.
Figure 7.4 illustrates the use of these statements.

    No Z co-ordinate for the centre of the circle is given because a circle applies at all Z heights. In the second instruction, the centre of the circle is defined by a symbolic name for a point. The Z co-ordinate of this point is ignored. The actual Z value is assigned by a ZSURF/ statement just before the circle definition is used later in the part program.

Point patterns

Point patterns can be described in number of different patterns which correspond to engineering design practice.

Points on straight lines

(a) Equal separation

$$\text{Symbol} = \text{PATERN/LINEAR}, p_1, p_2, n$$

where p_1 = the initial point of a point pattern

p_2 = the final point of a point pattern

n = number of points (including p_1 and p_2).

(b) Variable separation

$$\text{Symbol} = \text{PATERN/LINEAR}, p_1, \text{ATANGL}, a, \$$$
$$\text{INCR}, n_1, \text{AT}, d_1$$

where p_1 = the initial point of a pattern

a = the angle between the X axis and the line joining the holes

n_1 = a number of increments

d_1 = the length of the increments (positive).

The information 'n, AT, d' can be repeated as often as desired. With a single instruction it is possible to define a sequence of points with varying point separation. Figure 7.5 shows how these statements are applied.

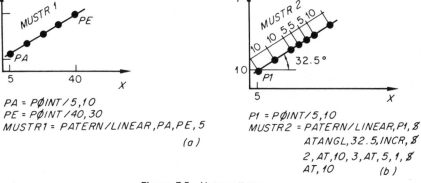

PA = PØINT/5,10
PE = PØINT/40,30
MUSTR1 = PATERN/LINEAR,PA,PE,5
(a)

P1 = PØINT/5,10
MUSTR2 = PATERN/LINEAR,P1,$
ATANGL,32.5,INCR,$
2,AT,10,3,AT,5,1,$
AT,10   (b)

Figure 7.5 Linear patterns.

Points on pitch circles

(a) Equal spacing

$$\text{Symbol} = \text{PATERN/ARC}, c_1, a, \frac{\text{CLW}}{\text{CCLW}}, n$$

where c_1 = symbol of a circle

a = starting angle

n = number of points in the point pattern.

The symbol 'a' gives the angle from the positive direction of the X axis to the straight line joining the centre of the circle to the first point of the pattern. The vocabulary words CLW or CCLW (clockwise or counterclockwise) indicate that the point sequence is numbered in either clockwise or counterclockwise direction.

(b) Variable spacing

$$\text{Symbol} = \text{PATERN/ARC}, c_1, a, \frac{\text{CLW}}{\text{CCLW}}, \text{INCR}, \$$$
$$\text{n, AT, da}$$

where c_1 = symbol of a circle
        a = starting angle
        n = number of increments (positive whole number)
        da = incremental angle (positive).

The information 'n, AT, da' can be repeated as often as necessary.

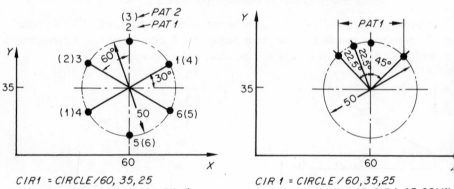

CIR1 = CIRCLE/60, 35, 25
PAT1 = PATERN/ARC, CIR 1, 30, $
        CCLW, 6
PAT2 = PATERN/ARC, CIR1, 210, $
        CLW, 6

CIR 1 = CIRCLE/60, 35, 25
PAT 1 = PATERN/ARC, CIR1, 45, CCLW,
        INCR, 1, AT, 45, 2, AT, 22.5

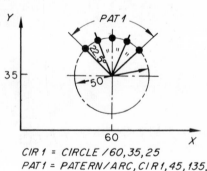

CIR 1 = CIRCLE/60, 35, 25
PAT1 = PATERN/ARC, CIR1, 45, 135, $
        CCLW, 5

Figure 7.6   Circular patterns.

(c) Equal spacing, sector of circle

$$\text{Symbol} = \text{PATERN/ARC, } c_1, a_1, a_2, \frac{\text{CLW}}{\text{CCLW}}, n$$

where c_1 = symbol of a circle
   a_1 = starting angle
   a_2 = final angle ($a_2 \neq a_1$)
   n = number of points (positive whole number).

Figure 7.6 shows how the statements are applied.

Transformation of patterns

(a) Pure translation

$$\text{Symbol} = \text{PATERN/TRAFO, } pat_1, pat_2$$

where pat_1, pat_2 = symbol of a point pattern or a single point.

The symbols pat_1 and pat_2 are two previously defined point sequences and/or single points. The instruction means that the points of the point sequence pat_1 are fixed in space. At each point of that sequence, the whole of the point sequence pat_2 is attached. The result is a point sequence containing points corresponding to the product of the number of points of each point sequence. The index numbers of the points in the new sequence can be seen from Fig. 7.7.

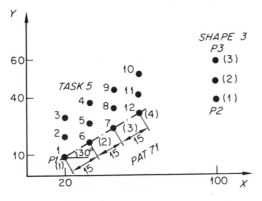

```
P1 = PØINT / 20, 10
P2 = PØINT / 100, 40
P3 = PØINT / 100, 60
PAT 71 = PATERN / LINEAR, P1, AT ANGL, Ø
        30, INCR, 3, AT, 15
SHAPE 3 = PATERN / LINEAR, P2, P3, 3
TASK 5 = PATERN / TRAFØ, PAT 71, Ø
        SHAPE 3
```

Figure 7.7   Transformed pattern.

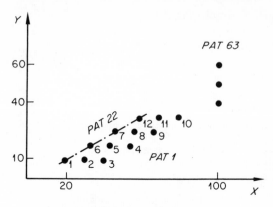

PAT 1 = PATERN / TRAFØ, PAT 22, 8
PAT 63, XYRØT, -90

Figure 7.8   Transformed pattern.

(b) Translation and rotation

$$\text{Symbol} = \text{PATERN/TRAFO}, \text{pat}_1, \text{pat}_2, \text{XYROT}, \text{da}$$

where $\text{pat}_1, \text{pat}_2$ = Symbol of a point pattern or a single point
           da = rotation angle of the second pattern (rotation about the
                  first point of pattern 2).

If the first of the two patterns to be joined is only a single point, the instruction indicates translation of the second pattern. Figure 7.8 shows how these statements are applied. Note the numbering order of the points in the final patterns.

Reflection of point patterns

$$\text{Symbol} = \text{PATERN/MIRROR}, l_1, \text{pat}_1$$

where l_1 = symbol of a straight line (mirror axis)
           pat_1 = symbol of a point pattern or a single point.

The defined pattern consists of the reflection of the pattern pat_1 about the straight line l_1. The use of MIRROR is shown in Fig. 7.9.

Linking point patterns

$$\text{Symbol} = \text{PATERN/RANDOM}, \text{pat}_1, \text{pat}_2, \text{pat}_3, \text{etc.}$$

where $\text{pat}_1, \text{pat}_2$, etc. = symbol of a single point or a point pattern.

   Any number of point patterns and single points with the same Z co-ordinate can be brought together in one name.

The RANDOM modifier can be used to group all the points of Fig. 7.9 into one pattern which can then be transformed, mirrored, or otherwise operated on as a group.

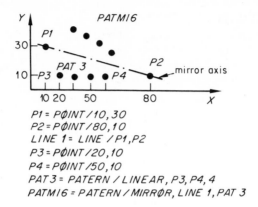

P1 = PØINT / 10, 30
P2 = PØINT / 80, 10
LINE 1 = LINE / P1, P2
P3 = PØINT / 20, 10
P4 = PØINT / 50, 10
PAT 3 = PATERN / LINEAR, P3, P4, 4
PATMI6 = PATERN / MIRRØR, LINE 1, PAT 3

Figure 7.9   Mirror and random patterns.

The statement should be

$$\text{PATAL1} = \text{PATERN/RANDOM}, p_1, \text{pat}_3, p_2, \text{pat}_6$$

Selecting a point from a point pattern

$$\text{Symbol} = \text{POINT/pat}_1, n$$

where pat_1 = symbol for a point pattern
          n = index number of desired point in the pattern.

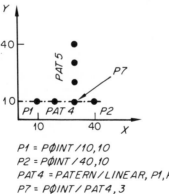

P1 = PØINT / 10, 10
P2 = PØINT / 40, 10
PAT 4 = PATERN / LINEAR, P1, P2, 4
P7 = PØINT / PAT 4, 3
PAT 5 = PATERN / LINEAR, P7, 8
             ATANGL, 90, INCR, 3, AT, 10

Figure 7.10   Point from a pattern.

Figure 7.10 shows the application of this statement. In this example, the third point in the pattern PAT4 is selected and given the symbol P7. The point P7 is then incorporated in a further point pattern, PAT5.

Modification of point patterns

When an instruction can call up a point pattern by a symbolic name, the following modifiers may be used:

RETAIN: Retains those points whose index numbers follow RETAIN whilst all other points are omitted.

OMIT: Omits those points whose index numbers follow OMIT whilst all other points are retained.

INVERS: Effects an inversion of the numbering sequence; the last point is given the index number 1.

The index numbers of the points must be given in ascending order.

$$\ldots, \text{pat}, \frac{\text{RETAIN}}{\text{OMIT}}, \text{pi1}, \text{pi2}, \ldots, \text{pin}$$

where pat = Symbol of a point pattern
pi1, pi2, pin = point indices

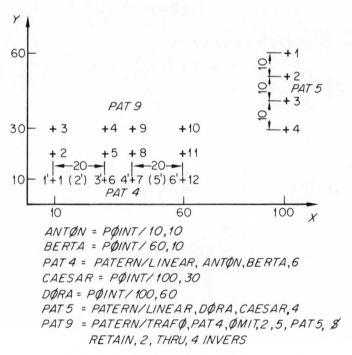

```
ANTØN = PØINT/10,10
BERTA = PØINT/60,10
PAT4 = PATERN/LINEAR, ANTØN,BERTA,6
CAESAR = PØINT/100,30
DØRA = PØINT/100,60
PAT5 = PATERN/LINEAR,DØRA,CAESAR,4
PAT9 = PATERN/TRAFØ,PAT4,ØMIT,2,5,PAT5, Ø
        RETAIN,2,THRU,4 INVERS
```

Figure 7.11 Modification of a point pattern.

If successive points of a pattern are to be retained or omitted, an abbreviated form of writing can be used:

$$\ldots, \text{pat}, \frac{\text{RETAIN}}{\text{OMIT}}, \text{pi1, THRU, pin}$$

All points from pi1 to final pin are retained or omitted. The modifier INVERS can be appended to the above expressions or, if RETAIN and OMIT are not present, can follow directly after the name of the point pattern. OMIT and RETAIN cannot both apply to one pattern. The modifiers do not change the original defined sequence but operate solely during the call-up.

In Fig. 7.11, the use of the OMIT and RETAIN modifiers is illustrated in the definition of PAT9. PAT9 is based on transforming the inverted PAT5 on each point of PAT4. In calling up PAT4, the second and fifth points are omitted, while the first point is omitted from PAT5 by retaining 2 to 4.

Technological definitions

The technological part of EXAPT 1 comprises instructions to describe the workpiece and machining which should be carried out at positions that have been described by geometrical instructions.

With these instructions, it is possible either to specify all the information necessary for manufacture or to have the data determined by the processor according to predetermined rules. In the language, provision is made for all automatically selected data to be directly stated, and for it to be adapted to a firm's experience or the varying manufacturing requirements.

The following are automatically selectable:

tool motion;
feeds and cutting speeds;
spindle return strokes when deep drilling;
tools;
work sequences.

Moreover, calculations are carried out by the data processor to avoid possible collisions by the tool tip. At the same time, the tool length is checked to ensure that it is large enough for the programmed work, and the tool changing position is determined.

In order to use the program, three card indexes are essential: a tool card index, a material card index, and a machining card index. By entering its own values in these indexes, cutting conditions can be suited to a firm's specific requirements. All the tools which may be required in part programs must be described in the tool card index.

However, the statement

NEWTOL, n

where n is the number of tool cards which follow, can be used to augment the tool card index for the particular part program in which it occurs.

In the programming language, there are four possible ways to describe the machining operations:

individually programmed tool motion with special instructions for feed and cutting speed;
single machining operations;
work cycles;
sequential operations.

The following description gives examples of data which can be determined automatically and data which must be specified by the part programmer.

Instructions for workpiece description

Before the first machining definition, the following information about the workpiece must be given:

material;
initial condition (optional);
initial surface finish (optional);
cutting value corrections (optional).

The workpiece description is given by the following definition:

PART/modifiers

If, in particular working instructions, a modifier of the principal word **PART** changes, then PART/... with the changed modifier can be given. Only the named modifiers are interpreted as changes to the previous information. The position of the particular modifiers is discretionary. The following modifiers are permitted:

Material modifier   The instruction reads

MATERL, b

where   b = material number.

Initial condition   The initial condition of the material where a hole is to be machined need only be defined when the program uses work cycles. If no initial condition is specified, solid material is assumed by the program.
The modifiers read:

UNMACH   unmachined
SEMI   premachined so that the final condition at the
            machining point can be achieved in one stage
CORED   hole cored or precast

Initial surface condition   If work cycles are called up in the working instructions, then information on the initial surface condition can be given. If no initial surface is defined, the condition SMOOTH is assumed. The modifiers read:

SMOOTH   smooth surface
ROUGH   rough surface

An additional operation to the normal sequence produced by a work cycle is caused by ROUGH. A preliminary machining operation is scheduled with a spot facer.

Cutting value corrections   When feed and cutting speed are determined automatically by the processor, no allowance is made for the rigidity of the part. If it is thought that the feed and/or cutting speed will not be suitable, then a corresponding change can be programmed by means of the cutting value correction. The change is given as a percentage of the value determined automatically, thus:

$$80 = 80 \text{ per cent of the determined value}$$
or $\qquad 140 = 140$ per cent of the determined value.

The instruction is given by

CORREC, f, s

where   f = percentage of the feed
           s = percentage of the cutting speed.

Clearance distance   The instruction reads:

CLDIST/tz

where tz = clearance distance.
    When positioning, the point of the tool proceeds to a distance tz above the workpiece surface. During the working stroke in the negative Z direction, at a distance tz before reaching the programmed position, the machine is switched over from rapid traverse to work feedrate.

Tool-change position   The instruction reads:

SAFPOS/x, y, z
or $\qquad$ SAFPOS/z

During tool changing, no tool will ever go to a Z value smaller than that given by these instructions. SAFPOS/x, y, z also determines a definite tool-change position. The instruction remains valid until the next **SAFPOS** instruction. The instruction

SAFPOS/NOMORE

cancels the tool-change position.

Machining operation definitions　The types of machining operation are specified by major words; modifiers describe these more closely. The order of the modifiers is immaterial.

Either single machining operations or work cycles can be specified. In the case of single machining operations, only the type of machining operation described is carried out. Premachining—for example, predrilling in the case of thread tapping—must be programmed separately.

In the case of work cycles, only the final machining operation is described. The required premachining operations are determined automatically.

Major words and modifiers for defining machining operations

The major words are shown in Fig. 7.12. Where an x is shown in the figure, the corresponding major word applies to the operation. Thus, MILL only applies to a single operation and not to a work cycle.

Operation	Major word	Single operation	Work cycle
Centre drilling	CDRILL	x	
Drilling	DRILL	x	x
Reaming	REAM	x	x
Core drilling	SISINK	x	x
Spot facing or counterboring	SINK	x	x
Countersinking	COSINK	x	
Tapping	TAP	x	x
Boring	BORE	x	
Recessing with special tool	RECESS	x	
Milling	MILL	x	
Other machining	MAKE	x	

Figure 7.12　Major words for machining definitions.

Several modifiers are associated with these major words, and these modifiers are shown in Fig. 7.13. Not every modifier can be used with each major word. The relationship between modifiers and major words is shown later in Fig. 7.15.

Coding of spindle return motion　Following the modifier SPIRET, a two-figure code has to be quoted. Figure 7.14 shows the standard codes available.

Thus, if SPIRET, 33 is programmed, the spindle will be stopped in its aligned position, will dwell, and will then return at the feedrate rather than at rapid.

Meaning	Modifier	Example	SØ	Work cycle
Diameter	DIAMET, d d = diameter		X	X
Machining depth	DEPTH, t t = depth of hole	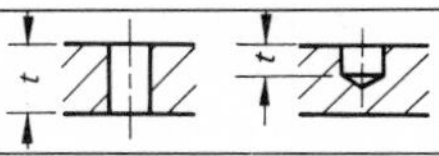	X	X
Tool	TØØL, e TØØL, e, f, e = tool number f = turret number		X	
Feed	FEED, s s = feedrate		X	
Cutting speed	SPEED, v v = cutting speed		X	
Spindle return motion	SPIRET, g g = spindle return code		X	
Drilling blind holes	BLIND, i i = code for bottom of hole i = 1 arbitrary shape i = 2 flat bottom		X	X
Countersinking	DIABEV, j, ANBEV, m j = diameter m = angle	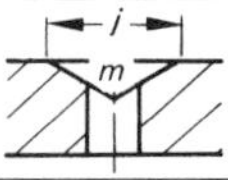	X	X
Thread type	TAT, p p = code for thread type p = 1 standard metric thread p = 2 fine metric thread p = 3 Whitworth thread		X	X
Thread pitch	PITCH, h h = pitch of thread		X	X
Single machining operation	SØ		X	

Figure 7.13   Modifiers for machining definitions.

	Return stroke at rapid traverse rate		Return stroke at feedrate	
	immediate	delayed	immediate	delayed
Same turning direction	0	1	2	3
Reverse turning direction	10	11	12	13
Stationary spindle	20	21	22	23
Aligned spindle	30	31	32	33

Figure 7.14   Spindle return codes.

Major word	Modifiers	SØ	DIAMET d	DEPTH t	TØØL e,f	FEED s	SPEED v	SPIRET g	NØREV	TØLPØ	BLIND i	BEVEL i	DIABEV i	ANBEV m	TAT p	PITCH h
Single operations																
Centre drilling	CDRILL	●	●(1)	○	●(2)	○	○	—	—	—	—	—	—	—	—	—
Drilling	DRILL	●	●	●	○	○	○	○	○	—	—	—	—	—	—	—
Reaming	REAM	●	●	●	○	○	○	○	—	—	○	—	—	—	—	—
Core drilling	SISINK	●	●	●	○	○	○	○	—	—	—	—	—	—	—	—
Spot facing or counterboring	SINK	●	●	●	○	○	○	○	—	—	—	—	—	—	—	—
Countersinking	CØSINK	●	●(1)	—	●(2)	○	○	○	—	—	—	—	●	●	—	—
Tapping	TAP	●	●	●	●(1)	—	○	○	—	—	—	—	—	—	●(2)	●(2)*
Boring	BØRE	●	●	●	●	●	●	○	—	—	—	—	—	—	—	—
Recessing	RECESS	●			●												
Milling	MILL	●	●	●	●	●	●	—	—	—	—	—	—	—	—	—
Other machining operations	MAKE	●		●	●												
Work cycles																
Drilling	DRILL	—	●	●	—	—	—	—		○	—	○(1)	○(2)	○(2)	—	—
Reaming	REAM	—	●	●	—	—	—	—		○	○	○(1)	○(2)	○(2)	—	—
Core drilling	SISINK	—	●	●	—	—	—	—		○	—	○(1)	○(2)	○(2)	—	—
Spot facing or counterboring	SINK	—	●	●	—	—	—	—		○	○	○(1)	○(2)	○(2)	—	—
Tapping	TAP	—	●	●	—	—	—	—		○	○	○(1)	○(2)	○(2)	●*	●*

● must be used
○ may be used
— may not be used

Pecking instruction Usually, when drilling deep holes, automatic retraction to clear the chips takes place and feedrate reduction is also applied. By using the modifier NOREV, reversal and feed reduction can be suppressed (only for a single machining operation).

Hole positioning accuracy The modifier TOLPO indicates that there is a special requirement for accurately positioning the hole. When TOLPO is programmed, a centre-drilling operation precedes drilling; if TOLPO is not programmed, there is no centre-drilling operation.

Operation	Result	Definition
Centre Drill	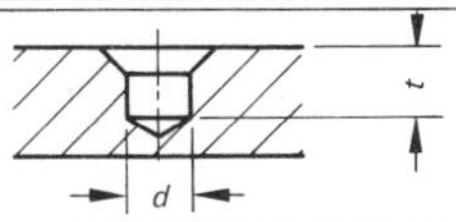	Symbol = *CDRILL/SO, DIAMET, d, DEPTH, t, $ FEED, s, SPEED, v*
Drill	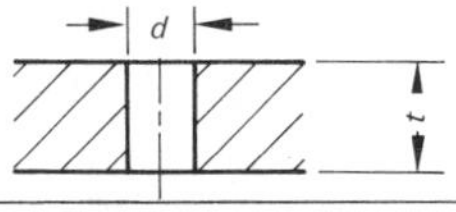	Symbol = *DRILL/SO, DIAMET, d, DEPTH, t, $ TOOL, e, f*
Ream	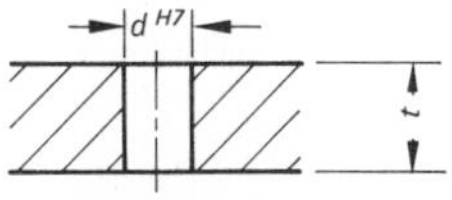	Symbol = *REAM/SO, DIAMET, d, DEPTH, t, $ SPIRET, g*
Spot face or counterbore	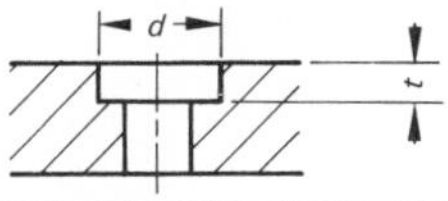	Symbol = *SINK/SO, DIAMET, d, DEPTH, t, $ TOOL, e, f*
Countersink	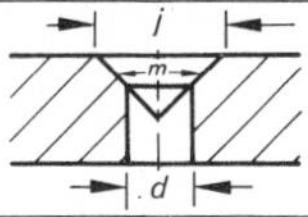	Symbol = *COSINK/SO, DIABEV, j,ANBEV, m, $ DIAMET, d*
Tap	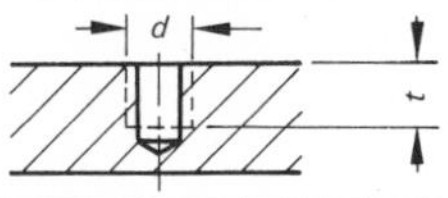	Symbol = *TAP/SO, DIAMET, d, DEPTH, t, $ TOOL, e, f*
Bore	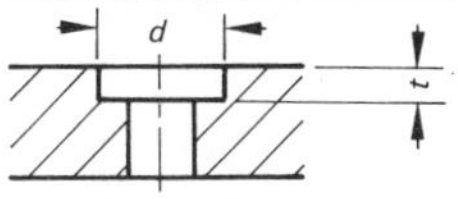	Symbol = *BORE/SO, DIAMET, d, DEPTH, t, $ TOOL, e, f, FEED, s, SPEED, v*
Mill	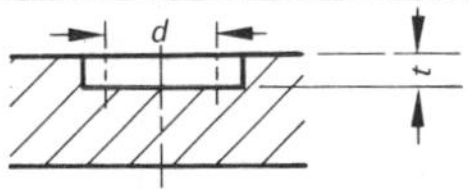	Symbol = *MILL/SO, DIAMET, d, DEPTH, t, $ TOOL, e, f, FEED, s, SPEED, v*

Figure 7.16 Single operation examples.

Single machining operations Figure 7.15 shows the permitted single machining operation definitions. The general form of the definition reads:

$$\text{Symbol} = \text{major word/SO, modifiers}$$

where SO, signifies a single operation.

The modifiers marked with a 1 or 2 in Fig. 7.15 cannot be used in the same statement. Those modifiers marked with a 2 in a given type of definition must be used together in that definition.

Figure 7.16 illustrates a number of typical SO definitions, and the work they define.

A number of the diagrams look similar, but the differences arise from the final results. Thus, the spot face, bore, and mill produce similar results. Bore would be used rather than spot face when an accurate diameter, d, is required. With the milling operation, d is the diameter of the circular path which the centre of the milling cutter follows.

Work cycles The permitted work cycles are also given in Fig. 7.15. When describing a work cycle, only the final state of a machining position is given. The necessary premachining and the machining sequence are automatically determined by the processor.

In the case of work cycles, tools and cutting conditions are determined automatically. The modifiers SO, TOOL, e (or e, f), FEED, s, SPEED, v, NOREV, SPIRET, g are not permitted.

The modifiers DIABEV, j and ANBEV, m describe the defined countersinking. If DIABEV, j and ANBEV, m are given, then the modifier BEVEL is inadmissible. If the modifier TOLPO is given, a centre-drilling operation is added to the machining cycle.

The modifier SPIRET, g applies only to that part of a work cycle named in the major word. Thus in a TAP/... cycle, only the tapping operation would be affected.

Figure 7.17 shows a number of work cycles and their effects. Thus the TAP/ definition illustrated calls up operations to drill, chamfer, and tap using the necessary tools and the feeds and speeds built into the processor.

Execution instructions

Motion instruction

The motion instructions define the final position of a motion. At the end of the motion, the tool point is at this specified position. When positioning in the X–Y plane, the actual path taken is not laid down. If the Z co-ordinate of the final point is greater than that of the starting point, then the tool first moves in the Z direction and is then positioned in the X and Y directions. If the Z co-ordinate of the final point is smaller than that of the starting point, movement in the Z direction follows positioning in the X and Y directions.

Operation	Result	Definition	Work sequence
Drill	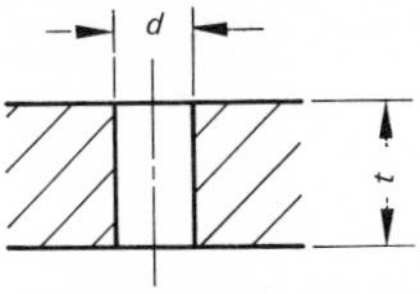	Symbol = $DRILL/DIAMET,d,\mathcal{S}$ $DEPTH ,t$	1. Predrill for $25 \leq d < 55$ one predrilling for $55 \leq d$ two predrillings
Drill with countersink	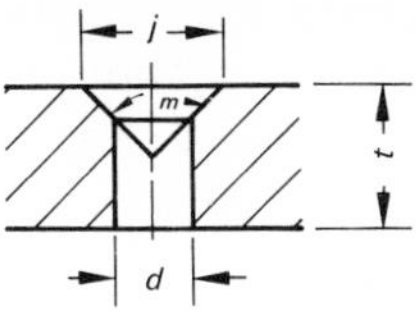	Symbol = $DRILL/DIAMET, d,\mathcal{S}$ $DEPTH , t, DIABEV,\mathcal{S}$ $j, ANBEV, m$	1. Predrill 2. Drill 3. Countersink
Ream	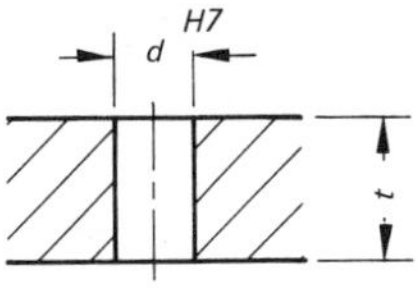	Symbol = $REAM/DIAMET,d,\mathcal{S}$ $DEPTH, t, T\emptyset LP\emptyset$	1. Centre drill 2. Drill 3. Core drill 4. Ream
Ream and countersink a blind hole	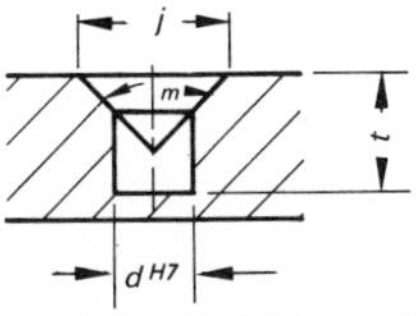	Symbol = $REAM/DIAMET,d,\mathcal{S}$ $DEPTH,t,T\emptyset LP\emptyset,\mathcal{S}$ $DIABEV,j,ANBEV,\mathcal{S}$ $m, BLIND,j$	1. Centre drill 2. Drill 3. Core drill 4. Countersink 5. Ream
Core drill	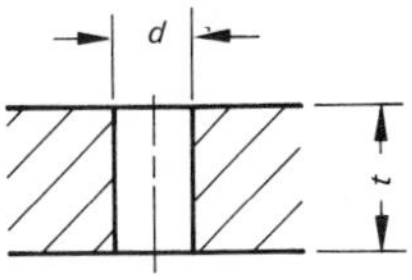	Symbol = $SISINK/DIAMET, \mathcal{S}$ $d,DEPTH,t$	1. Drill 2. Core drill
Counterbore and countersink	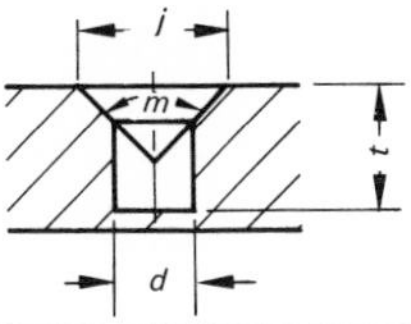	Symbol = $SINK/DIAMET,d,\mathcal{S}$ $DEPTH,t,T\emptyset LP\emptyset,\mathcal{S}$ $DIABEV,j,ANBEV,\mathcal{S}$ m	1. Centre drill 2. Drill 3. Core drill 4. Countersink
Tap and countersink	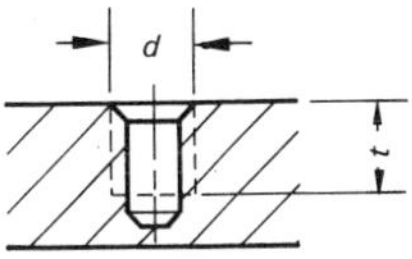	Symbol = $TAP/DIAMET,d,\mathcal{S}$ $DEPTH,t,TAT,p,\mathcal{S}$ $PITCH,h,BLIND,\mathcal{S}$ $i, BEVEL$	1. Drill 2. Countersink 3. Tap

Figure 7.17 Work cycle examples.

The effect of a motion instruction depends on its position in the part program. A distinction is made between motion instructions within the scope of a machining call-up statement and the motion called for by the 'additional instructions' described later.

If a motion instruction is given under the control of a machining call-up statement, the machining will be carried out at each programmed point.

Starting point for motion   Before the first motion instruction in a part program, the starting point must be defined by a FROM instruction. At this position there is no machining operation. If the motion instructions which follow are to be under the control of a machining call-up, a machining statement must appear before the FROM/ statement.
The instruction reads:

FROM/x, y, z

or   FROM/sp

where sp = symbol of a single point.

Before starting machining, the operator must set the point of the tool at the point defined in the FROM instruction.

Motion instruction to a specified point

GOTO/x, y, z

or   GOTO/sp

where sp = symbol for a single point or for a point pattern.

After the symbol for a point pattern, the modifiers RETAIN, OMIT, INVERS may be used as described earlier.

Incremental motion instructions

GODLTA/dx, dy, dz

or   GODLTA/dz

where   dx = increment in X direction
        dy = increment in Y direction
        dz = increment in Z direction.

The instruction describes the motion to a point that differs from the current position of the tool by the increments dx, dy, and dz.

Modification of motion instructions

Avoidance of obstacles   If obstacles arise during motion from one point in a pattern to the next, making it necessary to lift the tool in the Z direction before

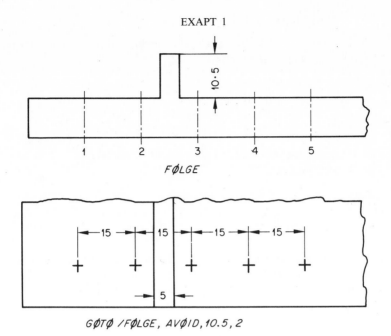

FØLGE

GØTØ /FØLGE, AVØID,10.5,2

Figure 7.18   The use of AVOID.

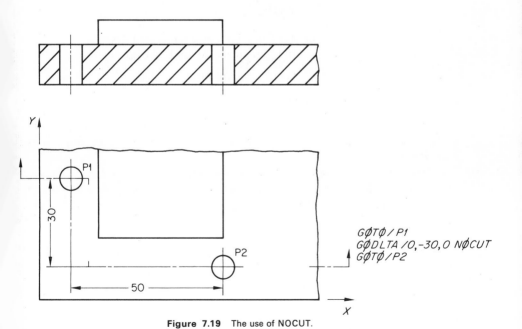

GØTØ / P1
GØDLTA /0,-30,0 NØCUT
GØTØ / P2

Figure 7.19   The use of NOCUT.

actually positioning in the X–Y plane, this can be specified by the modifier word AVOID in a GOTO/ instruction.

Thus ...AVOID, dz, pi
or AVOID, dz, pi1, pi2, ... , pin
or AVOID, dz, pi1, THRU, pin
or AVOID, dz1, pi1, AVOID, dz2, pi2

where dz = increment in Z direction (height of obstacle)
pi = point index.

This modifier follows the other modifiers normally used for point pattern. The point index refers to the original numbering of the point pattern. The obstacle occurs after the specified point. Figure 7.18 shows how AVOID should be used.

Positioning without subsequent machining The modifier NOCUT included in the motion instruction prevents machining at the specified position. This instruction can be used to circumvent obstacles and can be appended to the GOTO/ and motion instructions. See Fig. 7.19.

Work instruction

Machining call-up The machining definitions are called up by a machining instruction. This reads:

WORK/md1, md2, ...

where md1, etc., are symbols of previous machining definitions.
A WORK/... definition remains in force until cancelled by

WORK/NOMORE

or by another WORK/... definition.

The machining is carried out in the sequence of the listed symbols; therefore it is possible to describe the desired sequences of single machining operations and/or work cycles.

By the statement DEPTH, t after the symbol of the machining definition, the previously defined depth can be modified. The modification of the depth does not permanently change the machining definition, that is, it works only for that particular machining call-up.

Sequence of motions The machining operations are carried out at the positions given by the motion instructions. Different tool paths can be prescribed according to the type of machining.

(a) For all machining operations other than milling the tool point is brought to the desired position (tool not in contact with the work during positioning).

Next comes the machining in the negative Z direction followed by the return stroke of the tool. Figure 7.20 illustrates the tool path for drilling.

(b) In the case of milling, the tool point goes to the first position and then machines to the given depth. The tool then moves to further positions at the work feedrate with the tool still cutting. The return stroke of the tool only

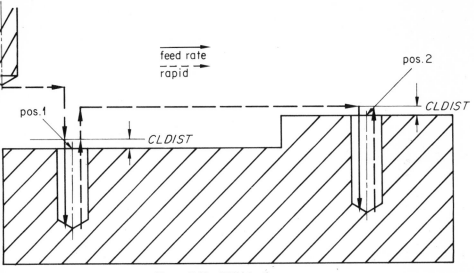

Figure 7.20   DRILL/ action.

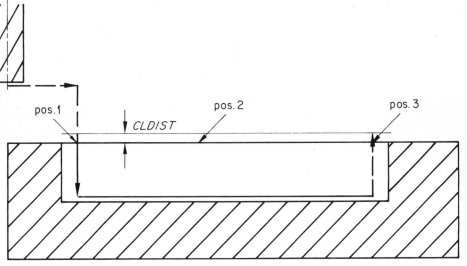

Figure 7.21   MILL/ action.

follows after reaching the last position. A simple MILL/ sequence would produce the tool path shown in Fig. 7.21.

(c) Any positions specified after a WORK/NOMORE instruction are taken up by the tool point of the last specified tool, but there is no cutting motion into the metal. This is illustrated in Fig. 7.22.

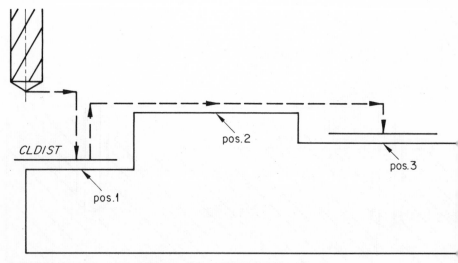

Figure 7.22   WORK/NOMORE action.

Tool changing   If the same machining operation is to be carried out at several positions, it is necessary to state whether the sequence of operations in a work cycle is to be carried out at each hole in turn or whether one operation is to occur at all the holes before the next operation in a cycle is similarly treated.

Normally, a tool-change occurs after a tool has been used at all the points referenced in the motion statement. However, use of the modifier PH (for per hole) permits a complete machining sequence to be carried out at each point in turn, involving several tool changes at each position. The modifier PH must be given directly after the major word WORK/ before the first machining definition symbol.

Additional instructions

To allow arbitrary work sequences which cannot be achieved by single machining operations or work cycles, the tool feed and cutting speed can be defined by special single instructions. These single instructions have the same effect as in APT.

These statements are:

Cutting fluid	COOLNT/...
Definition and call-up of auxiliary functions	AUXFUN/...
Table rotation with simultaneous rotation of the workpiece co-ordinate system	ROTABL/...
Machine stop	STOP
Beginning of a part program	PARTNO/...
Remarks	REMARK/...
Post processor printout	PPRINT/...
Post processor call-up	MACHIN/...
Printout of intermediate results	CLPRNT
Definition of synonyms for vocabulary words	SYN/...
Definition and call-up of post-processor functions	PPFUN/...
End of a part program	FINI

Subscripting, program loops, and subroutines

Subscripting

In a part program, symbolic names may not be defined twice unless they are arithmetic variables. In order to be able to use definition instructions with symbolic names within programming loops and subroutines, subscripted symbols must be employed. Without the subscript, the symbolic name would be re-defined each circuit round the loop. The subscript value must be a positive whole number greater than zero.

Before the subscripted symbol can be used, a RESERV/ instruction must be programmed. Thus

$$\text{RESERV/symbol, az}$$

where az = greatest subscript value.

If it is desired to have eight point symbols, P(1) to P(8), the instruction is

$$\text{RESERV/P, 8}$$

after which   P(1) = POINT/3, 3, 3
             P(7) = POINT/4, -2, 1
are legal statements. A symbol may only be shown once in a RESERV instruction.

Program loops

In a part program without program loops, the instructions are executed consecutively in the order in which they are programmed. Occasionally, it proves to be useful to be able to influence the order of the execution of instructions by building branches into the program, the branch selected being dependent on conditional statements in the part program. Branches of this kind, which interrupt the linear flow of the program, are initiated by jump commands.

In order to simplify the task of the processor, the instruction

LOOPST

must appear before the first instruction in a loop and

LOOPND

after the last instruction. A labelling system has to be provided to indicate the instruction to which a jump has to be made. The label, or address, is prefixed to the instruction:

11) Instruction

where 11 is a label, separated from the instruction by a right parenthesis).

Two instructions are available to initiate an instruction jump. These are:

JUMP/11

and $\qquad$ IF (arithmetic expression) 11, 12, 13

JUMP/ causes an unconditional jump to the instruction labelled 11.

The IF statement causes a jump to the statement labelled 11 if the value of the arithmetic expression is negative, to 12 if it is zero, and to 13 if it is positive.

All the instructions in a loop, other than LOOPST, may be labelled. Between LOOPST and LOOPND, the vocabulary words FINI, MACHIN, MACRO, PARTNO, and TERMAC must not appear.

Figure 7.23 shows a program loop used to define points P1 to P12 lying on the pattern illustrated.

Subroutines

A subroutine is a portion of a program, complete in itself, named with an unsubscripted symbol. Once programmed, it is stored in the computer and only introduced and worked on when called up by its symbol in the part program. The same subroutine can be called up as often as required within a part program.

Definition of a subroutine The definition of a subroutine begins with the instruction

Symbol = MACRO/fp1, fp2

or $\qquad$ Symbol = MACRO

where fp1, fp2 = formal parameters.

The conclusion of the subroutine is defined by the instruction

TERMAC

All language statements following the MACRO instruction up to and including TERMAC are stored as subroutines without the instructions being carried out.

Call-up of a subroutine A subroutine is called up by the following instruction:

CALL/su, fp1, fp2 = ap2

or   CALL/su

where su = symbol for a subroutine

        ap = actual parameter (symbol or value)

        fp = formal parameter (symbol).

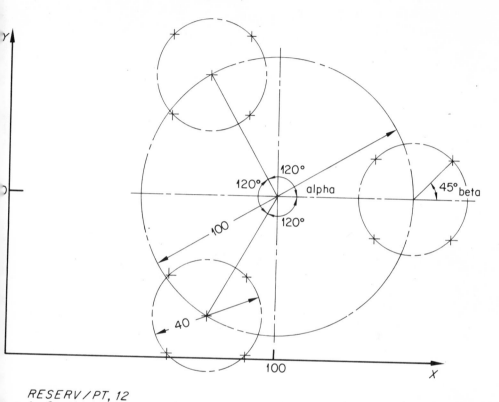

```
    RESERV/PT, 12
    I = O
    K = O
    LØØPST
A)  M = O
B)  I = I + 1
    PT(I) = PØINT /(100+50*CØS(K*120)+20 * CØS(45+M*90)),$
    (60 +50*SIN(K*12Ø) +20*SIN(45 +M*90))
    M = M +1
    IF (M-4) B,C,C
C)  K = K+1
    IF(K-3) A,D,D
D)  LØØPND
```

Figure 7.23 Program loop example.

Computer processing

For processing a part program, a digital computer and the required EXAPT processor program must be available. Although it runs like a single program the processor has two parts, followed by a post processor, corresponding to Fig. 7.24.

These parts are:

geometrical processor ⎱
technological processor ⎰ the processor

post processor (to suit a particular machine tool).

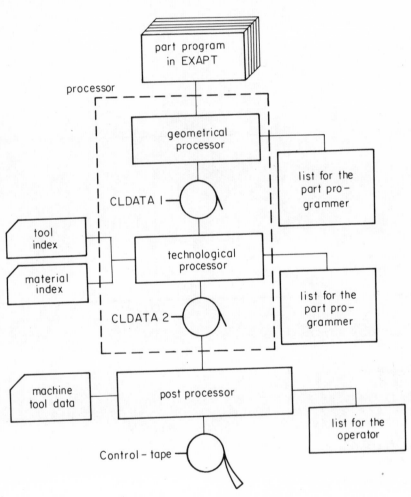

Figure 7.24   Processing with EXAPT 1.

The part program, punched on to punched cards, is processed by means of the geometrical and the technological processors independent of any particular machine tool. These parts of the program have been produced once and for all, and are suitable for all machine tools which perform drilling, tapping, reaming, etc.

In the geometric processor, the part program instructions are analysed and tested for errors, and the EXAPT language words are turned into code numbers. The information is converted to the unique format of the CL (cutter location) DATA1. This completes the geometric processing.

In the technological processor, the machining definitions are worked on, and the data which is not available for the carrying out of the operations is determined in the correct sequence. The required information is calculated for each operation at the machining point called up in the part program. The necessary tool movements are determined, and the output is transferred sentence by sentence to the CLDATA 2.

CLDATA 1 and CLDATA 2 represent input/output media for intermediate results which, according to the particular computer configuration available, may be on magnetic tape, disc, or drum store.

Following the processor program is the post processor, the program which adapts the standard output to suit a particular type of machine. The post processor is always developed for a particular NC machine and its control system, and it adapts the neutral CLDATA 2 to the particular technical requirements of this machine.

At this stage the CLDATA 2 forms the input. In its format, it is nearly compatible with the post-processor input for APT post processors so that an EXAPT CLDATA can be processed with an APT post processor with the minimum changes to the post processor.

In the post processor, the contents of the sections of the CLDATA 2 are converted into the format for the particular numerical control system. The workpiece co-ordinates are recalculated into the machine co-ordinates of the particular machine tool, and the technological data is translated into setting values and commands for the particular machine. Furthermore, the post processor can determine magazine-loading details and can printout lists for the operator and for tool preparation. Finally it will produce a punched tape.

The machine tool and control system manufacturers therefore require experienced technical experts who are in a position to produce the required post processors for the EXAPT system and to develop and maintain them. They must be in a position to produce suitable post processors for new combinations of machine tool and control systems and the EXAPT Association can help with the required training program.

Post processors are described in more detail in chapter 11.

Programming example with EXAPT 1

It is intended to machine the plate shown in Fig. 7.25. The 30 mm diameter hole must be reamed and requires premachining. The four threaded holes lie on a pitch circle and again must be premachined before final machining. The arrangement of six holes of 10 mm diameter can be treated as a point pattern. These holes can be produced in one machining sequence.

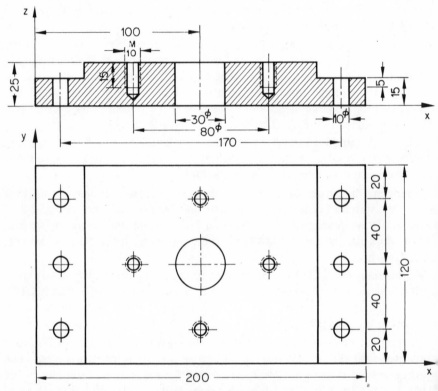

Figure 7.25   Drilled plate.

In instructions 5 to 13 of the part program shown in Fig. 7.26, the positions at which machining should take place are defined. Line 5 defines the co-ordinates of P1 (30 mm diameter hole). Line 6 gives the Z co-ordinates of the following holes. Line 7 defines the circle on which, in line 8, the positions of the threaded holes are indicated. The instructions 9 to 13 define the position of the holes of 10 mm diameter as point patterns L1 and L2.

The technological definitions describe the material (line 14) and the types of machining operations to be used. In each case, only the final machining operation is given. The instructions signify:

line 15: ream 30 mm diameter, depth 25 mm
line 16: thread tapping 10 mm diameter, depth 15 mm, metric, blind hole
line 17: drilling 10 mm diameter, depth 15 mm.

In the execution instructions (lines 18 to 27), the previously defined machining operations and machining positions are called up. By means of CLDIST (see line 18), the clearance distance for the approach to the workpiece surface is laid down. COOLNT/ON in line 19 is intended to switch on the coolant.

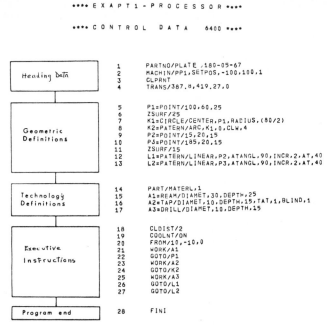

```
•••• E X A P T 1 - P R O C E S S O R ••••

•••• C O N T R O L   D A T A    6400 ••••
```

Heading Data	1 2 3 4	PARTNO/PLATE ,180-05-67 MACHIN/PP1,SETPOS,-100,100,1 CLPRNT TRANS/387.8,419.27,0
Geometric Definitions	5 6 7 8 9 10 11 12 13	P1=POINT/100,60,25 ZSURF/25 K1=CIRCLE/CENTER,P1,RADIUS,(80/2) K2=PATERN/ARC,K1,0,CLW,4 P2=POINT/15,20,15 P3=POINT/185,20,15 ZSURF/15 L1=PATERN/LINEAR,P2,ATANGL,90,INCR,2,AT,40 L2=PATERN/LINEAR,P3,ATANGL,90,INCR,2,AT,40
Technology Definitions	14 15 16 17	PART/MATERL,1 A1=REAM/DIAMET,30,DEPTH,25 A2=TAP/DIAMET,10,DEPTH,15,TAT,1,BLIND,1 A3=DRILL/DIAMET,10,DEPTH,15
Executive Instructions	18 19 20 21 22 23 24 25 26 27	CLDIST/2 COOLNT/ON FROM/10,-10,0 WORK/A1 GOTO/P1 WORK/A2 GOTO/K2 WORK/A3 GOTO/L1 GOTO/L2
Program end	28	FINI

Figure 7.26 Part program for plate.

By means of the FROM instruction, the starting point is given. In the WORK instruction (line 21), the machining instruction A1 is called up (see lines 21 and 15). Then follows the definition of the machining position in the GOTO/ call-up (line 22).

The instructions for the machining of the part are enclosed between the heading and the program-ending statement. The heading is contained in line 1, a general title with the name of the part. Line 2 gives the machine to be used. CLPRNT effects a printout of the computer results on the CLDATA. The TRANS instruction states to the post processor the relationship between the workpiece and machine tool co-ordinate system. FINI in line 28 indicates that the part program is finished.

After processing the part program shown in Fig. 7.26, the EXAPT 1 pro cessor causes the printout of CLDATA 1 at the end of the geometric processing The CLDATA 1 (see Fig. 7.27) contains after each numbered punched card the contents of the punched card. At the same time, the geometric processor has converted all geometric definitions (see lines 5 to 13) of the part program directly into GOTO instructions. Corresponding, for example, to the instruction GOTO/P1 (line 22 of the part program), CLDATA 1 shows the co-ordinates X = 100, Y = 60, Z = 25 which correspond to P1. The technological defini- tions are fully retained; for example, compare line 15 of the part program with lines 11 and 12 of CLDATA 1. The definition A1 = ..., etc. was converted into line 12 of CLDATA 1 where the modifiers had been given code numbers following the ream definition. The end of CLDATA 1 is again marked with the word FINI.

```
                PLATE ,180-05-67

PRINTOUT OF CLTAPE 1
   1  CARDNO    1      1
   2  PARTNO    2   1045PLATE      ,180-0     5-67
   3  CARDNO    1      2
   4  MACHIN    2   1015  PP1    SETPOS     -100.000    100.000    1.000
   5  CARDNO    1      3
   6  CLPRNT    2      3
   7  CARDNO    1      4
   8  TRANS     2   1037   387.800   419.270    0.000
   9  CARDNO    1     14
  10  PART     15   1501         3101   1.000
  11  CARDNO    1     15
  12  REAM     16      3  A1       3205   30.000   3202   25.000
  13  CARDNO    1     16
  14  TAP      16      7  A2       3205   10.000   3202   15.000   3214   1.000   3213   1.000
  15  CARDNO    1     17
  16  DRILL    16      2  A3       3205   10.000   3202   15.000
  17  CARDNO    1     18
  18  CLDIST    2   1071    2.000
  19  CARDNO    1     19
  20  COOLNT    2   1030         71
  21  CARDNO    1     20
  22  FROM      5      3       0        0   10.000  -10.000    0.000
  23  CARDNO    1     21
  24  WORK     15   1502  A1      -1.000
  25  CARDNO    1     22
  26  GOTO      5      5       P1        0  100.000   60.000   25.000
  27  CARDNO    1     23
  28  WORK     15   1502  A2      -1.000
  29  CARDNO    1     24
  30  GOTO      5      5  K2        1   140.000   60.000   25.000
  31  GOTO      5      5  K2        2   100.000   20.000   25.000
  32  GOTO      5      5  K2        3    60.000   60.000   25.000
  33  GOTO      5      5  K2        4   100.000  100.000   25.000
  34  CARDNO    1     25
  35  WORK     15   1502  A3      -1.000
  36  CARDNO    1     26
  37  GOTO      5      5  L1        1    15.000   20.000   15.000
  38  GOTO      5      5  L1        2    15.000   60.000   15.000
  39  GOTO      5      5  L1        3    15.000  100.000   15.000
  40  CARDNO    1     27
  41  GOTO      5      5  L2        1   185.000   20.000   15.000
  42  GOTO      5      5  L2        2   185.000   60.000   15.000
  43  GOTO      5      5  L2        3   185.000  100.000   15.000
  44  CARDNO    1     28
  45  FINI     14         END OF CLTAPE 1
           544  WORDS
```

Figure 7.27   CLTAPE 1 for plate.

In processing the technological definitions, the technological processor deter mines tool paths. Thus, for example, in CLDATA 2 (see Fig. 7.28) the definition of the ream-machining operation (part program line 15, CLDATA 1 line 12 has disappeared. In its place, the work sequence, the necessary tools, feeds, and rotational speeds were determined and—when called up (by part program line 21) the developed sequence of the machining operations—were given out a individual instructions on CLDATA 2 (see Fig. 7.28). Thus, for the machining operation A1, the first output is the tool for predrilling. Line 22 in Fig. 7.28 states in column 5 that the first tool in the tool list is to be used. In line 23, the calculated spindle speed is 332 rev/min. Details of the first tool are given in line 6 of Fig. 7.28 and summarized in line 1 of Fig. 7.29 and A1 (the corresponding machining operation) in Fig. 7.30.

Then the output gives the RAPID motion to a position above P1 with the Z co-ordinate 227·0 (see part program line 22, GOTO/P1, CLDATA 2 line 27).

```
PLATE ,180-05-67

PRINTOUT OF CLTAPE 2
  1  CARDNO  1    1
  2  PARTNO  2  1045    PLATE    ,180-0   5-A7
  3  CARDNO  1    2
  4  MACHIN  2  1015    PP1      SETPOS  -100.000   100.000   1.000
  5  CARDNO  1   17
  6  TOOLST  2  1061
                         1      24     1       0      2  200.000  160.000  24.900   8.400
               118.000   0     103     0   0.000      0    0.000    0.000   1.000   1.000
  7  TOOLST  2  1061     1      44     1  25.000
                 0.000   0             0             2  240.000  190.000  29.750   4.500
                         1     115     0   0.000      0    0.000    0.000   1.000   1.000
  8  TOOLST  2  1061     1      41     1  25.000
                 0.000   0             0             2  120.500   70.000  30.000   4.200
                         1     113     0   0.000      0    0.000    0.000    .900    .900
  9  TOOLST  2  1061     1      24     1  25.000
               118.000   0             0             2  100.000   75.000   8.400   3.100
                         1     111     0   0.000      0    0.000    0.000   1.000   1.000
 10  TOOLST  2  1061     1       7     0  20.600
                 0.000   0             0             2   50.000   26.000  10.000   5.000
                         1     122     0   1.500      0    6.000    6.000   1.000   1.000
 11  TOOLST  2  1061     1      24     1  15.000
               118.000   0             0             2  105.200   80.000  10.000   3.400
                         0     105     0   0.000      0    0.000    0.000   1.000   1.000
 12  CARDNO  1    3                          15.000
 13  CLPRNT  2    3
 14  CARDNO  1    4
 15  TRANS   2  1037   387.800   419.270   0.000
 16  CARDNO  1   18
 17  CLDIST  2  1071     2.000
 18  CARDNO  1   20
 19  RAPID   2    5
 20  FROM    5    3        0       0   10.000  -10.000   2.000
 21  CARDNO  1   21
 22  TOOLNO  2  1025     1.000
 23  SPINDL  2  1031   332.382    59
 24  CARDNO  1   22
 25  COOLNT  2  1030     1.000
 26  RAPID   2    5
 27  GOTO    5    5       P1
 28  FEDRAT  2  1009     .366     0   100.000   60.000   227.000
 29  GOTO    5    5        0   100.000   60.000   191.600
 30  RAPID   2    5
 31  GOTO    5    5        0   100.000   60.000   227.000
 32  CARDNO  1   23
 33  TOOLNO  2  1025     2.000
 34  SPINDL  2  1031   235.396    59
 35  CARDNO  1   22
 36  RAPID   2    5
```

Figure 7.28   CLTAPE 2 for plate.

FIRST RUN OF TECHNOLOGICAL PROCESSOR IN CONNECTION WITH MACHIN/   PP1

TOOL LIST   PLATE ,180-05-67

SER.NO.	IDENTITY NO.	MAG.-NO.	DIAMETER	L1	SYSTEM-NO.
1	103	0	24.90	160.00	124 10 2
2	115	0	29.75	190.00	144 10 2
3	113	0	30.00	70.00	141 10 2
4	111	0	8.40	75.00	124 10 2
5	122	0	10.00	26.00	1 7 10 2
6	105	0	10.00	80.00	124 10 2

Figure 7.29   Tool list for plate.

After converting to the working feed (see line 28, Fig. 7.28, FEDRAT), output of the position relative to the working depth (see line 27 GOTO), and finally return stroke rapid motion (see line 30 and 31, RAPID, GOTO), the pre-drilling has been completed. Now follows the output of instructions for counter-sinking with tool number 2 (line 33, Fig. 7.28), while in the remainder of CLDATA 2 printout, which is not reproduced, are the instructions for reaming with tool number 3 in the same position. The instructions for the other holes (thread cutting and drilling 10 mm diameter) at the corresponding positions then follow. The CLDATA 2 also finishes with the instruction FINI.

LIST OF MACHINING OPERATIONS   PLATE,180-05-67

NAME	MACH. OP.	TOOL NO.	DIAMETER	DMAX	DMIN	DEPTH	MAX.DEPTH
A1	DRILL	1	22.31	26.77	22.31	25.00	25.00
A1	SISINK	2	29.75	29.75	29.75	25.00	25.00
A1	REAM	3	30.00	30.00	30.00	25.00	25.00
A2	DRILL	4	8.30	8.40	8.20	20.60	20.60
A2	TAP	5	10.00	10.00	10.00	15.00	15.00
A3	DRILL	6	10.00	10.00	10.00	15.00	15.00

END OF 1 ST RUN

Figure 7.30   Machining operations for plate.

Apart from CLDATA 2, the computer also prints out the tool list shown in Fig. 7.29, and the machining operation list (Fig. 7.30), which serve the programmer for control as well as for tool preparation. The processing then passes to the post processor which produces the punched tape for the machine tool and a listing of the control punched tape with some additional information for the operator of the machine. For this particular part program, the processing time on the CDC 6400 was 28 s costing about £3.

The MACHIN/ statement determines whether CYCLE/... type records should be written on CLDATA 2 instead of writing the series of individual tool movements which machine tools require when they do not have control circuits which can deal with CYCLE/, or G81, etc. control-tape (see chapter 6) commands.

By writing out CYCLE/ records for this part program, the length of CLDATA 2 would be reduced to 50 per cent and the total processing time to 20 s.

This description of the processing of a part program does not explain how the tools and cutting conditions are calculated by the EXAPT 1 processor. Chapter 15 contains an example of cutting technology processing in EXAPT 1, dealing in detail with the selection of tools and the machining sequence for a REAM/ statement, and showing how the feeds and speeds are derived.

8. EXAPT 2

Dieter Reckziegel

For career summary see chapter 7.

The language EXAPT 2 is used for the automatic programming of NC lathes with line and continuous-path control. Just as in EXAPT 1, both geometrical and machining problems are described by simple, easily understood, free-format instructions. Drilling operations on NC lathes can be described with the instructions of EXAPT 1, so that, by combining the facilities of EXAPT 1 with the new facilities of EXAPT 2, production problems can be programmed for NC lathes. In contrast to the EXAPT 1 language, in which geometric descriptions are confined to definitions of points and point patterns, the geometric descriptions in EXAPT 2 had to be supplemented to describe the contours of both the raw material (or blank) and the finished part. It is necessary to describe contours so that the material, which has to be removed from the blank in order to leave the finished shape, can automatically be divided into a series of machineable cuts.

All technological information necessary for the description of the turning work can be described by technological instructions.

It is possible to determine automatically:

the division of the material to be cut into single cuts between the blank and the finished part;

the co-ordinates of the end points of the approach, working, withdrawal, and return motions;

the cutting depth, feed, and cutting speed.

In this chapter, reference will be made to the information on EXAPT 1 in chapter 7 and on the technological principles of EXAPT in chapter 15 where identical situations exist.

8

EXAPT 2 is briefly summarized below:

(a) It is suitable for programming lathe work.

(b) The information on the workshop drawing can be included in the part program without any recalculation.

(c) Simply constructed free-format instructions describe the blank, the finished part, and the machining process.

(d) The sequence of cuts, the tool paths, feeds, and rotational speeds are determined automatically. The initial data for this automatic determination is based on the particular values required by each user.

(e) Any data which can be automatically determined can also be directly specified in the part program. When this is done, the automatic determination is omitted.

(f) Both tool and material information is required to determine tool path and technological data. This information is made available to the processor program in the computer in the form of tool and material card indexes. The availability of such card indexes is a requirement for the introduction of EXAPT.

(g) The processing of part programs requires a data-processing installation with a suitable processor and also with a suitable post processor for the machine tool being used.

Language structure

The language structure of all parts of the EXAPT language is the same, the information given in chapter 7 applies equally to EXAPT 2.

Co-ordinate systems

Machine tool and workpiece co-ordinate systems

The part to be turned is defined in a right-handed cartesian co-ordinate system with axes x, y. The rotational axis is the x axis. If the part programmer has also been manually producing control tapes he must remember not to use zy axes normally used on the lathe (see chapter 6). The origin of the co-ordinate system is chosen by the programmer in such a way that the geometric description of the part is as simple as possible.

In this way the part program becomes independent of the machine tool on which the part is to be machined. To convert from workpiece to machine-tool co-ordinates, the relationship between the two co-ordinate systems must be given.

This relationship is described by two instructions which give explicitly:

the position of the chuck relative to the machine tool origin (CHUCK/...);

the position of the workpiece co-ordinate system (of the part) relative to the chuck (CLAMP/...).

Figure 8.1 illustrates the connection between the various positions.

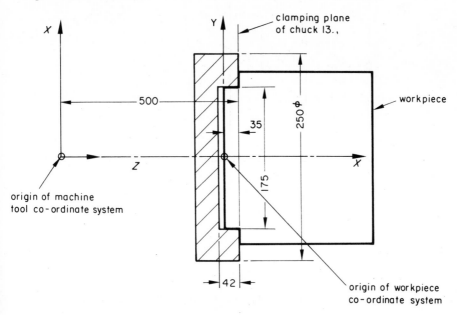

CHUCK /13, 500, 250, 0, 175, -42
CLAMP /35

Figure 8.1   The use of CHUCK/... and CLAMP/.

Clamping instructions   The main dimensions of the chuck and its position relative to the machine tool origin, are given in the chuck instruction

$$\text{CHUCK/nr, l, } d_a, l_a, d_i, l_i$$

where   nr = the number of the chuck
   l = the length, measured along the axis of rotation, between the origin of the machine tool co-ordinate system and the workpiece-locating face on the chuck (clamping plane)
   d_a = the outer diameter of the clamping device
   l_a = the distance on the chuck between its outer locating face and its workpiece locating face
   d_i = the inner diameter of the clamping device
   l_i = the distance on the chuck between its inner locating face and its workpiece locating face.

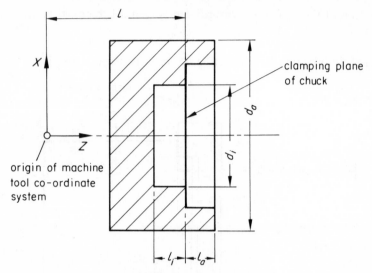

Figure 8.2 Dimensions of a chuck.

Clamping instructions The position of the turned part in the chuck is given by the statement

$$\text{CLAMP}/l_s$$

where l_s = co-ordinate of the clamping plane in the workpiece co-ordinate system.

If the turned part is not to be clamped in the position described but is to be rotated through 180°, this is done by adding the modifier INVERS in the clamp instruction

$$\text{CLAMP}/l_s, \text{INVERS}$$

Reference co-ordinate system The introduction of further reference co-ordinate systems, in addition to the workpiece co-ordinate system chosen at the beginning of the program, sometimes simplifies the programming task and sometimes is the only way to solve a problem (production of an eccentric bore with the spindle stationary).

Definitions

Geometric definitions

The contours of the blank and the finished part are defined by geometric descriptions in the workpiece co-ordinate system. Because of rotational symmetry, turned parts are completely defined if one half of the geometric cross-section is described.

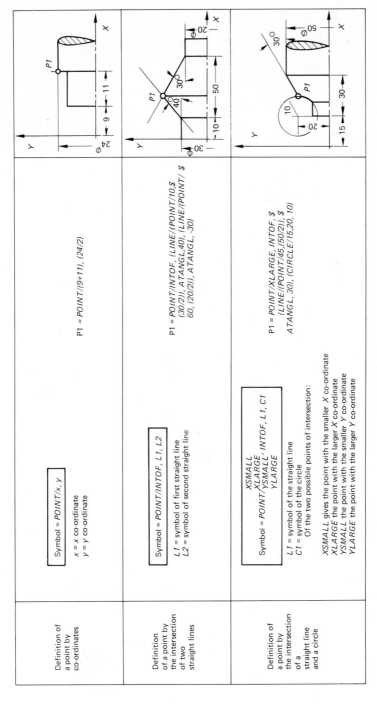

Figure 8.3 Point definitions.

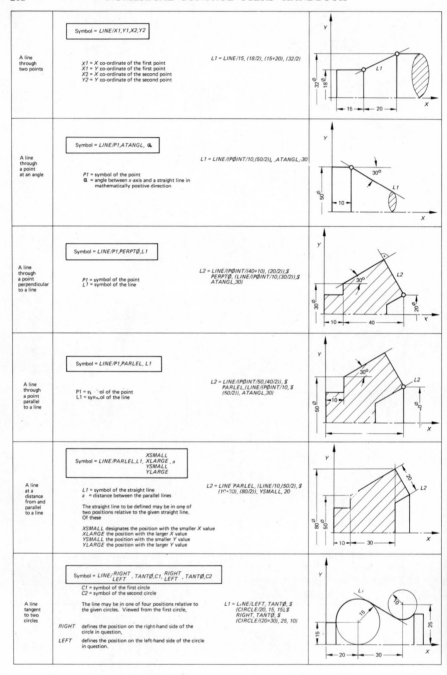

Figure 8.4   Straight-line definitions.

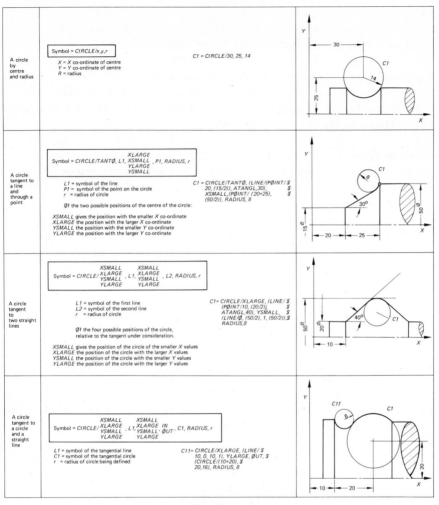

Figure 8.5 Circle definitions.

Geometric definitions are required for the geometric description of the blank and finished parts, and for the provision of tool paths in any desired machining process.

In the description of the blank and the finished part, the geometric definitions can be called up by means of their particular symbol or can be given in brackets. Cylindrical surfaces and plane faces, bevelled and rounded edges, can be described in a simplified form.

Point definitions The definition of a point may be required to give the tool-changing position in a work sequence, to determine the starting direction in the description of the contour, and to define straight lines and circles.

Figure 8.3 gives the permitted point definitions for EXAPT 2.

Definition of a straight line   The definition of a straight line may be required to give the tool path in a work sequence, for the description of contour elements, and for the definition of points and circles. In Fig. 8.4, the permitted definitions are illustrated by means of examples.

Definition of circles   The definition of a circle may be required for giving the tool paths in any particular work sequence, for describing contour elements, and for the definition of points and straight lines. The circle definitions permitted in EXAPT 2 are given in Fig. 8.5.

Description of contour

The perimeter of half of the longitudinal cross-section is described as a closed contour. The contour is always described in a clockwise direction to identify where the material lies.

The contour is described by a sequence of geometric elements, each instruction adding a geometric element to the part of the contour already described.

In order to avoid unnecessary repetition of the description of identical contour sections, simple instructions deal with shifting and forming mirror images of parts of contours.

If the shape of the raw material is given by means of an allowance (for example by a casting allowance relative to the finished part), then the description of the shape of the raw material can be left out if the allowance is given in the contour description of the finished part.

Form of the contour description   The contour is described between the words CONTUR/ and TERMCO. The modifiers BLANCO (raw material) and PARTCO (finished part) after the word CONTUR/ state whether the raw material or the finished part is being described.

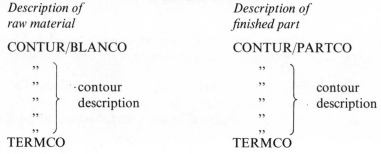

The contour is always closed in a clockwise direction.

Links between instructions   In the description of a contour, the contour elements are linked together in the sequence determined by the direction of description. Thus each element can be regarded as linked to the preceding element. By using this linking concept, geometric elements of infinite extent are converted to contour elements with the linking points as their boundaries. These linking points are considered to be the tangential or corner points of the contour, and can be given symbolic names by means of the contour mark feature.

Bevelling and rounding of corners may be defined as additions to contour elements by means of modifiers after the contour-element instructions. Similarly, the surface finish of contour elements (e.g., mating faces) may be defined by modifiers. Figure 8.6 shows the principle of the geometric descriptions.

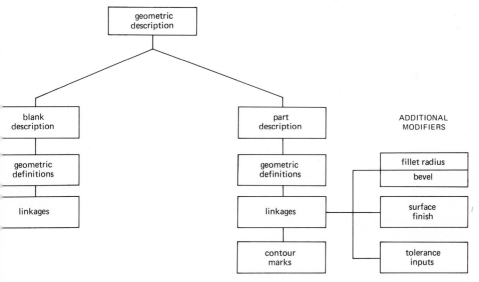

Figure 8.6   Describing geometry in EXAPT 2.

Each contour instruction takes its direction with reference to the direction of the previously defined element.

The starting point and direction of the contour description are given in the following way:

$$\text{BEGIN/P1, } \begin{matrix} \text{XSMALL} \\ \text{XLARGE} \\ \text{YSMALL} \\ \text{YLARGE} \end{matrix} \begin{matrix} \text{L1} \\ \text{C1} \end{matrix}$$

where   P1 = symbol of a contour point as starting point
        L1 = symbol of a straight line
        C1 = symbol of a circle

From the starting point the direction of the contour description is:

for   XSMALL ...   in the negative x direction,

         XLARGE ...   in the positive x direction,

         YSMALL ...   in the negative y direction,

         YLARGE ...   in the positive y direction.

In place of the symbolic name for a point its two co-ordinates may be given.

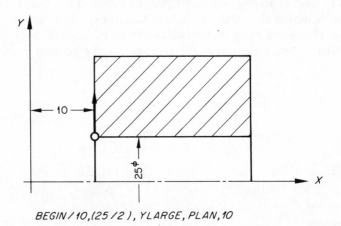

BEGIN / 10, (25 / 2) , YLARGE , PLAN, 10

Figure 8.7   The use of a BEGIN/ instruction.

The direction of each subsequent contour element is referred to the previously defined element by the major words:

         RGT/   ...   right

         LFT/   ...   left

         FWD/ ...   forwards

         BACK/ ...   reverse

These major words may be followed by the following modifiers:

(a) *For geometric elements*   The symbolic name of a straight line or circle. In addition, simplified definitions of contour elements for

               cylinder   ... /DIA, d

or   plane surfaces ... /PLAN, x

where   d = diameter

        x = co-ordinate of the plane surface.

(b) *To define end points of contour elements*   The specific definition of the end point of a contour element can be left out if the end point is built up from the following element in a unique manner.

When the following contour element is produced from the same geometric element, then the end point must be defined on a line intersecting the geometric element. This end point then becomes the link point. After the geometric element the end point is defined thus:

$$.../ \ldots, \text{ON, L3}$$

where L3 is the symbol of an intersecting line.

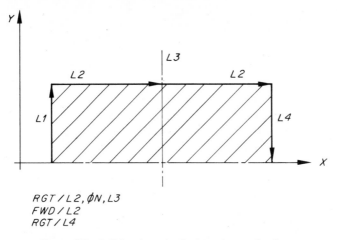

RGT / L2, ØN, L3
FWD / L2
RGT / L4

Figure 8.8   Defining the end-point by an intersecting line.

To indicate which of two possible end points is intended   Where there are two points of intersection with the following contour element and where the second point of intersection is to be the end of a contour element, the modifiers 2, INTOF are used thus:

$$.../ \ldots, \text{ON, 2, INTOF, C2}$$

where 2 identifies the second point of intersection and C2 is the symbol of the following geometrical element (see Fig. 8.9).

For additional contour transitions   Bevelled or rounded edges can be specified as transitional elements in the contour element. The statement refers to a contour transition between that contour element and the following one. It is assumed that both contour elements intersect. The modifier has the following form:

Bevelled edge: .../..., BEVEL, b

where b = length of each element removed by bevel.

Rounded edge: .../..., ROUND, r

where r is the radius of circle tangent to both elements, as in Fig. 8.10.

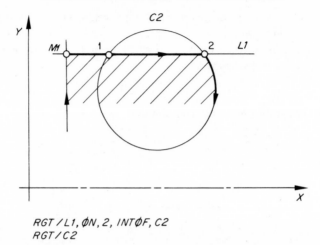

```
RGT/L1,ØN,2,INTØF,C2
RGT/C2
```

Figure 8.9   Indicating the second point of intersection.

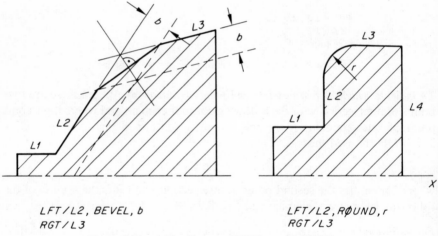

```
LFT/L2, BEVEL, b              LFT/L2, RØUND, r
RGT/L3                        RGT/L3
```

Figure 8.10a and b   The use of BEVEL... and ROUND....

To add technological information   Modifiers can be added to the contour element instruction to indicate:

surface finish;
tolerances, or unmachined surfaces, respectively;
dimensional correction statements.

These are dealt with below, under technological definitions.

For contour marks In order to apply subsequent instructions to parts of contours, mark symbols can be placed before each contour element. They are separated from the statements by commas and refer to the starting points of the respective contour elements.

A statement which contains no modifiers to define a transition can only be given one mark symbol. If a transition (bevel or rounded edge) is defined, then a second mark can be added to the statement. The second mark symbol refers to the starting point of the transition. Since a contour element statement may not contain more than one geometrical element and one transition, it can only be given two markings at the most.

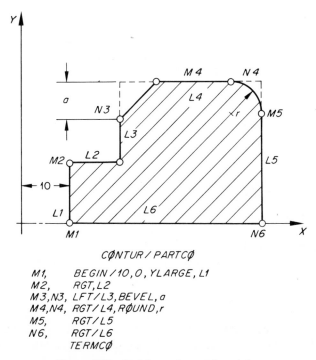

CØNTUR / PARTCØ

```
M1,     BEGIN /10,0,YLARGE, L1
M2,     RGT,L2
M3,N3, LFT/L3,BEVEL,a
M4,N4, RGT/L4,RØUND,r
M5,     RGT/L5
N6,     RGT/L6
        TERMCØ
```

Figure 8.11 Applying contour-mark symbols.

Simple description of the shape of the raw material by an excess metal dimension
The shape of the raw material can be defined by means of an excess dimension relative to the finished part by means of the major word OVCONT/

OVCONT/a

where 'a' is the excess dimension of the raw material relative to the finished part.

The statement specifies a constant excess dimension of the raw material relative to the elements of the following finished part contour until a new OVCONT statement is given.

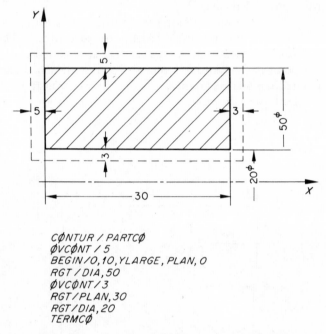

```
CØNTUR / PARTCØ
ØVCØNT / 5
BEGIN/0,10,YLARGE, PLAN, 0
RGT / DIA, 50
ØVCØNT / 3
RGT / PLAN, 30
RGT / DIA, 20
TERMCØ
```

Figure 8.12   Describing shape of raw material by OVCONT/.

Technological definitions

The technological part of EXAPT 2 contains statements for describing the workpiece technology and the cutting sequence. These instructions contain all the necessary information for turning work; they contain information about the workpiece, the machining operation, the machine tool, and the tools.

Part of the information must be given; other parts can be either automatically determined or given in advance. The following can be automatically determined:

Turning	Drilling
Tool paths (division of the region to be machined into single cuts)	Tool paths
	Spindle return motions
Feeds and cutting speeds	
Cutting depth	

The machining definitions available for formulating the machining work are summarized in Figs. 8.13 and 8.14. Machining definitions for concentric internal holes correspond to the definitions of EXAPT 1. In addition, there are the turning definitions which define the various basic modes of machining on lathes with their corresponding major words. By adding the modifiers, LONG, CROSS, or ATANGL, the desired feed direction is specified. The identity number of the desired tool and its setting angle must also be given in the machining definition, since, in contrast to EXAPT 1, the present stage of development of EXAPT 2 does not contain any automatic choice of tools. To specify the desired quality of machining, a statement must also indicate whether a roughing, finishing, or fine-finishing operation is intended. Internal, or external, machining with turning tools is determined by the processor program from the position of the volume of material to be removed, but the part programmer

Major words			SO	LONG	CROSS	ATANGL	TOOL	SETANG	DEPTH	FEED	SPEED	ROUGH	FIN	FINE	TAT	PITCH	OSETNO	SPIRET
(Modifiers)							e,f	t		s	v				p	h	no	g
Single machining operations	turn	TURN	●	①	①	①	●	●	○	○	○	②	②	②	—	—	○	○
	contour turn	CONT	●	①	①	①	●	●	○	○	○	②	②	②	—	—	○	○
	recess	GROOV	●	①	①	①	●	●	○	○	○	②	②	②	—	—	○	○
	thread	THREAD	●	①	①	①	●	●	○	—	●	②	②	②	●	●	○	○

Those modifiers characterized by 1, exclude each other mutually.
The same applies for modifiers characterized by 2.

● Modifier must be given
○ Modifier can be given

Figure 8.13　Turning definitions.

Major words		SO	DIAMET d	DEPTH t	TOOL e,f	FEED s	SPEED v	SPIRET g	NOREV	DIABEV i	ANBEV m
(Modifiers)											
Single machining operations	centre drill	CDRILL	●	○	○	●	●	○	—	—	—	—
	drill	DRILL	●	●	●	●	○	○	○	○	—	—
	ream	REAM	●	●	●	●	○	○	○	—	—	—
	spiral sink	SISINK	●	●	●	●	○	○	○	—	—	—
	sink	SINK	●	●	●	●	○	○	○	—	—	—
	tap	TAP	●	●	●	●	—	○	○	—	—	—
	counter sink	COSINK	●	○	—	●	○	○	○	—	●	●
	boring bar	BORE	●	●	●	●	●	●	○	—	—	—

● Modifier must be given
○ Modifier can be given

Figure 8.14　Drilling definitions.

must select an appropriate tool. In addition, the data-processing installation will carry out some collision-course calculations. A check is also made to see whether the length of the tool is adequate for the programmed machining operation and the tool-changing position is determined.

The part programmer is assisted by two card indexes dealing with tool cards and material cards. These card indexes contain his own firm's particular metal-cutting values entered in the special manner described in more detail in chapter 15. The cutting conditions, selected automatically by the processor, can thus be suited to the specific requirements of the firm.

For machining, the following descriptions can be used:

(a) Single programmed tool motions with separate instructions for feed and cutting speed.

(b) Single machining operations.

(c) Sequential operations.

In each case, some of the machining data can be determined automatically. The following description of the program indicates where the data is determined automatically and where it must be specified by the part programmer.

Technological statements concerning workpiece   Technological statements may apply either to the whole part, or only to a particular contour element.

Technological statements which contain data referring to the whole workpiece are instructions which are modal. The modal effect means that each statement made in the minor part of the instruction applies until it is replaced by a corresponding statement following the same major word.

Before the first machining statement, the following statements about the workpiece must be made:

Tool statement
Cutting value correction (if required)
Machining allowance (if required)
Surface finish (if required)

These statements are described below:

Description of workpiece   The description of the workpiece is carried out by the statement:

PART/modifier

To alter these modifiers for a subsequent machining instruction, PART/... is used with the changed modifiers and only those modifiers named in such a PART/... instruction are interpreted as changes to the previous statements. The position of the various modifiers is immaterial. The following modifiers are permitted:

(a) Material modifiers have the form

MATERL, b

where b = the material number (see chapter 16).

(b) Cutting value corrections. In the automatic determination of feed and cutting speed, the strength and stability of the part is not taken into account. If it is not desired to use the feed and/or cutting speed determined, then a cutting value correction can be programmed. The change is given as a percentage of the determined value, e.g.,

80 = 80 per cent of the determined value
140 = 140 per cent of the determined value

The information is given in the form

CORREC, ts, tv

where   ts = percentage of the feed
        tv = percentage of the cutting speed.

Machining allowances   Machining allowances can be made during the automatic subdivision of the cuts. These may be given using the major word OVSIZE/...

Thus $OVSIZE/FIN, c_1, FINE, c_2$

where   FIN = finished machining.
        FINE = precision finished machining.
        c_1, c_2 = equidistant allowances in mm.

The modifiers FIN and FINE can also be given individually.

If no allowance is given, then 0·5 mm is assumed for both finishing and precision finishing.

The allowance used in calculating a tool path will be determined by the type of machining which is derived from the particular machining definition used and from the final surface finish for any particular contour element.

Surface finish   If the contour elements have a predominantly uniform surface finish, it can be given by the following statement:

ROUGH
SURFIN/FIN, r1
FINE, r2

where   ROUGH = rough machined
        FIN   = finish machined
        FINE = precision finish machined
and   r1, r2 = the permitted cusp heights in mm.

ROUGH signifies contours which will be finished to size by roughing cuts.

Any surface finishes departing from that specified in a SURFIN/ statement for a particular contour element can be specified with the corresponding modifier in a contour-element instruction.

Contour elements which are given with either FIN or FINE are produced by rough machining, a finishing cut and, in the latter case, by an additional precision finishing cut. At the same time, the programmed allowances remain for subsequent machining operations.

For the automatic determination of feed, roughness values in mm can be given after the modifiers FIN and FINE. If no values are given, feed calculations are carried out with a peak-to-peak roughness given by the German DIN 3141, series 2, machining triangles:

0·016 mm for finish machining;
0·004 mm for precision finish machining.

Clearance distance   With the following statement, it is possible to program a clearance distance from the actual contour. This will be taken into account while lifting off the tool on the return stroke, and while bringing the tool to the work:

$$CLDIST/tz$$

where tz is the clearance distance in mm. If no clearance distance instruction is given, then tz = 1 mm is assumed.

Tool-changing position   The tool-changing position can be given by the following statement:

$$SAFPOS/x, y$$

where x, y = co-ordinates of the tool-changing position.
The definition is modal and is superseded by a new SAFPOS/ instruction or by

$$SAFPOS/NOMORE$$

Technological information in link statements   Technological data which concerns individual contour elements can be given by modifiers to the corresponding contour element instructions. The following information can be given:

Surface finish
Mating faces
Dimensional corrections by means of correction switches.

Surface finishes   If, on a particular contour element, the surface finish is different from that given in the SURFIN statement, it must be given in the corresponding contour element instruction by one of the following modifiers:

> ..., ROUGH, ...
> ..., FIN, rt1, ...
> ..., FINE, rt2, ...

where rt1, rt2 are the permitted surface finishes in mm. If no SURFIN instruction has been given, then the surface finish must be given in each contour element statement.

Mating faces   By using the modifier

> ..., FIT, hm, ...

where hm is a dimension in mm, contour elements are defined as mating faces. Depending upon the capability of the control system, different control commands can be generated in the post processor (for example, optional halt, withdrawal cycles, etc.).

The significance of the dimension hm depends on the post-processor operation. It can, for example, give a radial dimension which has to be held, or a cutting length at the point of measurement.

Dimensional corrections   In order to change the position of the tool relative to particular contour element by means of correction switches (e.g., for mating faces, or faces with grinding allowance), it is possible to give the number of the correction switch to be used in an instruction for a link contour element.

Thus   ..., OSETNO, n, ...

where n = number of the correction switch. This correction only becomes effective during the last cut which produces the finished contour and is only valid for the particular contour element.

Machining definitions   Machining definitions are definition statements to which a symbol is assigned, and specify the type of machining operation.

A major word gives a basic type of machining. Modifiers describe the conditions under which the basic machining operation is to be carried out. The sequence of modifiers is entirely optional.

Drilling   If drills are used for concentric internal machining, the limited number of EXAPT 1 definitions summarized in Fig. 8.14 are currently permissible. The major words and modifiers have been explained in chapter 7.

Turning

Basic types of machining   For internal and external machining with turning tools, the basic types of machining given in Fig. 8.15 can be used and these are distinguished by the various methods of automatically dividing up the cuts and cutting the threads.

TURN
The major word TURN/ defines the basic machining operation of turning in which the action between the beginning of the cut and the lift-off position is carried out by straight-line feed motions. If the direction of feed does not follow the required contour, then the required contour cannot be created with the basic machining process TURN.

CONT
The major word CONT/ defines the basic machining operation of contour turning. Turned parts can be thus produced in so far as tool geometry permits.

   If no particular feed direction is given by means of a modifier, the feed motion is in the direction of the finished part contour displaced in the main cutting direction. If a particular feed direction is stated, the tool moves in this direction until the desired contour is reached and then follows that contour. When the permitted cutting depth is again reached, the chip removal is carried out again in the previously given direction.

GROOV
The major word GROOV defines the basic machine process of grooving. The feed motion is linear in the predetermined direction.

THREAD
The major word THREAD defines the process of thread-turning. Both work sequence and subdivision of cuts depend on the machine tool. The data required for the determination of the control commands can be given by means of modifiers.

   The meaning of the modifiers permitted in the machining definitions can be seen from Fig. 8.16.

Special information for turning work
Constant cutting speed   To avoid large variations in cutting speeds when doing facing, taper turning, and radial operations on machines with variable speeds, one can use the modal instruction

$$CSRAT/tp$$

where tp = percentage cutting speed ratio = $(v_{min}/v)100$ can be programmed or determined automatically. When facing a large area, the necessary speeds and switch-over points are determined automatically.

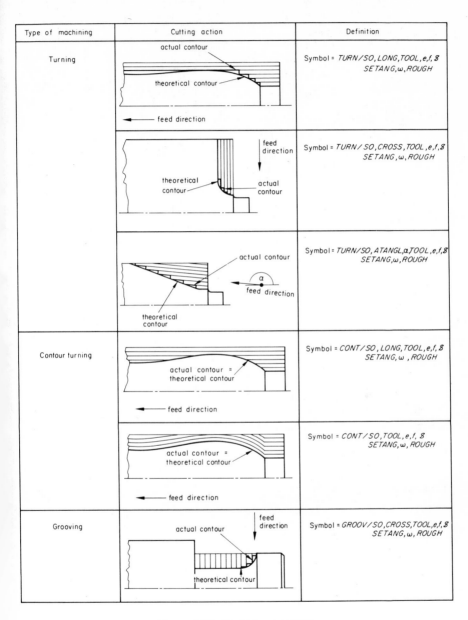

Figure 8.15 Machining definitions.

	Meaning	Modifiers	Example	Turning operation	Boring operation
1.	Machining depth	DEPTH, t t = machining depth		X	X
2.	Countersink	DIABEV, j, ANBEV, m j = diameter m = angle		X	X
3.	Diameter	DIAMET, d d = diameter			X
4.	Feed	FEED, s s = feed value		X	X
5.	Feed direction longitudinal transverse with angle ∝	LØNG CRØSS ATANGL ∝ ∝ = angle with rotational axis		X	
6.	Boring without reversal	NØREV			X
7.	Tool correction	ØSETNØ, no no = number of the correction switch		X	
8.	Machining qualities Roughing _ _ _ _ _ _ _ _ _ _ Finish machining _ _ _ _ _ _ Precision Finish machining _ _ _ _ _ _ _	RØUGH _ _ _ _ _ _ _ _ _ _ _ _ _ _ _ FIN _ _ _ _ _ _ _ _ _ _ _ _ _ _ _ _ _ _ FINE _ _ _ _ _ _ _ _ _ _ _ _ _ _ _ _ _		X	
9.	Setting angle	SETANG W W = angle in degrees		X	
10.	Single operation	SØ		X	X
11.	Cutting speed	SPEED, v v = cutting speed		X	X
12.	Spindle return	SPIRET, g g = spindle return code			X
13.	Type of thread Thread pitch.	TAT, p, PITCH, h p = type of thread h = pitch		X	X
14.	Tool	TØØL, e,f. e = tool identification number f = magazine number		X	X

Figure 8.16 Machining modifiers.

Torque limitations   It is possible to limit the automatically determined cutting depth and feeds by programming a torque limitation. The statement reads:

$$TORLIM/a$$

where 'a' is the torque in mkp. This is a modal instruction and is superseded by a new torque limitation or by the instruction

$$TORLIM/NOMORE$$

If no TORLIM statement is made, then the torque information from the inbuilt machine data is used.

The use of TURN   Figure 8.17 illustrates the sub-division of cuts using TURN/ for a simple component defined as raw material and finished part.

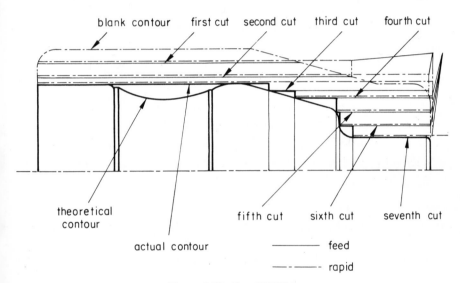

Figure 8.17   Use of TURN/.

Starting from the maximum diameter of the blank, the first cut is taken with the maximum cutting depth. The first point of intersection in the direction of feed with the contour of the blank gives the starting point of the cutting motion at feed velocity. The tool is moved to this point with rapid motion on a collision-free course, taking into account the programmed clearance distance. The cutting motion along the first cutting line finishes where the motion leaves the blank. There the tool lifts off by the amount of the clearance distance. It then moves with rapid motion, in the opposite direction to the feed, to a position from which it can be brought, on a collision-free course in the main cutting direction, to the starting point of the next cut. The example shows that, for the third cut after reaching the contour of the finished part, the cutting line is raised to the position of the finished part diameter still to be machined, the tool being lifted off rapidly and set to the new cutting path. Here the cutting depth is less than its maximum value so that the program determining the cutting values can raise the cutting speed while maintaining the required tool life.

In cuts 4 to 6, the finished part contour is reached and the new starting point approached with rapid motion. A test is made when calculating each cut and if necessary a correction can be made to the cut and the following one so that the cut will never be less than the minimum cutting depth a_{min}. After the last cut in a particular section of a part the tool is moved in an axial direction to a new part section where the subdivision of cuts begins again. After finish-machining the last section of the part, the tool moves back to the tool-change position.

Executive statements

Work call-up

The machining definitions are called up by means of a WORK/ statement. This work call-up statement reads

> WORK/symbol of a machining definition, $
> symbol of a machining definition, ...

or   WORK/NOMORE

Machining takes place in the sequence of the symbols used. Thus it is possible to describe any desired sequence of operations.

Machining position call-up

After each work call-up, it is necessary to state where the machining operation is to be carried out.

The major word for the machining position call-up is

$$CUT/...$$

Drilling position call-up   In the case of drilling operations, this is followed by the modifier

$$CENTER, x$$

where x = x co-ordinate of the hole depth.

Thus CUT/CENTER, x is the form used.

Turning position call-up   In the case of turning operations, the area to be cut is specified by quoting two marks after the CUT/ instruction. Thus the form used is:

$$CUT/M1, \; {TO \over RE}, \; M2$$

where M1 is the mark for the starting position
       M2 is the mark for the finished position.

The CUT statement thus specifies the section of the finished part contour on which machining is to be carried out. The contour section is bounded by the marks which were given in the contour element instructions.
    The modifiers TO and RE define the machining directions as follows:

TO defines the contour section in the direction of the description (clockwise);
RE defines the contour section in the opposite direction of the description.

The sequence of marks, taken in conjunction with the feed modifiers given in the called-up machining definition, lays down the feed direction. In order to determine the feed direction unambiguously, the marks quoted in the CUT statement must be such that the straight line joining them deviates by at least $2°$ from the perpendicular to the direction of feed.

Additional statements

Position of the machining point relative to the lathe axis Although the machining positions have been described on half of the finished part, the actual machining position on a lathe can be either in front of, or behind, the lathe axis. The actual machining position, for the CUT/ statement for which it is valid, is given by the modal statement

$$CUTLOC/{BEFORE \over BEHIND}$$

where BEFORE indicates in front of the lathe axis
       BEHIND indicates behind the lathe axis.

Limitation of tool movements Tool movements which are determined automatically by the processing of a CUT statement can be limited by plane and/or cylindrical surfaces.
    The statement for feed limitation has the following construction

FDSTOP/PLAN, x, DIA, d

where x = x co-ordinate of the plane surface
       d = diameter of the cylinder.

The information in the auxiliary part of the statement can also relate only to a plane or only to a cylinder. The PLAN and DIA information is valid until replaced by corresponding information in a further FDSTOP/ statement or until cancelled by the instruction FDSTOP/NOMORE or by a new CLAMP/ statement.

Further instructions

Special work sequence

Technological information There may be work sequences which cannot be programmed by using machining definitions and machining position call-up. Alternatively, the part programmer may wish to specify a speed, feed, cutting sequence, or turret position for a tool instead of having these provided automatically. Single statements are available for this purpose but these statements must not be used between a WORK/ and a WORK/NOMORE statement. The statements available are:

TOOLNO/e, f

where e = tool identification number
f = turret position.

TOOLNO/e, f, OSETNO, n

where n = number of the correction switch assigned to the tool.

STAN/w

where w = tool setting angle for the turning tool.
This must be given when using TOOLNO.

OFFSET/a

where a = number of the correction switch, modal acting on all tools until the next OFFSET/a or OFFSET/NOMORE appears.

FEDRAT/s

where s = feed value per revolution.

FEDRAT/s, PERMIN

where s = feed value per minute.

$$\text{SPINDL/n, } {\text{CLW} \atop \text{CCLW}}$$
SPINDL/OFF

where CLW is clockwise spindle rotation
CCLW is anticlockwise spindle rotation
n = revolutions per minute
OFF is spindle at rest.

DELAY/a
DELAY/b, REV

where a = delay time in seconds
       b = delay time in number of revolutions
       REV is modifier to indicate revolutions.

$$\text{CSPEED}/v, \genfrac{}{}{0pt}{}{\text{CLW}}{\text{CCLW}}$$

$$\text{CSPEED}/\text{OFF}$$

where v = cutting speed
       CLW is clockwise spindle rotation
       CCLW is anticlockwise spindle rotation
       OFF is spindle at rest.

RAPID

The major word RAPID specifies rapid feedrate until countermanded by a subsequent FEDRAT/ or WORK/ instruction.

Tool motions   While the machining position call-up results in the automatic determination of tool paths, it is possible to move the tool in a desired direction with motion instructions. In that case, description of the blank and finished part are not required.

The motion instructions refer to the cutting reference point of the tool. They are expressed by the modal major word

TLON

which must be used in the part program as the statement preceding the motion statements to indicate that the *tool* reference point is to be on the cutting position. For each tool, the initial point of the motion is a tool-change position. The motion instructions are:

$$\text{FROM}/x, y$$

where x, y are the co-ordinates of the starting point

$$\text{GOTO}/x, y$$
or $$\text{GOTO}/P$$

where x, y are the co-ordinates, or P is the symbol, of the point to which the reference point on the tool is to be moved.

$$\text{GODLTA}/dx, dy$$

where dx, dy are x and y increments of tool motion.

The instructions for tool motion along geometric elements must always be preceded by a FROM/ and a GOTO/ statement to establish a sense of direction.

The instructions permitted in EXAPT 2 are:

GO/ON, E

where E is the symbol of a previously defined contour element (circle or line)

This statement calls for the tool reference point to be moved by the shortest path to the element.

GOLFT/E, GORGT/E, GOFWD/E, GOBACK/E

The major words GOLFT, GORGT, GOFWD, GOBACK correspond in their usage to the major words LFT, RGT, FWD, BACK in the link statements, and are the correct APT instructions (see chapters 9 and 10).

The final position of a motion is given in exactly the same manner as a limitation of a geometric element in the CONTUR/ statements. Statement can have, for example, the following constructions:

GOLFT/L1, ON, 2, INTOF, C2

Go left on L1 to the second intersection of L1 with C2

GORGT/L1, ON, C2

Go right on L1 to the first intersection of L1 with C2

GOFWD/L1

Go forward on L1 to the first intersection of L1 with the geometric element defined in next motion statement

GOBACK/C2

Go back on C2 to the first intersection with the geometric element defined in the next motion statement.

Cutting fluid

COOLNT/q

COOLNT/$^{ON}_{OFF}$

The first instruction calls for coolant type 'q' to be turned on. COOLNT/ON calls for the coolant corresponding to the type of work and material to be turned on. This coolant is specified on the material index card. COOLNT/ is a modal instruction cancelled by another COOLNT/ instruction.

Auxiliary functions

AUXFUN/tr

AUXFUN/tr, OFF

where tr = the number of the auxiliary function.

This instruction is dealt with by the post processor which should generate a miscellaneous function code on the control tape. Thus AUXFUN/71 should cause an M71 to be punched on the control tape, and a subsequent AUXFUN/71, OFF will cancel the M71.

To stop machine   The statement STOP brings the machine to a halt. The exact effects can be derived from the description of the post processor. An optional stop can be programmed with OPSTOP.

Arithmetic statements   The arithmetic statements correspond to those in EXAPT 1 described in chapter 7. The functions are supplemented by

ABS (arg)   absolute value of arg

Program technical statements   The following statements have been described in chapter 7 and are also used in EXAPT 2.

Start of a part program
Commentary statement
Text in the post-processor output
Call-up of the post processor
Printout of intermediate results
Definition of synonyms for vocabulary words
End of part program.

Direct control commands for the machine tool   If functions occur in the control system of a machine tool which cannot be called up by the vocabulary words available in EXAPT 2, then it is possible to program direct control-tape commands in the format required by the control system. The statement has the following form:

INSERT/control command for machine tool

The programmed statement to the right of the oblique line will be inserted by the post processor at the programmed place, thus

INSERT/01S102X00000Y00000H9

will generate 01 S102 X00000 Y00000 H9 on the control tape. The continuation of the control command beyond the end of the card is only possible by repeating the statement INSERT/.

Instructions to the post processor   If the post processor called up by the MACHIN/... statement is required to carry out any special functions then this can be programmed at the appropriate place in the part program by the statement

PPFUN/list

The list may contain any desired elements of the language (symbols, vocabulary words, and numbers) suitably separated by commas. The post processor will interpret and process the list elements.

Printout of defined quantities   The values of any of the defined quantities can be printed in the output. The instruction reads:

PRINT/S1, S2, S3
or   PRINT/ALL

where S1, S2, S3 are the symbols which have been assigned to the desired outputs.

The modifier ALL indicates that all defined quantities are to be printed.

If the printout of the parameters is to start on a new sheet, this can be indicated in the PRINT statement by an 0 which must follow directly after the oblique line:

PRINT/0, ALL

Indexing, program loops, subroutine   The use of indexing, program loops, and subroutines have been described in chapter 7.

Figure 8.18 shows how a program loop can be used to deal with a series of equally spaced grooves. The value of I is 1 the first time the loop is traversed and this takes the tool to the STRTPT(1) with $x = 21$, $y = 21 \cdot 5$ and then it grooves using GROOVE(1) at $x = 21$, $y = 16$.

Next time round, the loop $I = 2$ and the groove is now centred on $x = 41$ and so on until $I = 5$ and 5 grooves have been cut.

Example of a subroutine for grooving   The problem represented in Fig. 8.18 (several grooves in one diameter) can also be solved in general terms by means of a subroutine which can then be used for any shaft of similar shape. For this purpose the following formal parameters are chosen:

I	Name for the number of the groove
DSHAFT	Diameter of the shaft
DGROOV	Diameter of the groove
CNTRS	Distance between two neighbouring grooves
START	Distance of the L.H. plane of the first groove from the y axis
PERIPH	Cutting speed
FEDIN	Feed for grooving
FEDOUT	Feed on the return stroke out of the groove
NUMBER	The number of the grooves

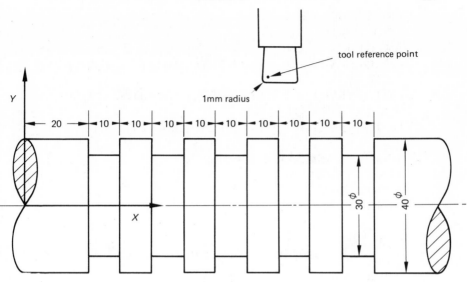

```
             RESERV/STRTPT, 5
             RESERV/GRØØVE, 5
             FRØM/0,50,0
             CSPEED/20
             TØØL NØ./1724,5 $$ STRAIGHT GRØØVING TØØL, WIDTH 10MM
             CØØLNT/ØN
             I=0
             LØØPST
       A)    I=I+1
             STRTPT(I)=PØINT/(X=1 + 20*I), 2I.5
             GRØØVE(I) = PØINT/X,16
             RAPID
             GØTØ/STRTPT(I)
             FEDRAT/.1
             GØTØ/GRØØVE(I)
             FEDRAT/1
             GØTØ/STRTPT(I)
             IF(I-5)A,B,B
       B)    LØØPND
```

Figure 8.18 Using a program loop in EXAPT 2.

The subroutine assumes that the required grooving tool has been brought into the working position by the statement TOOLNO/. . . and is in a position which permits an unhindered approach to the first groove.

The subroutine then has the following form:

GRUCUT = MACRO/I, DSHAFT, DGROOV, START, NUMBER, $
 PERIPH = 20, FEDIN = .1, FEDOUT = .3
I = 0

CSPEED/PERIPH
COOLNT/ON
A) RAPID
   GOTO/(START + 1 + I * CNTRS), (DSHAFT/2 + 1.5), 0
   FEDRAT/FEDIN
   GOTO/(START + 1 + I * CNTRS), (DGROOV/2 + 1), 0
   FEDRAT/FEDOUT
   GOTO/(START + 1 + I * CNTRS), (DSHAFT/2 + 1.5), 0
   I = I + 1
   IF (I − NUMBER)A, B, B
   COOLNT/OFF
   SPINDL/OFF
   TERMAC

A complete program in EXAPT 2

The shaft shown in Fig. 8.19 is to be machined. In order to save space, the programming in this example is to be confined to the roughing cut (i.e., to the premachining of the shaft). Before starting programming, the co-ordinate system with its x and y axis was put onto the drawings. The origin of co-ordinates was chosen on the centre-line of the shaft at the point M0. After naming

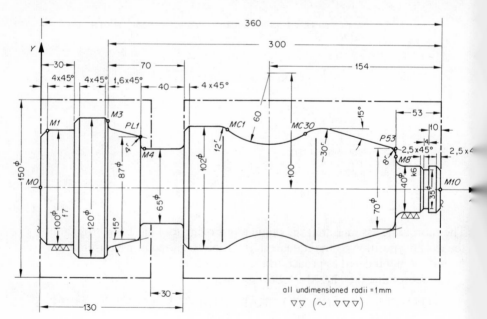

Figure 8.19 Shaft.

the part program in line 1 of Fig. 8.20, TURNING A SHAFT, there follows information about two machine tools intended for carrying out this information having the names ZEISIG (line 4) and EX2PP (line 5).

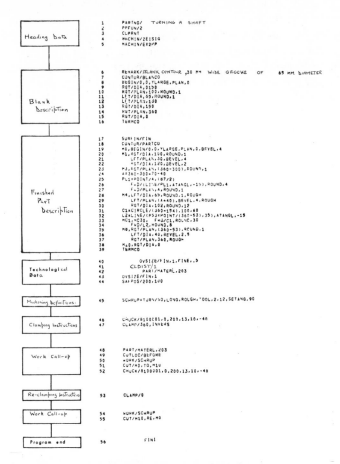

Figure 8.20 Part program for shaft.

Following these heading cards comes the description of contour of the raw material. The description begins with the statement CONTUR BLANCO (contour of blank) and ends at line 16 with the statement TERMCO. The actual contour description begins in line 8 at point 0, 0 with a straight line which runs in the direction of increasing Y parallel to the Y axis and at a distance of 0 mm (see line 8). This is followed in the right-hand direction by another straight line parallel to the X axis which limits the workpiece to a diameter of 150 mm. At a distance of 100 mm in the X direction (see line 10, PLAN/100)

9

this cylindrical region is bounded by a line parallel to the Y axis. The external contour can be followed in this way until it is closed at line 15 by the statement RGT/DIA, 0. The blank shape is shown chain dotted in Fig. 8.19.

The raw material contour is followed by the finished part contour. Special attention should be given here to the definition of the marks M0, M1, etc., which are shown in Fig. 8.19. To supplement the normal description of the contour, contour transitions for example line 21 with BEVEL = 4, are defined. This statement indicates that, between the geometric elements described by line 21 and line 22, a bevelling of 4 mm is defined. Apart from the simplified definitions of straight lines, using the words DIA and PLAN, a line has been defined in line 26, but has not been given a symbol, as it is only used once. The straight line L2 (see line 32) was separately defined to improve understandability and was given the symbolic name L2. In line 34 the straight line L2 is linked with the previous contour element. Line 39 TERMCO ends the description of the finished part.

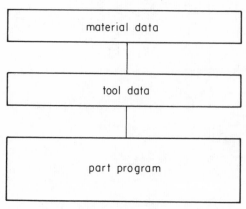

Figure 8.21 Data set for processor.

In line 45, the sole machining operation on this part is given. For future reference, it is given the name SCHRUP. It indicates that a rough machining operation is to take place. The feed direction is defined parallel to X (LONG). The surface finish is to correspond to a roughing operation (ROUGH). Tool No. 2 is to be used and this has the setting angle of 90° to the X axis. Before calling up the machining operation it is necessary to describe the connection between the workpiece and the machine tool co-ordinate system. This is done by the chuck statement (line 46) and by the clamp instruction (line 47). This states: clamp the workpiece INVERS (i.e., turned through 180°). The origin of the co-ordinate system is positioned at a distance of 360 mm from the clamping face of the chuck. The workpiece is then to be machined corresponding to the machining call-up (line 50, WORK/SCHRUP). The region to be machined is given in line 51. It indicates: machine from Mark M0 to Mark M10, i.e.,

```
******************************************************************************************************
* CODE-NUMMER * WERKSTOFF *   KS11  * EXPON. 1-Z *  FINK  *  CROSSK  *  GROOVK  *   RKAPK  *SPITZENWINKEL*
*             *           *         *            *        *          *          *          *             *
*    203      *   C 45    *   180   *    .80     *  .50   *   .50    *   .30    *   .70    *   118.00    *
*             *           *         *            *        *          *          *          *             *
******************************************************************************************************
```

DREH WERKSTOFFE
****

```
**********************************************************************************************************************************
*CUT MAT *-CODE*KUEHLUNG* B/H *B/H.MIN* V.MAX* V.MIN* A.MIN*E.MITTEL*F.MITTEL*G.MITTEL*H.MITTEL* TLIFE*B.EMPF*H.EMPF*
*   SS   *  1  *   1    * 8.0 *  .1   *  80  * 10  * .20 * -.2389 * .2807 * -.1485 * .4300 * 120.0 * 5.00 * .36 *
*  P 10  *  5  *   0    *14.0 *  .1   * 250  *100  * .10 * -.2367 * .1078 * -.2094 * .1753 * 307.0 * 2.00 * .20 *
*  P 20  *  6  *   0    *12.0 *  .1   * 200  * 73  * .10 * -.2723 * .1572 * -.2732 * .3387 * 230.0 * 2.13 * .16 *
*  P 25  *  7  *   0    *10.0 *  .1   * 180  * 60  * .20 * -.1981 * .1430 * -.1836 * .2498 * 228.0 * 4.62 * .24 *
*  P 30  *  8  *   0    *10.0 *  .1   * 150  * 55  * .20 * -.3497 * .1064 * -.1528 * .4134 * 201.0 * 5.56 * .37 *
*  P 40  *  9  *   0    *10.0 *  .1   * 120  * 45  * .30 * -.1807 * .1000 * -.1260 * .3929 * 63.0 * 5.56 * .56 *
*  P 50  * 10  *   0    *10.0 *  .1   * 110  * 30  * .30 * -.1732 * .1000 * -.1260 * .3929 * 61.0 * 5.56 * .37 *
*  I 10  * 11  *   0    *12.0 *  .1   * 160  * 70  * .10 * -.1806 * .0985 * -.2143 * .2320 * 320.0 * 2.00 * .20 *
**********************************************************************************************************************************
```

BOHR WERKSTOFFE
****

```
**********************************************************************************************************************************
*OPERATION *-CODE*SCHNSTOFF*-CODE*KUEHLUNG*  V   *  A1  *  B1  *  A2  *  B2  *  A3  *  B3  *
* CDRILL  * 1 *  SS  * 1 * 1 * 22.0 * .93 * .0005 * .73 * .0008 * .17 * .0025 *
* DRILL   * 2 *  SS  * 1 * 0 *  7.2 *1.93 * .0015 *1.27 * .0025 * .82 * .0038 *
* REAM    * 3 *  SS  * 1 * 1 * 20.0 *1.10 * .0013 * .67 * .0020 * .28 * .0031 *
* SISINK  * 4 *  SS  * 1 * 1 * 12.6 * .76 * .0008 * .54 * .0012 * 0 00 * 0 0000 *
* SINK    * 5 *  SS  * 1 * 1 * 18.0 * .97 * .0010 * .40 * .0019 * .19 * .0025 *
* CISINK  * 6 *  SS  * 1 * 3 * 18.0 * .90 * .0010 * .40 * .0010 * .10 * .0020 *
* TAP     * 7 *      * 0 * 0 *  0.0 *0 00 *0 0000 *0 00 *0 0000 *0 00 *0 0000 *
**********************************************************************************************************************************
```

Figure 8.22a   Material card information.

```
*******************************************************************************************************
* BEZEICHNUNG      * ABK. * 1.WERKZEUG * 2.WERKZEUG * 3.WERKZEUG * 4.WERKZEUG * 5.WERKZEUG *
* IDENTNUMMER      *      *            *            *            *            *            *
* TOOL IDENTNUMMER *      *   153001   *   363001   *   363010   *   403002   *   433001   *
* SCHAFTIDENTNUMMER*      *    -0      *    -0      *    -0      *    -0      *    -0      *
* SYSTEMNUMMER     *      *            *            *            *            *            *
* FERTIGUNGSVERFAHREN*    *     4      *     4      *     4      *     4      *     4      *
* SCHNEIDENGEOMETRIE *    *    31      *    11      *    39      *    41      *    41      *
* SCHAFTART        *      *     3      *     3      *     7      *     7      *     3      *
* SCHNEIDSTOFF     *      *     6      *     6      *     6      *     6      *     6      *
* SPANNMITTEL      *      *     1      *     1      *     1      *     1      *     1      *
* EINSPANNUNG      *      *     0      *     0      *     0      *     0      *     0      *
* WINKELPLANWAREN  *      *            *            *            *            *            *
* KORREKTURWINKEL  * ETA  *  90.00   *  90.00   *  90.00   *  90.00   *  90.00   *
* HAUPTSCHNEIDENWINKEL* PSI*275.00   * 347.00   * 267.00   *  0.00    * 347.00   *
* NEBENSCHNEIDENWINKEL* TAU*  0.00   *  0.00    *  0.00    *  0.00    *  0.00    *
* SPITZENWINKEL    *EPSILON* 60.00  * -45.00   * -36.00   * -30.00   * -45.00   *
* ABMESSUNGEN      *      *            *            *            *            *            *
* EINSTELLAENGE XS * XS   * 202.20   * 202.00   * 194.00   * 200.50   * 202.50   *
* EINSTELLAENGE YS * YS   * -55.20   * -55.00   * -87.00   * -62.50   * -55.50   *
* SPITZENRADIUS    * RS   *   .80    *  1.00    *  1.00    *  2.50    *   .50    *
* EINSTECHBREITE   * I    * -0.00    * -0.00    * -0.00    * -0.00    * -0.00    *
* EINSTECHTIEFE    * TMAX * -0.00    * -0.00    * -0.00    * -0.00    * -0.00    *
* MINDEST-TIMEUDURCHM.* DMIN* -0.00  * -0.00    * -0.00    * -0.00    * -0.00    *
* FREIMASS RECHTS  * E    *  28.00   *  29.00   *  60.00   *  38.00   *  30.00   *
* EINSATZLAENGE RECHTS* L1 * 35.00   *  35.00   *  25.00   *  35.00   *  35.00   *
* FREIMASS LINKS   * F    *   .80    *  1.00    *  1.00    *  2.50    *   .50    *
* EINSATZLAENGE LINKS* L2 * 500.00   * 500.00   * 500.00   * 500.00   * 500.00   *
* FREIMASS VORN    * G    *   .80    *  1.00    *  9.00    *  2.50    *   .50    *
* ACHSABSTAND      * K    * -42.00   * -42.00   * -44.00   * -43.00   * -42.00   *
* SCHAFTBREITE     * H    *  25.00   *  25.00   *  32.00   *  32.00   *  25.00   *
* FREIMASS RECHTS  * E1   * 500.00   * 500.00   * 500.00   * 500.00   * 500.00   *
* FREIMASS LINKS   * F1   *  0.00    *  0.00    *  0.00    *  0.00    *  0.00    *
* SCHAFTHOEHE      * H    *  32.00   *  40.00   *  40.00   *  40.00   *  40.00   *
* EINSATZBEDINGUNGEN*     *            *            *            *            *            *
* ZUL. SCHNITTBREITE* AZUL*  8.00    *  0.00    *  3.00    *  6.00    *  3.00    *
* ZUL. SCHNITTKRAFT* PZUL *1500.00   *1500.00   * 500.00   * 700.00   * 500.00   *
* SPANFORM-FAKTOR  *B/WREL*  1.00    *  1.00    *  1.00    *  1.00    *  1.00    *
* VERSCHLEISSMARKENBR.* VB*   .60    *   .60    *   .60    *   .40    *   .40    *
* ZUL. STANDZEIT   * TZUL *  20.00   *  20.00   *  40.00   *  40.00   *  40.00   *
* SCHNITTTIEFEN-FAKTOR*AMINREL* 1.00 *  1.00    *  1.00    *  1.00    *  1.00    *
* MAX. SPANDICKE   * HZUL *   .45    *   .50    *   .70    *   .20    *   .30    *
* MIN. SPANDICKE   * HMIN * -0.00    * -0.00    * -0.00    * -0.00    * -0.00    *
* UNIT MM ODER INCH *     *     0    *     0    *     0    *     0    *     0    *
*******************************************************************************************************
```

Figure 8.22b   Tool card information.

the complete part is to be machined. After the completion of this machining operation, the workpiece is to be re-clamped corresponding to instruction 53 CLAMP/0 and is to be so clamped in the chuck that the origin of the co-ordinate system of the workpiece lies at the clamping face of the chuck. Machining is then to take place from M10 to M0. Statement 56 ends the part program.

For computer processing, it is also necessary, as shown in Fig. 8.21, to give tool data and material data in addition to the part program. The type of data to be put in can be seen in Fig. 8.22. Each user fills in data of this type once for each combination of tool and material which he wishes to use (see chapter 15).

```
CLTAPE 1
  1   1     1
  2   2  1045                      TOR   NI   N   GA  SH   AFT
  3   1    .2
  4   2  1079        2.000
  5   1    .3
  6   2  1081
  7   1    .4
  8   2  1015  ZEISIG
  9   1    .5
 10   2  1015  EX2PP
 11   1    16
 12  30     1     360.000    75.000    0.000     0.000       10
 13  30     2       0.000     0.000    0.000     1.000    0.000
 14  30     2       0.000    75.000   -0.000    -0.000   -1.000
 15  30     2     100.000    75.000   -0.000    -1.000   -0.000
 16  30     2     100.000    33.500    1.000   101.000   33.500
 17  30     2     101.000    32.500   -0.000    -0.000   -1.000
 18  30     2     129.000    32.500    1.000   129.000   33.500
 19  30     2     130.000    33.500    0.000     1.000    0.000
 20  30     2     130.000    75.000   -0.000    -0.000   -1.000
 21  30     2     360.000    75.000   -0.000    -1.000   -0.000
 22  30     2     360.000     0.000    0.000     0.000    1.000
 23   1    39
 24  30     3     360.000    60.000    0.000     0.000       32
 25  30     4       0.000     0.000    0.000     1.000    0.000      3222      .100    -1.000    -1.000
  :
  :
  :
  :
  :
 77   1    50
 78  15  1502  SCHRUP
 79   1    51
 80  19     1        1       4003       32
 81   1    52
 82   2  1073  •00001.000    0.000   200.000    13.000    10.000   -40.000
 83   1    53
 84   2  1051       0.000
 85   1    54
 86  15  1502  SCHRUP
 87   1    55
 88  19     1       32       4004        1
 89   1    56

 90  14     END OF CLTAPE 1
```

Figure 8.23 CLDATA 1 output.

The processing in the computer is done in exactly the same way as for EXAPT 1 in the geometric and the technological passes. The intermediate results are given as CLDATA 1 and CLDATA 2 data respectively.

Figure 8.23 shows a section of CLDATA 1. Lines 1 to 10 give the information from the heading data. Following this comes the representation of the blank determined by the geometric processor in the so-called canonical form of geometric definitions. From line 23 onwards follows the representation of the finished part contour. From line 77 onwards, it is possible to see some information from the machining call-ups and the machining definitions.

The CLDATA 1 becomes the input information for the technological processor and is processed into CLDATA 2 (see Fig. 8.24). First information on the CLDATA 2 after the PARTNO statement is a listing of the tools required in the part program with their dimensions (see line 2). From line 14 to line 24

```
CLTAPE 2

 0    1   CARDNO     1
 1    2   PARTNO  1045                               - TOR    NI     N      G      A       SHA      FT
 2    2   TOOLST  1061      4..0     24.00    5.00   6.00    0.00    0.00     2.00     0.00    114.00
                          144.00     90.00    3.00   0.00   55.00    1.00    -0.00    -0.00    100.0
                            1.00    115.00   28.00  65.00    1.00  128.00    25.00    10.00   1500.0
                            1.00       .40   60.00   1.00  187.00   32.00     1.00              .63

 3    1   CARDNO     2
 4    2   PPFUN   1079      2.00
 5    1   CARDNO     4
 6    2   MACHIN  1015     ZEISIG
 7    1   CARDNO    41
 8    2   CLDIST  1071      1.00
 9    1   CARDNO    44
10    2   SAFPOS  1070    200.00    100.00
11    1   CARDNO    47
12    2   CHUCK   1073  *00001.00     0.00   200.00  13.00   10.00   -40.00
13    2   CLAMP   1051    360.00       6
14   30   BLANC1     1    360.00     75.00
15   30   BLANCO     2    360.00      0.00    0.00   0.00    1.00
16   30   BLANCO     2      0.00      0.00   -0.00   1.00   -0.00
17   30   BLANCO     2      0.00     75.00   -0.00   0.00   -1.00
18   30   BLANCO     2    230.00     75.00    0.00  -1.00    0.00
19   30   BLANCO     2    230.00     33.50    1.00 231.00   33.50
20   30   BLANCO     2    231.00     32.50   -0.00   0.00   -1.00
21   30   BLANCO     2    259.00     32.50    1.00 259.00   33.50
22   30   BLANCO     2    260.00     33.50   -0.00   0.00   -0.00
23   30   BLANCO     2    260.00     75.00   -0.00   0.00   -1.00
24   30   BLANCO     2    360.00     75.00    0.00  -1.00    0.00
25   30   PARTC1     3
26   30   PARTCO     4    360.00      0.00    0.00   0.00    1.00   .3222    .10   -1.00   -1.00
27   30   PARTCO     4      0.00      0.00   -0.00   1.20   -0.00    3221    .10   -1.00   -1.00
28   30   PARTCO     4      0.00     17.50    0.00    .71    -.71    3222    .10   -1.00   -1.00
29   30   PARTCO     4      2.50     20.00   -0.00  -1.00   -1.00    3222    .10   -1.00   -1.00
30   30   PARTCO     4     52.00     20.00    1.00  52.00   21.00    3222    .10   -1.00   -1.00
31   30   PARTCO     4     53.00     21.00   -0.00   1.00   -0.00    3222    .10   -1.00   -1.00
32   30   PARTCO     4     53.00     28.86    -.13  61.00   28.86    3222    .10   -1.00   -1.00


58    1   CARDNO    50
59    2   TOOLNO  1025      1.00   O12.00
60    2   STAN    1080     90.00
61    1   CARDNO    51
62    2   RAPID      5
63    5   GOTO       5                      506.52  -191.00
64    5   GOTO       5                      506.52  -190.00
65    5   GOTO       5                      506.00  -180.01
66    2   FEDRAT  1009       .00
67    2   SPINDL  1031    103.67    60
68    5   GOTO       5                      403.00  -184.01
69    2   RAPID      5
70    5   GOTO       5                      376.00  -180.01
71    2   FEDRAT  1009       .00
72    2   SPINDL  1031    103.67    60
73    5   GOTO       5                      158.00  -180.01
74    5   GOTO       5                      159.05  -181.01
75    2   RAPID      5
76    5   GOTO       5                      506.58  -181.01
77    5   GOTO       5                      506.00  -170.03


107   1   CARDNO    53
108   2   CHUCK   1073  *00001.00     0.00   200.00  13.00   10.00   -40.00
109   2   CLAMP   1051     0.00
110  30   BLANC1     1    360.00     75.00
111  30   BLANCO     2    360.00     75.00   -0.00  -1.00   -0.00
112  30   BLANCO     2    360.00      0.00    0.00   0.00    1.00
113  30   BLANCO     2      0.00      0.00    0.00   1.00    0.00


237   2   FEDRAT  1009       .00
238   2   SPINDL  1031    262.53    60
239   5   GOTO       5                      453.00  -140.07
240   5   GOTO       5                      454.05  -141.07
241   2   RAPID      5
242   5   GOTO       5                      506.27  -141.07
243   5   GOTO       5                      506.00  -136.00
244   2   FEDRAT  1009       .00
245   2   SPINDL  1031    311.08    60
246   5   GOTO       5                      453.00  -136.00
247   5   GOTO       5                      454.05  -137.00
248   1   CARDNO    56
249  14      END OF CLTAPE 2
```

Figure 8.24   CLDATA 2 output.

is the description of the blank for use in the post processor to check for collision paths (in the actual clamped position). From line 25 begins the information about the finished part contour. In lines 58 to 77, one can see the beginning of the machining called up by the first CUT call-up and this information includes the automatically determined cutting speeds and feeds as well as the tool positions calculated by the computer. Lines 107 to 109 give information about reclamping. After the re-clamping instruction, it is normal to repeat the information about the blank and finished part contours, since the re-clamping of

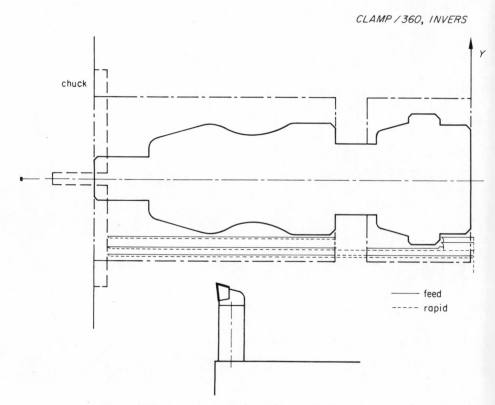

Figure 8.25   Cut sequence developed after line 47, CLAMP/360, INVERS.

the workpiece causes changed geometric conditions for the collision-path calculations in the following post processor. Following the blank and finished contour, there come again the calculated tool positions as well as the information about feed and cutting speed values. The sequence of these commands is repeated again from RECORD 237. The statement FINI in line 247 (Code of FINI = 14) finishes CLDATA 2.

Figure 8.25 shows the machining operations in the first clamping position.

The initial blank contour and the finished part contour have been drawn out as well as the path of the tool during feed and rapid motion. Before beginning the operation, the chuck was related to the workpiece by the processor so that the chuck was taken into account in checking for collision with the tool during the division of cuts. Figure 8.26 shows the re-clamped part with the actual blank contour after the machining operations in the first clamping position were finished. It shows the path of the tool during the division of the cuts.

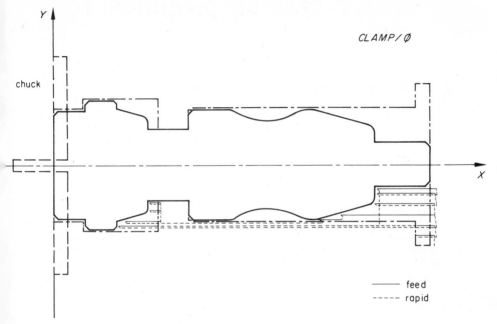

Figure 8.26   Cut sequence developed after line 53, CLAMP/0.

Following on these two machining operations, it would be necessary to make a further call-up which would produce the desired finished contour by one contouring cut.

The total computer time required for the processing of this part program was 35·2 seconds of CDC 6400 central processor time which would cost £4.

9. NEL computer programs for NC

J. F. McWaters and D. S. Welch

J. F. McWaters, MSc, BSc, CEng, MIMechE, is a Principal Scientific Officer, responsible for the continuing development of the 2C,L numerical control program, at the National Engineering Laboratory, East Kilbride, Scotland. He obtained his BSc in mechanical engineering at the University of Strathclyde, Scotland, and his MSc degree in thermodynamics and related studies at the University of Birmingham, England. After serving an apprenticeship with British Polar Engines, Glasgow, he joined the English Electric Nelson Engineering Laboratories, where he was engaged in research into the characteristics of hydrostatic oil bearings. He joined the NEL in 1962, and in 1965 transferred to the newly formed Numerical Control Division.

D. S. Welch, FBCS, BSc, is Senior Systems Programmer, Ferranti Ltd, Dalkeith, Scotland, now Plessey N.C Ltd. He obtained his BSc with first class honours in Physics from King's College, Newcastle, England (now Newcastle University), and entered industry as a programmer with the C. A. Parsons Nuclear Research Centre. Before taking up his present position he was Chief Programmer with Scottish Brewers Ltd. At Ferranti, he designed and controlled the writing of the prototype version of 2C,L.

The need for a standard language, with which parts to be manufactured by NC machine tools could be described, was seen in 1965 by an expert committee set up by the National Engineering Laboratory Steering Committee. The committee, after a study of the metal-cutting industry in Great Britain, reported that there was a great need for a nationally available standard part-programming language and an associated computer processor which was independent of particular machine tools or control systems or computers.

At that time, the greatest need for such a standard system was in the area of two-dimensional contour milling and, towards the end of 1965, NEL set about the task of specifying the part programming language for this work (2C,L). In conjunction with the Numerical Control Division of Ferranti Limited at Dalkeith, a computer processor was to be provided to produce machine control tapes from instructions written in the part-programming language. Since that time, however, the rapid adoption in industry of NC lathes and drilling machines has produced a demand for processors which will allow computers to be used in the preparation of control tapes for these and similar types of machine tool.

To meet this demand, the original 2C,L processor, which had been designed with such contingencies in mind, has been expanded to allow parts which have to be machined on lathes, drilling machines, punching machines, contour milling machines, and burning machines, to be described within the scope of one language. Such part programs can be processed by a common computer program—named the NELNC processor. The integrated system also allows parts to be planned for manufacture on machining centres.

The choice of standard language

When work began on the NELNC processor, none of the programming languages available at that time fully met the needs found in British industry, but the American APT language best met the requirements of the basis of a standard for the following main reasons.

(a) APT is an English-like language; this allows part programs to be easily read and understood by users.

(b) APT was well established in the USA; this fact meant that there APT was approaching being a standard.

(c) The APT language allows programming to be carried out at all levels up to and including multi-axis machining; this means that part programmers could be trained in the simpler point-to-point or two-dimensional contouring work and, provided they had the ability, they could graduate to more complex work without difficulty and without a change of programming language.

(d) APT is capable of expansion to meet future needs; being like English and having scope for addition to its vocabulary, it allows the APT type of statement to be used to describe any type of situation.

(e) APT is descriptive without being too verbose; the use of English-like words make the part program written in APT easily readable, while at the same time the use of words which do not add to the description is avoided.

(f) Most of the features which had to be provided in 2C,L already had descriptions in APT.

Those features which were not already available in APT were studied closely and APT-like descriptions devised for them. To ensure wide acceptance of the form of these descriptions, the statements were presented to two groups of APT users for approval. First of all to the APT Language Standards Committee in Europe for discussion, and after modification (where this was found to be desirable) to the APT Vocabulary Review Committee in the USA for approval. At the same time, these Committees were also studying the extensions to the APT language required by EXAPT 1 (chapter 7) and, as a result, a great deal of uniformity between these part programming languages has been achieved.

The full language has also been presented by the British Standards Institution (BSI) to the International Standards Organization (ISO) as one of a series of candidate languages for international consideration.

The 2C,L subset—the part-programming language

Where a component has to be produced on an NC 2C,L contour milling machine, or a control tape has to be produced to operate some similar device, the bits of information which have to be provided to describe the desired tool path will normally run into hundreds of thousands. Furthermore, in contour milling, because the information normally required by the control system refers to the cutter centre, calculations have to be made to generate a cutter path which is offset from the desired part shape by a cutter radius. The language of 2C,L allows a planner to take the geometry of a part from a drawing and, by means of a few English-like statements, describe the path to be taken by a cutter to produce the part, leaving the computer to generate the vast amount of tool offset and other information required.

The English-like statements referred to are composed of a combination of vocabulary words, symbolic names, and data (normally obtained from a drawing), these being separated by punctuation, exactly as described in chapter 7.

The vocabulary words are English-like words of up to six alphabetic characters in length. These words have a precise meaning in the language; for example, INTOF means 'the intersection of' and may not be used in any other context. Many of the more commonly used words have been allocated terse alternatives so that a part programmer may, if he so desires, write shorthand versions of a part program. Examples of terse alternatives are:

C – the terse form for CIRCLE
INT – the terse form for INTOF
GB – the terse form for GOBACK

In writing the part program shown in Fig. 9.1, terse alternatives were used extensively but, so that the computer listing of the program has the full words included, the instruction FULIST/ON has been programmed (Seq. No. 4). Thus, to generate Seq. No. 25, the part programmer wrote

L5 = L/PAR, L9, XL, .1875

and the printout listed it as

L5 = LINE/PARLEL, L9, XLARGE, .1875

Within the system, provision has been made to allow every vocabulary word to have one or more alternatives. However, initially only 50 of the more commonly used words have been allocated alternative forms. In this respect, the NELNC language differs from EXAPT (described in chapters 7 and 8) and APT (described in chapter 10), neither of which has this facility. It reduces writing to a minimum for the more experienced part programmers without a loss of readability at the check-out stage. Every item which has to be defined and referred to later in a part program is allocated a symbolic name at the time of definition. For instance, a point on the drawing whose co-ordinates are x = 3 in. y = 2 in. could well be given the name P10 and defined P10 = POINT/3, 2. The symbolic name allocated at this time is the name which is used whenever reference has to be made to the element in either a subsequent

Figure 9.1 Part-program for mould shown in Fig. 9.10 (page 245).

definition or a tool motion command. If, in the same part program, there was a need to take a tool to the above point the part programmer would write GOTO/Pl0. Symbolic names may be up to six characters long and composed of any of the alphanumeric characters, provided that the first character is alphabetic. So that there is no danger of inadvertently using a vocabulary word for a symbolic name, it is recommended that every symbolic name should contain at least one numeric character. Vocabulary words do not have any numeric characters. In the language, spaces do not act as word separators as they do in English, but are everywhere ignored. Every item in a statement must be separated from the one which follows by punctuation. For this, the special characters = / and , are provided. Their use is fully described in chapter 7.

The part programming statements of the language lie in six groups:

(a) Geometric
(b) Motion
(c) Post processor
(d) Computer
(e) Arithmetic
(f) Statements associated with the special programming features.

Geometric

Within the system, a variety of methods of defining geometric features is available so that it should never be necessary for a part programmer to become in-

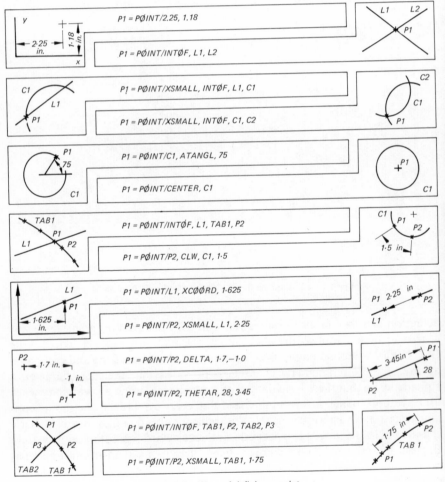

Figure 9.2 Ways of defining a point.

volved in complex geometric calculations. The part programmer has available 14 ways of defining a point, 10 ways of defining a line and 19 ways of defining a circle. In addition, there are facilities for defining planes, vectors, matrices, empirical curves. The full range is shown in Figs. 9.2 to 9.9. Although 2C,L is aimed mainly at contour milling, limited facilities are available for the definition of patterns of holes. Where point-to-point work is found in conjunction with contour milling in the one component, then this can be handled within one part program. A 2C,L component is shown in Fig. 9.10 and the part program

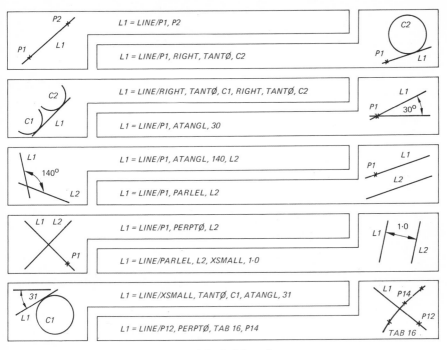

Figure 9.3 Defining a line.

shown in Fig. 9.1 is the one written to clear the area within the outer profile leaving the cores required on the mould to produce the ventilation holes and the terminal recess. The component machined to this stage is shown in Fig. 9.11. The example uses a few of the geometric definitions detailed in Figs. 9.2 to 9.9 and shows how these are put together in a part program.

When part programming, it is normally convenient to describe the geometric features of a component in the order and in the way in which they are constructed in the engineering drawing. In the example shown and with an origin chosen at the centre of the mould, the definition of the geometry can readily begin with

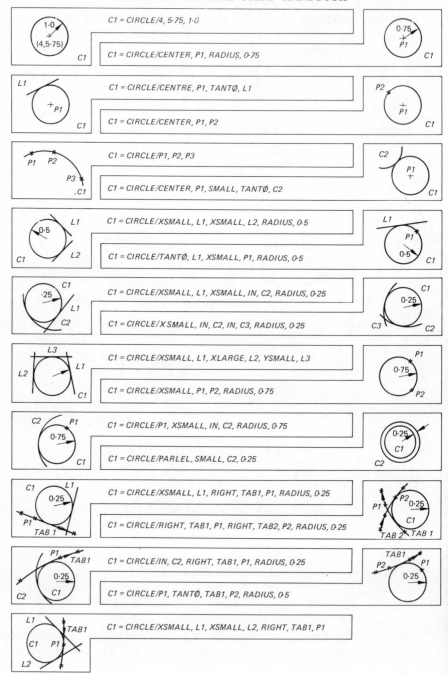

C1 = CIRCLE/4, 5·75, 1·0

C1 = CIRCLE/CENTER, P1, RADIUS, 0·75

C1 = CIRCLE/CENTRE, P1, TANTØ, L1

C1 = CIRCLE/CENTER, P1, P2

C1 = CIRCLE/P1, P2, P3

C1 = CIRCLE/CENTER, P1, SMALL, TANTØ, C2

C1 = CIRCLE/XSMALL, L1, XSMALL, L2, RADIUS, 0·5

C1 = CIRCLE/TANTØ, L1, XSMALL, P1, RADIUS, 0·5

C1 = CIRCLE/XSMALL, L1, XSMALL, IN, C2, RADIUS, 0·25

C1 = CIRCLE/XSMALL, IN, C2, IN, C3, RADIUS, 0·25

C1 = CIRCLE/XSMALL, L1, XLARGE, L2, YSMALL, L3

C1 = CIRCLE/XSMALL, P1, P2, RADIUS, 0·75

C1 = CIRCLE/P1, XSMALL, IN, C2, RADIUS, 0·75

C1 = CIRCLE/PARLEL, SMALL, C2, 0·25

C1 = CIRCLE/XSMALL, L1, RIGHT, TAB1, P1, RADIUS, 0·25

C1 = CIRCLE/RIGHT, TAB1, P1, RIGHT, TAB2, P2, RADIUS, 0·25

C1 = CIRCLE/IN, C2, RIGHT, TAB1, P1, RADIUS, 0·25

C1 = CIRCLE/P1, TANTØ, TAB1, P2, RADIUS, 0·5

C1 = CIRCLE/XSMALL, L1, XSMALL, L2, RIGHT, TAB1, P1

Figure 9.4   Defining a circle. (All dimensions in inches)

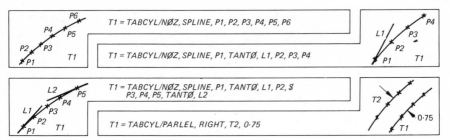

Figure 9.5 Defining a tabcyl.

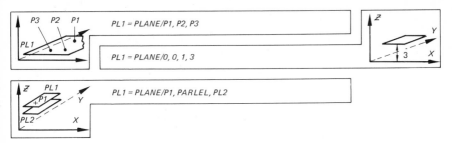

Figure 9.6 Defining a plane.

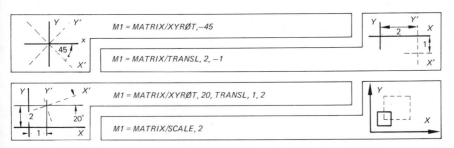

Figure 9.7 Defining a matrix. (All dimensions in inches)

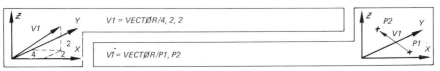

Figure 9.8 Defining a vector. (All dimensions in inches)

the 7 in. radius circles of the mould profile. C1 which has its centre at -4 in. in X and 0 in Y is then defined as

$$C1 = CIRCLE/-4, 0, 7 \quad \text{(Seq. No. 7)}$$

and in a similar way circles C2, C3, and C4 are defined. The part programmer can then define the smaller circles of the profile using another of the 2C,L circle

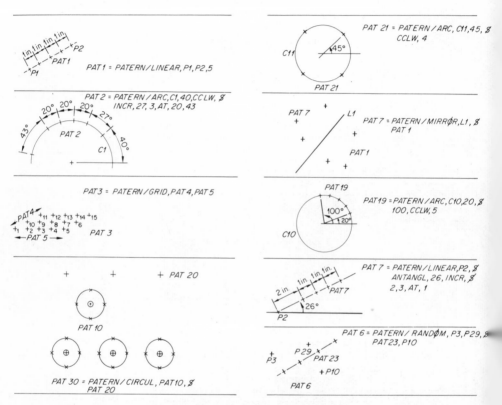

Figure 9.9   Defining a pattern.

definition forms. For example, in statement Seq. No. 11, C5, being tangent to previously defined circles C1 and C2, is defined as

$$C5 = CIRCLE/YLARGE, IN, C1, IN, C2, RADIUS, 1$$

The words YLARGE and IN, known as modifiers, must be included by the part programmer to indicate which of the 8 possible circles he requires (Fig. 9.12). From this it will be seen that the required circle is one of a group which lie YLARGE with respect to the other group which is YSMALL. Hence the modifier YLARGE is chosen for inclusion in this definition. For this particular

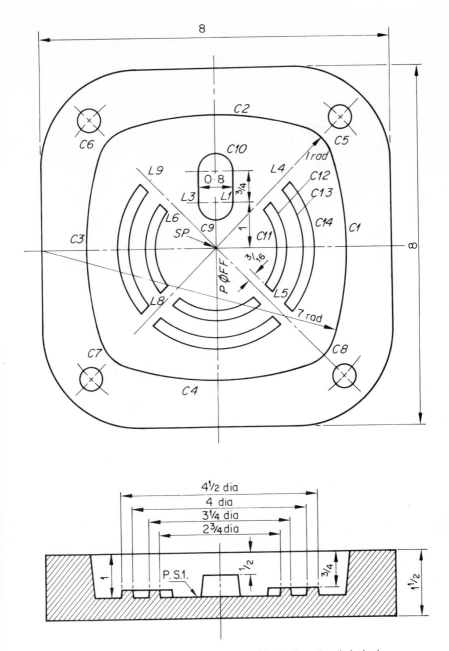

Figure 9.10 Drawing of end cover mould. (All dimensions in inches)

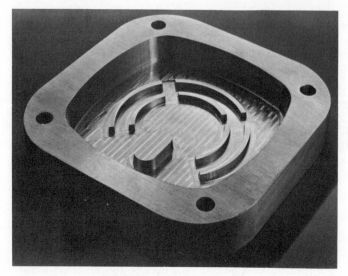

Figure 9.11 End cover mould.

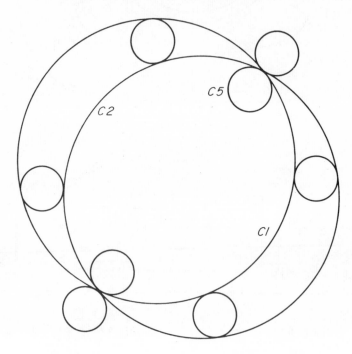

Figure 9.12 C5 and other 1 in. circles tangent to C1 and C2.

definition, XLARGE could alternatively be used since the desired group is also XLARGE relative to the undesired group. Of the four possibilities in this group, two lie inside the circle C1 and two lie outside C1. The required circle is one of those lying inside therefore the modifier IN is added in front of C1. Similarly, of these two remaining possibilities, the one required is the one which is also inside C2 and therefore the modifier IN is applied to C2. Modifiers are used in this way throughout the language. Usually each modifier deals with one of two possibilities. Two modifiers can then indicate which one of four possibilities is intended while three modifiers can deal similarly with eight possibilities.

The above process of definition is continued until the part geometry is complete. Normally, in addition to the geometry of the profiles to be cut, it is necessary to define additional geometric features for use during the cutting process; for example, the point at which the tool is to be initially set must be defined (in this case named SP in Seq. No. 6) and the horizontal plane on which the end of the tool has to run while cutting must be defined (named PS1 in Seq. No. 59).

Motion

The motion statements in 2C,L allow the part programmer to isolate those parts of the geometry, already defined, which form the contour to be produced. The first movement of the tool must be introduced by a FROM statement (here FROM/SP in Seq. No. 61) to establish a sense of direction. Motion statements describe where the tool has to go and the route to be taken. Statements of this type are used in the example in statements Seq. Nos. 68 to 78. (These are enclosed in a MACRO. This will be described later and for the moment can be ignored.)

The purpose of this group of statements is to finish round the cores for the ventilation holes lying between lines L7 and L5 in Fig. 9.10. The cutter, after doing some clearing work, has been taken to the point POFF by the instruction GOSAFE/POFF (Seq. No. 65). The next motion instruction

GO/TO, C11, TO, PS1, TO, L5, 15   (Seq. No. 68)

causes the tool to move to a position which makes it tangent to C11 and L5 and at the same time takes the end of the tool on to the plane PS1 and thus establishes the height at which the tool will run during subsequent motion. In making this move, a sense of direction in the XY plane is established so that, from now on, the part programmer may imagine the tool being driven round the contour as if it were a car, in that a change of direction is described relative to the existing motion. The next cut required is along C11 and therefore the instruction GOLFT/C11, PAST, L7 (Seq. No. 69) is programmed. This takes the tool left along C11 until it is just past line L7 (the check surface) when a move to the right along L7 is required so that the following statement is

GORGT/L7, PAST, C12

Similar instructions are added to the program until the tool has been taken round the desired contour. Since statements of the type described above are aimed at generating information relating to the cutter centre path required to produce the desired profile, it is necessary to specify a cutter diameter before embarking on cutter motion. This can be done in one of two ways either by defining the diameter in a MIL/ statement which is used in conjunction with the technology of area clearance (described later), or, in simple continuous-path work, more conveniently in a cutter statement, viz.:

$$\text{CUTTER}/0.284 \quad (\text{Seq. No. 80})$$

which states that the offsets are to be calculated on the basis that the cutter to be used is 0·284 in. diameter. The part programmer should also specify the feed-rate to be used. This can either be done by means of a feedrate instruction

$$\text{FEDRAT}/15$$

specifying 15 in./min or by tagging the feedrate to the end of a motion statement

$$\text{GO/TO, C11, TO, PS1, TO, L5, 15} \quad (\text{Seq. No. 68})$$

Any feedrate given remains in operation until a new value is specified.

Post processor

Within 2C,L, part programming instructions exist to allow data to be specified to initiate a variety of machine tool functions. The information contained in this type of statement is generally very much machine-tool-dependent, and within the 2C,L processor instructions of this type are only checked to ensure that they are of acceptable form before being passed to output for processing in the post processor. A typical example of such a statement is COOLNT/ON which, on a suitable machine tool, would cause the coolant to be turned on. Again, provided the machine tool had the facilities available, the word ON could be replaced by AIR or FLOOD or MIST.

Computer

Every 2C,L part program must begin with the instruction PARTNO which indicates to the system that it has begun to read a part program. The word PARTNO may be followed by any alphanumeric string of characters by which the part programmer wishes to identify his program. Similarly, an indication should be given that a part program is complete. To do this, the instruction FINI should be the last line in the part program.

In addition, certain optional computer instructions allow the part program-mer some control over the type of output which he will obtain. The effect of these instructions is dealt with in the description of the output from the pro-cessor.

Arithmetic

A part programmer may if he wishes perform arithmetic calculations within a part program. The facilities available are

+	addition
−	subtraction
/	division
*	multiplication
**	exponentiation

SIN	To find the sine of an angle
COS	To find the cosine of an angle
ATAN	To find the angle from a given tangent
ASIN	To find the angle from a given sine
ACOS	To find the angle from a given cosine
ABS	To find the value of a number without regard to sign
EXP	To find the value of 'e' raised to a power
LOG	To find the natural logarithm of a number
SQRT	To find the square root of a number.

Arithmetic variables when defined are given symbolic names similar to those given to geometric variables and may be referenced by name at any time.

The principal use of the arithmetic facilities is to permit drawing dimensions to be stated directly without mental arithmetic. Where a program provides full geometric facilities (as in 2C,L), it should not normally be necessary to use trigonometric or square and square-root functions. However, with these facilities available, unusual problems can be tackled with a minimum of inconvenience. The computation facilities also allow some design calculations to be carried out.

Special programming features

These features have been incorporated in the 2C,L system to reduce the amount of repetitive programming which would otherwise be necessary. The features available are MACRO, TRACUT, and AREA CLEARANCE.

MACRO and TRACUT   A macro is a series of part programming statements written between the words MACRO and TERMAC and allocated a symbolic name, viz.:

$$MAC1 = MACRO$$

Statements in the macro

TERMAC

This is a definition of a macro and, in processing the part program, no action is taken on the above piece of part program unless a call to the macro is made.

In the example shown in Fig. 9.1, a macro MAC1 is defined in statements Seq. Nos. 67 to 79 and calls to this macro are made in statements Seq. Nos. 81, 85, and 87. The macro facility in this example has been used in conjunction with another special feature TRACUT. This is a facility which allows the part programmer to transform, using a previously defined matrix, cut vectors (or cutter path) generated for any motion statements written between the statements

TRACUT/name of a matrix

and   TRACUT/NOMORE

In this particular example, the motion statements required to form the cores for the ventilation holes to the right of the part are written in the macro MAC1. A call to the macro statement Seq. No. 81 is made to generate the cutter paths for these cores. A further call to the same macro is made in statement Seq. No. 85 but this time preceded by the statement

TRACUT/M1

where M1 is a previously defined matrix calling for a 90° rotation about the origin with the result that the cutter paths produced by the same cut sequence are now rotated 90° to produce the cores at the bottom. A third call to the same macro, again preceded by a TRACUT but using the other matrix M2, produces the cutter paths to form the cores to the left.

An alternative way of programming this would be with the use of macro variables and no TRACUT. A study of Fig. 9.10 and the cut sequence for each group of cores shows that the variables are L7 which becomes L6 and then L8, and L5 which becomes L7 and then L6. The macro is then written in general terms with the variable A, say, in place of L7 or L6 or L8, and B, say, in place of L5 or L7 or L6 and, when calling the macro, the real values are inserted. The part programming necessary would then be

L6 = L/PAR, L9, XS, .375
L8 = L/PAR, L4, XS, .375
MAC1 = MACRO/ A, B
GO/TO, C11, TO, PS1, TO, B15
GL/C11, PAST, A
GR/A, PAST, C12
GR/C12, PAST, B
GL/B, TO, C13
GL/C13, PAST, A
GR/A, PAST, C14
GR/C14, PAST, B
GR/B, PAST, C11
GD/1
GT/POFF

TERMAC
CALL/MAC1, A = L7, B = L5
CALL/MAC1, A = L6, B = L7
CALL/MAC1, A = L8, B = L6

Area clearance   The area clearance facility allows a part programmer to program for the clearance of an area without having to specify all the necessary cuts, provided that the area to be cleared has a closed boundary which can be specified as a combination of parts of lines and circles. If islands occur, they can be avoided as long as they also can be described as closed contours composed of lines and circles. It should be noted that any area which has to be cleared must be completely bounded, otherwise cuts of infinite length and time would be involved.

To use the area-clearance facility, three part programming stages are required:

definitions of contours;
definition of cutting conditions;
execution of clearing process.

Contours are defined by a series of statements as seen in Fig. 9.1 (statements Seq. Nos. 29 to 56) where the contours CON1, CON2, and CON3 are defined. The shape of the contours is sketched out in Fig. 9.13. The first line of the contour definition is always

$$name = CONTUR$$

and the last line always

TERCON

The second line of the definition is the specification of a previously defined horizontal plane to which the end of the tool will be taken at any time that it requires to cross over the area within the contour if not cutting. Any convenient point on the contour is then chosen and the motion statements are written to describe the path taken to draw around the contour until the first element mentioned is again reached. The contour description is then closed by a statement TERCON.

The technological data relating to the cutting process is defined in a statement of the form

$$Name = MIL/SO, string of couplets$$

an example of which is given in part program statement Seq. No. 57. The data is defined as a series of couplets consisting of a word describing the type of data followed by the value. These can be given in any order and the following are allowed:

DIAMET   The diameter (in.) of the cutter

SPEED   The spindle speed (rev/min)

CUTFED   The feedrate (in./min) used to clear an area

PLGFED   The feedrate (in./min) at which the cutting tool will plunge into a pocket area

FINFED   The feedrate (in./min) used for the final or finishing cuts round the contour

SWATH   The incremental distance between tool centres on successive cuts in the area, specified as a decimal fraction of the cutter diameter.

The statements in the part programming sequence which will execute the clearing process are bracketed between the statements ACT/ and ACT/NO-MORE.

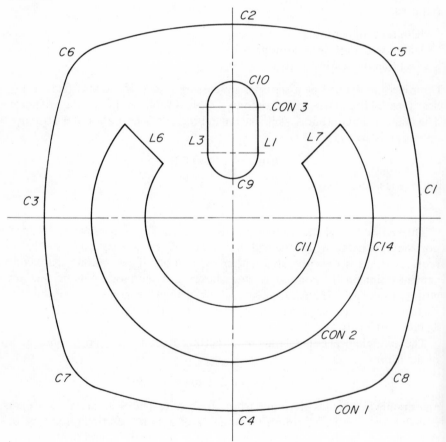

Figure 9.13   Geometry of the three contours.

The use of these is seen in statements Seq. Nos. 60 to 66. The statement ACT/TECH1 indicates that the following operations are to be carried out using the cutting conditions described in TECH1. These conditions will apply until the next ACT statement which may be either a statement of new conditions or ACT/NOMORE (indicating that this type of cutting instruction no longer applies). The start-up in area clearance is the standard form. The cutter is taken to a position where it is in contact with one of the component parts of the contour (modified, as described below, by a FINCUT allowance which shifts the contour) and must be in a suitable position for the next command, i.e., if the next instruction is GOZIG/IN, contour, then the start-up must have positioned the cutter inside rather than outside the contour.

The statement used to clear an area is GOZIG/ followed by the name of the contour inside which metal has to be removed. This can be followed by the names of contours describing any islands to be left. In the part programming example shown, one stage of the machining process requires metal to be cleared within the area surrounded by the outer contour CON1 leaving the horse shoe shaped island CON2 and the terminal recess core CON3. The part programming statement for this operation is statement Seq. No. 64 in which the part programmer has also made use of the FINCUT facility whereby a finishing allowance may be specified on any contour. This material can then be removed separately by GORND statements of the form

GORND/IN, CON1
GORND/OUT, CON2

In the GOZIG/ routine, the direction of the cuts generated to clear the area are at right-angles to the normal to the drive surface specified in the start-up, so that, if the surface is a straight line, they are parallel to the line and, if a circle, they are parallel to the tangent at the start-up point. During the clearing process, the cutter is taken to a distance of SWATH + FINCUT from each contour, and at the end of the process automatically taken round each contour twice— once to clear off any scallops left and on the second time round to clear the SWATH which was left. The FINCUT is left for a part programmed GORND.

The 2C,L subset processor

System design

In order that the 2C,L language processor could be made generally available as a national program, two major requirements were placed on the system design. Firstly, the program must be capable of being implemented on as wide a range of computers as possible; secondly, it must be simple to maintain and expand.

The first requirement, general availability, led to the specification of the minimum configuration on which the 2C,L system could reasonably be implemented. This was determined to be a computer with

28,000 24-bit words of available core store (a nominal 32K Computer)
4 magnetic tape units and preferably disc or drum store
1 card or paper-tape reader
1 line printer
1 paper tape punch

An example of this sort of computer would be an ICT 1904 with 32K words of core store of which about 4K would be occupied by the EXECUTIVE program. It was assumed that computers of at least this size would be generally available throughout the engineering industry.

To enable the program to run on this size of computer, it was designed to operate as four independent sections which could overlay each other in the core store, data being transmitted from one section to the next via magnetic tape, disc, or drum files.

The second requirement, ease of maintenance and extension of the program, was satisfied by writing the program in a highly modular form. This form was also of assistance in constructing the overlay structures of the program for the different sizes of computer.

Both requirements dictated the choice of a higher level language as the programming language for the system. Fortran 4 and Algol being the most widely available languages, the choice was restricted to these two. A survey of the programming facilities offered in common on a number of computers by these two general-purpose languages indicated that Fortran 4 with only one or two restrictions on the USASI subset would be the most useful. Similar decisions were made about the same time by the originators of APT 4 and EXAPT.

The 2C,L program

As previously stated, the processor has been written as four independent sections, viz., input, decoding, geometry, and motion. These sections may be run under the control of a master program as a chained program, as in the UNIVAC 1108 implementation, or each section may be run as an independent program, as in the ICT 1904 implementation.

In either form, the system is a batch-processor, i.e., a batch of part programs may be processed by a single run of the 2C,L system. In the case of the UNIVAC 1108 version, where high-speed drum stores are available, each part program is processed as far as possible—i.e., until an error is detected or until all specified post processing is complete—before the next part program is read-in via the input section. In the other form of the program (four independent sections), the whole of the part-program data is processed through each section in turn to minimize tape-rewinding times. Jobs discovered to be in error are excluded from the data file used as input data to the next section in sequence.

The difference between the two forms of the program is connected with the type of back-up memory of the particular configurations of the computers concerned. In the case of the ICT 1904 version aimed at the minimal form without disc stores, the time penalty involved in overlaying is too great to allow separate processing of individual part programs right through the system.

A block diagram for the 2C,L system is shown in Fig. 9.14.

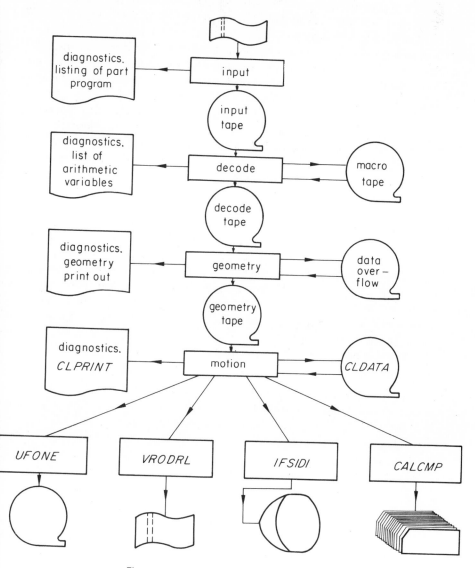

Figure 9.14   Processing stages in 2C,L.

Input section   This is the first section of the program. Its function is to read the part-program data from punched cards or paper tape, either a job at a time or a complete batch, depending on the version of the program under consideration. Output from this section consists of a listing of the part program and a data file which becomes the input file to the decoding section.

During input, 2C,L vocabulary words are recognized, as are terse forms of vocabulary words, and converted to an internal code on the output file. Terse forms appear as the equivalent full forms in the part-program listing, provided this has been specified by means of a FULIST/ON statement.

Scalar quantities are converted to floating-point numbers for internal use and symbols such as $+$, $-$, (,), etc. are also converted to an internal code. Alphanumeric items other than vocabulary words are flagged as being undefined but otherwise are unaltered.

Certain rudimentary checks are performed on the data by the input section, e.g., not more than six characters are allowed in an alphanumeric item. Any errors detected are flagged by letters in front of the offending statements on the listing of the part program. These errors are considered to be not necessarily fatal and processing is allowed to proceed to the decoding section whether or not a job is error-free.

Decoding section   This is the second section of the program. Its function is to analyse the part program data as read by the input section. Output from this section consists of a decoded data file to be used as the input file for the geometry section, and a listing of error diagnostic numbers (if any) and symbolic geometric variable names which have been redefined, i.e., appear more than once on the left-hand side of a geometric assignment statement. If printing of arithmetic variables has been specified in a PRINT/ statement, these values will also appear on the listing.

During the decoding section, the following actions are performed.

(a) Nested expressions. Nested expressions, i.e., expressions enclosed in parentheses, are analysed as separate statements starting with the lowest level of nesting and working outwards. The maximum depth of nesting allowed is level 10.

(b) Arithmetic statements. All arithmetic statements, i.e., statements using the Fortran-like computational facilities of the 2C,L language, are evaluated during the decoding section.

(c) Macro logic. All macro logic is performed during the decoding section. Macro definitions are copied from the input file to a macro file which contains any system macros inherent in the system being used. A CALL to a macro causes the input to switch from the input file to the appropriate place on the macro file which is used as the input file until the TERMAC statement is detected, when input reverts to the original input file.

(d) Decoding geometric definitions. Geometric definitions statements, i.e., statements containing POINT/, LINE/, CIRCLE/, TABCYL/, MATRIX/, VECTOR/, PLANE/, or PATERN/, are checked for format. On recognizing a definition format, the statement is written on the output file with two code integers, one identifying the type of definition and the other identifying the set of modifier words used.

(e) Decoding motion statements. Motion statements such as GO/, GOTO/, etc., are checked for format. On recognizing a format, the statement is written on the output file with a code integer identifying the format. For statements such as GOFWD/L1, TO, L2, a special modifier is inserted before the surface name not preceded by a modifier, e.g., L1 in this example.

(f) Other statements. Other statements, such as post-processor statements, which are not specifically processed by the decoding section, are copied from the input to the output files with a minimum of checking, but substituting scalar values of arithmetic variables where appropriate.

(g) Errors. All statements in error are flagged by printing the sequential statement number followed by an error diagnostic number identifying the type of error. Once a job is in error, no more output occurs and the job is not processed beyond the decoding section, but it is examined for more decoding errors.

Various lists or dictionaries are maintained within the decoding section, any of which will cause an error if an attempt is made to exceed its capacity. The restrictions on a part program as a result of this are:

Number of symbolic geometric names	256
Number of symbolic arithmetic names	50
Number of macros	30
Number of macro variables per macro	20
Depth of nesting of statements	10
Depth of nesting macros	4

Geometric and arithmetic variables may be redefined any number of times. A part program could have a thousand geometric definitions named providing that no more than 256 actual names were employed. This degree of freedom does not exist in APT and EXAPT. Where a symbolic name is redefined, a warning diagnostic is printed out.

Geometry section   This is the third section of the program. Its function is to calculate the canonical forms of all geometric definitions as specified on the decoded data file. Output from this section consists of a motion-data file to be used as the input file for the motion section, and a listing of error diagnostic numbers (if any). If printing of geometric variables has been specified in a PRINT/ statement, the canonical forms of these variables will also appear on the listing, as will any output required from the TABCYL routines.

During the Geometry section, the following actions are performed.

(a) Geometric definitions. This type of definition code is used to determine the type of geometric surface to be defined and which particular definition sub-routine is required. The latter calculates the canonical form of the definition and stores the result in a buffer in the core store. If the canonical form data exceeds the buffer size, the buffer is written on an auxiliary file, the geometry file, whence it may be read back when required and the buffer reused.

The size of the canonical form depends on the type of surface, e.g., for a point the canonical form is the x, y, and z co-ordinates of the point, whereas for a matrix it consists of the twelve elements of the (3, 4) matrix.

(b) Other statements. All other statements not explicitly processed by the geometry section are copied directly to the output file, except that all geometric variables referenced have their canonical forms inserted in the output statements.

(c) Errors. All statements in error are flagged by printing the sequential statement number followed by an error diagnostic number identifying the error. Once the job is in error, no more output occurs and the job is not processed beyond the geometry section, but any remaining geometry is checked.

(d) Printing. Canonical forms of geometric elements may be listed using the PRINT/ statement. Automatic printing of TABCYL canonical forms and other data such as slopes, radii of curvature, etc., occurs unless a TABPRT/OFF statement is encountered. Only in this latter condition will a PRINT/ statement cause printing of a TABCYL canonical form, thus avoiding duplication of information.

Motion section   This is the fourth and last section of the program. Its function is to calculate the tool movements or cut vectors required to machine the part as specified in the motion statements on the motion file. Output from this section consists of a CLDATA file to APT standards and any post processing which has been specified by the appropriate MACHIN/ statements. A listing of the CLDATA will also be produced if a CLPRNT statement appears in the part program.

During the motion section, the following actions are performed.

(a) Motion statements. The format code inserted by the decoding section is used to assist in analysing the motion required. Motions are analysed in a similar fashion to the APT system, in particular, circles and tabcyls are approximated to by a series of straight-line cut vectors lying within the specified tolerance bands.

For certain post processors, e.g., for a Ferranti control system, circles do not need to be approximated and tabcyls are best approximated by cubics. If a post processor of this type is requested, special records are included in the CLDATA file giving the end points and canonical forms for circular and curved arcs.

(b) Area clearance statements. These statements are processed by a sub-section of the program. The contours which have been defined are stored on a contour file whose format resembles the CLDATA file. These are then referenced via the GOZIG/ and GORND/ statements.

Data file maintenance

The purpose of the maintenance program is to allow standard macros to be added to the system and to create the vocabulary file and the definition file required by the input and decode sections respectively. In the extensions to the program for drilling and turning, the maintenance program is also used to create the tool files required.

The files are created once for any implementation, and the program is only used again when new standard macros, words, or definitions are being added.

Where standard macros are being included in a system, then the macros are written in standard part-programming language and all put between the words PARTNO/ and FINI. This part program is then read as data on execution of the maintenance program and results in the macros being processed to a semi-decoded state and stored on the system tape. The part program containing the standard macros is immediately followed by the card FINISH and by 18 files, the first of which is a series of cards each containing one vocabulary word and its alternatives; the second is a card giving the names of standard macros, while the remaining 16 files contain specimen geometric definitions and successor lists for geometric definitions.

The 2C,L subset—its output

The output from 2C,L is in two forms:

(a) CLDATA is numerical data, written on magnetic tape or a drum or disc file, which is finally handed over to a post processor to be matched to the requirements of the input to a particular control system.

(b) Printed output is produced at various stages of processing in 2C,L for use by the part programmer in interpreting the results of what he has programmed.

Output on CLDATA

The CLDATA, so-called because the bulk of the information which it carries refers to cutter locations, is the data interface between 2C,L and its post processors. (Post processors are described in chapter 11.) The CLDATA produced by 2C,L is exactly compatible with the APT 3 CLDATA so that the output from 2C,L is directly acceptable to APT post processors.

The form of the physical records on the CLDATA is not important from the point of view of standardization, since the tape-reading routine used by the post processors is one of the utilities supplied in 2C,L and, provided that in any given computer implementation arrangements are made to match the tape-writing and -reading routines, no difficulty exists. However, what is important from the point of view of standardization is that the logical records should conform to some standard, and in 2C,L it is those records which comply with the standards laid down for APT 3.

The APT 3 records put out by 2C,L and the special records put out for use by time-based control systems are fully described in chapter 11.

Printed output

Line-printer output is obtained at each stage of processing by the 2C,L processor. In addition to the normal output, the part programmer may by means of special part-programming statements induce additional output.

Every page is headed by identification of the processor and the part program. In Fig. 9.15a, a page of output is shown. At the top of each page, a system identification extends across the full 72-column field. Depending on the installation, this would normally give an indication of the version of the processor and possibly the date on which that version was put together. The examples shown here have all been run under the author's development version headed

2CL SYSTEM FOR ACCEPTANCE TESTS COPY

J. F. McWATERS UTIL FILE—X503ZC

On the following line, the name of the section of the processor in which the output was generated is given together with the date on which the processing took place, for example

2C,L—INPUT SECTION 7 JAN 69

On the following line of print, the PARTNO statement for the part program is given together with the number of the page. The printout from each section of the processor is closed with the message END OF JOB and this is followed by the number of errors (if any detected by that section of the processor) and by the time in minutes and seconds taken by that section of the processor to process the particular part program.

Line-printer output from 2C,L is restricted to columns 1 to 72 so that the output can, where conditions allow, be put out on teletype without modification to the print format.

Output from the input section The output which results from this section of the processor is a listing of the part program and an indication of any errors detected during the simple checks made at this stage.

Normally, the listing of the program is a printout of the statements in the form in which they were put in, but if FULIST/ON has been programmed, then terse forms of vocabulary words are replaced by the full form on the printout.

Errors detected by this section are indicated on the line on which the error occurs by alphabetic flags printed in the left-hand column of the listing. An example is shown in statement Seq. No. 10 of the part program in Fig. 9.15a where, owing to a missing comma, a word longer than six characters has been detected (RADIUS 1).

Output from the decode section   The output obtained from the decode section is simple and consists of a list of the numbers of those statements which are in

```
<2CL SYSTEM FOR ACCEPTANCE TESTS COPY    J.F.MCWATERS UTIL FILE - X503ZC>

2CL - INPUT SECTION  7 JAN  69

PARTNO BLOCK                                          PAGE    1

ERROR    SEQ.
FLAGS    NO.    INPUT STATEMENT

          1     PARTNO BLOCK
          2     MACHIN/UFONE,1
          3     CLPRNT
          4     CUTTER/0.5
          5     TOLER/.001
          6     PC1=P/0,0,1
          7     L1=L/3,0,3,1
          8     L2=L/0,3,1,3
          9     L3=L/0,0,0,1
         10     C1=CIRCLE/XLARGE,L1,YSMALL,L2,RADIUS1
         11     PRINT/3,L1,C1
         12     TAB1=TABCYL/NOZ,SPLINE,0,0,1,1,1.5,1.3,2,1,3,0
         13     PP1=PLANE/0,0,1,0
         14     PRINT/3,ALL
         15     FROM/PC1
         16     GO/PAST,L1,TO,PL1
         17     GL/L1,TT,C1
         18     GF/C1,TT,L2
         19     GF/L2,PAST,L3
         20     GOTO/PC1
         21     FINI

END OF JOB :-  1 ERROR FLAG  : INPUT SECTION TIME  =   0: 1.
```

```
<2CL SYSTEM FOR ACCEPTANCE TESTS COPY    J.F.MCWATERS UTIL FILE - X503ZC>

2CL - DECODING SECTION  7 JAN  69

PARTNO BLOCK                                          PAGE    1

        SEQ.     STAT.
        NO.      LABEL

         10               ERROR NO.   15
         16               ERROR NO.   33

END OF JOB :-  2 DECODING ERRORS : DECODING SECTION TIME =   0: 0.
```

```
<2CL SYSTEM FOR ACCEPTANCE TESTS COPY    J.F.MCWATERS UTIL FILE - X503ZC>

2CL - GEOMETRY SECTION  7 JAN  69

PARTNO BLOCK                                          PAGE    1

STATEMENT NO     11    PRINT OF CANONICAL FORMS

NAME      TYPE      CANONICAL FORM

L1       LINE       1.000000    -.000000    .000000    3.000000

C1       CIRCLE     4.000000     2.000000    .000000    .000000
                     .000000     1.000000   1.000000
```

Figure 9.15   Output from processor: (a) from input section, (b) from decode section, (c) from geometry section.

10

error together with an indication of the type of error (Fig. 9.15b). In the example shown in Fig. 9.15, two errors have occurred, one due to the mis-spelling already detected during input and the other due to an error in the motion statement Seq. No. 16 where the plane PP1 was referenced as PL1.

Output from the geometry section   Error diagnostic messages from this section appear on the printed output in a similar form to those in decode. However, at this stage, the part programmer may induce additional output by the inclusion of a PRINT/3 statement in his part program. This statement can have the forms:

(a) PRINT/3, list of geometric names, in which case the canonical form of each element named in the list is printed out provided that it has been defined earlier in the program. An example is shown in Fig. 9.15c.

(b) PRINT/3, ALL, which causes the printout of the values in the canonical form of every geometric element which has been defined up to the position of the print statement in the part program.

When tabcyls are being processed, printed output relating to the curve is produced. The output is in four sections.

(a) A table (Fig. 9.16a) giving information relating to the tabcyl input data. For each point, the cartesian and polar co-ordinates are given. These are followed by the segment length, which is the distance between the point and the next one in the list. The segment angle listed is the angle which the line joining the point with the next makes with the positive x axis. The extension angle is the angle between the segment line produced and the next segment line.

(b) Tabulated information on the spline (Fig. 9.16b). For each of the defining points, the following information is given.

SLOPE:	The slope of the spline, i.e., dy/dx at the defining point.
NORMAL:	The angle which the normal to the curve at a defining point makes with the positive x axis.
ALPHA:	The angle between the normal to the curve at the defining point and the line joining the defining point to the origin.
TANGENT A:	The tangent of the angle between the slope at the defining point and the segment line, i.e., the line joining the defining point to the next defining point. At the last point the value is set to zero.
TANGENT B:	The tangent of the angle between the slope at the next defining point and the current segment line. At the last point the value is set to zero.
CURV A:	The curvature at the defining point of the cubic fitted over the current segment. At the last point the value is set to zero.

DELTA: The difference in curvature between CURV A and the curvature, at the defining point, of the cubic fitted over the previous segment. At the first and last points, the value is set to zero.

```
<2CL SYSTEM FOR ACCEPTANCE TESTS COPY    J.F.MCWATERS UTIL FILE - X503ZC>

2CL - GEOMETRY SECTION  7 JAN 69

PARTNO BLOCK                                          PAGE   2


SPLINE OUTPUT FOR    TAB1

NO  X CO-ORD   Y CO-ORD    RADIUS     THETA    SEG LEN   SEG ANG    EXT ANG
 1   .000000    .000000    .000000    .0000   1.414214   45.0000     .0000
 2  1.000000   1.000000   1.414214   45.0000    .583095   30.9638  -14.0362
 3  1.500000   1.300000   1.994943   40.9144    .583095  -30.9638  -61.9275
 4  2.000000   1.000000   2.236068   26.5651   1.414214  -45.0000  -14.0363
 5  3.000000    .000000   3.000000    .0000     .000000     .0000     .0000
```

(a)

```
<2CL SYSTEM FOR ACCEPTANCE TESTS COPY    J.F.MCWATERS UTIL FILE - X503ZC>

2CL - GEOMETRY SECTION  7 JAN 69

PARTNO BLOCK                                          PAGE   3


SPLINE OUTPUT FOR    TAB1

NO   SLOPE     NORMAL     ALPHA      TAN A      TAN B     CURV A    DELTA
 1   .9264   -47.1889   -47.1889   -.038222    .038222    .0539     .0000
 2  1.0795   -42.8111   -87.8112    .291002   -.600000    .0546     .0007
 3   .0000   -90.0000  -130.9144    .600000   -.291073  -1.9657     .0002
 4  -1.0796   42.8074    16.2424   -.038286    .038286    .0540    -.0002
 5  -.9263    47.1926    47.1926    .000000    .000000    .0000     .0000
```

(b)

```
<2CL SYSTEM FOR ACCEPTANCE TESTS COPY    J.F.MCWATERS UTIL FILE - X503ZC>

2CL - GEOMETRY SECTION  7 JAN 69

PARTNO BLOCK                                          PAGE   4


SPLINE OUTPUT FOR    TAB1

               -2.05       -1.50        -.95        -.40        .15
    CURVATURE .+.............+.............+.............+...........+.+.
 1   .053935  .                                                 *  *.
 2   .054639  .                                                 *   .
 3 -1.965673  .  *                                                  .
 4   .054026  .                                                 *   .
 5   .000000  .                                                  *  .
    CURVATURE .+.............+.............+.............+.......+....+.
```

(c)

```
<2CL SYSTEM FOR ACCEPTANCE TESTS COPY    J.F.MCWATERS UTIL FILE - X503ZC>

2CL - GEOMETRY SECTION  7 JAN 69

PARTNO BLOCK                                          PAGE   5


CANONICAL FORM OF    TAB1

NUMBER OF POINTS

   7
      U          V          A          B      LENGTH     MAX       MIN
  -7.3360    -6.7958     .0000      .0000   10.0000     .0000     .0000
   .0000      .0000      .0000      .0270    1.4142     .0000    -.0096
  1.0000     1.0000     -.9088      .0309     .5831     .1145     .0000
  1.5000     1.3000      .9386    -1.5588     .5831     .1146     .0000
  2.0000     1.0000      .0000      .0271    1.4142     .0000    -.0096
  3.0000      .0000      .0000      .0000   10.0000     .0000     .0000
 10.3364    -6.7954
```

(d)

Figure 9.16 Continuation of Fig. 9.15c—TABCYL output.

(c) A plot of curvature against defining point number (Fig. 9.16c). The line printer is used to plot the values of curvature obtained. During computation, the values of curvature which have been calculated are scanned and a scale is chosen so that the diagram will fill the width of the print field. The plot allows a quick check to be made of the curvature values but is not a highly accurate plot. The actual numerical values of curvature are printed alongside the point numbers at the left of the diagram. In Fig. 9.17, the curve fitted is shown with

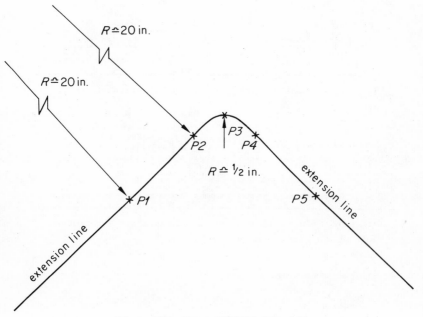

Figure 9.17   Shape of TABCYL of Fig. 9.16.

the approximate radii of curvature added. From this it will be seen that, because of the way in which the curve fitting mechanism operates, the curvature may be opposite to what would be expected where the curve is very close to a straight line. In a case where a satisfactory fit could not be made, the following message is printed below the diagram

WARNING, CURVE SMOOTHING UNSATISFACTORY.
EXAMINE TABCYL INPUT DATA

This message is given out only when

$$\text{DELTA} > 0{\cdot}001$$

(d) The canonical form of the tabcyl (Fig. 9.16d). Following a print of the name of the tabcyl is the number of points in the tabcyl. This is always two greater than the number of defining points, since the canonical form includes the points at the ends of the extension lines. This is followed by a table giving the seven elements for each point in the canonical form.

U and V are the cartesian co-ordinates of the defining points of the tabcyl; if the tabcyl has been transformed, then they are the co-ordinates in the new co-ordinate system and not in the co-ordinates system from which the defining points were given.

The values of A and B are the coefficients of the cubed and squared terms respectively of the equation of the cubic fitted over the segment, i.e., the segment between the defining point being considered and the next defining point.

MAX and MIN relate to the maximum deviation of the curve over the segment from the straight line joining the defining points at the ends of the segment. If the tabcyl-defining points are considered as being strung out from left to right in the order in which they were defined, then the values of MAX refer to the height of hills and the value of MIN to the depth of valleys. In some circumstances, a MAX and MIN value may occur within one segment.

Printed output from tool offsets section   The normal output from this section is similar to that obtained from decoding; that is, if errors are detected, the sequence number of the statement in error and the error type are printed out. If the part programmer includes CLPRINT in his program, then a listing of information in each CLDATA record is put out. The exact form of the output depends on the information contained in the record. The part program sequence number and line number is given followed by the CLDATA record number. If the record is the type which carries information on cutter size or feedrate, then the part programming word followed by the value is given. These are self-explanatory and examples can be seen in Fig. 9.18. The printout of cutter positions is given in four ways.

(a) When a start up is given. The name of the point from which motion will start is given. This is followed by the co-ordinates of the start point. An example is shown in Fig. 9.18 (CLDATA REC. No. 11).

(b) Where a move is made to a point via a GOTO/ statement, the printout is of the form DSIS/ followed by the name of the point, line, or circle to which the tool is taken and the co-ordinates of the tool centre after the move is made.

(c) In the case of straight-line motion, the printout is, as shown in Fig. 9.18, CLDATA REC. No. 15 which gives the drive-surface name followed by the co-ordinates of the point at the end of that particular motion.

(d) Where motion is round a circle, the information is contained in two records (Fig. 9.18, CLDATA REC. Nos. 17 and 18). The printout of the first gives the circle name followed by the values of the canonical form of the circle. The second record again gives the name of the circle as the drive surface followed by the co-ordinates of a series of points which are the points at the end of small straight-line cuts made to approximate to the circle.

Where area-clearance work is being done, records are put out which differ slightly from those described in (c) and (d), in that drive surfaces are not given since, when the cutter is away from the boundary of the contour, the drive surface is not specified but is generated internally in the processor.

```
NCLL SYSTEM FOR ACCEPTANCE TESTS COPY    J.F.MCWATERS UTIL FILE - X503ZC>

NCL - TOOL OFFSETS/CUT VECTORS SECTION   7 JAN 69

PARTNO BLOCK                                                    PAGE  1

SEQ.  STAT.  CLTAPE
NO.   LABEL  REC.NO.


  1           2   PARTNO      BLOCK


  4           6   CUTTER /    .5000000

  5           8   INTOL /     .0000000   .0000000   .0000000

  5           9   OUTTOL /    .0010000   .0010000   .0010000

 15          11   FROM /        PC1
                              X          Y          Z
                              .0000000   .0000000   1.0000000

 16          13   US IS /       L1
                              X          Y          Z
                              3.2500000   .0000000   .0000000

 17          15   US IS /       L1
                              X          Y          Z
                              3.2500000   2.0000000   .0000000

 18          17   C1 = CIRCLE / CANON

                              4.0000     2.0000     .0000     1.0000

 18          18   US IS /       C1
                              X          Y          Z
                              3.2539974   2.0773313   .0000000
                              3.2651185   2.1498293  -.0000000
                              3.2832674   2.2208945  -.0000000
                              3.3002712   2.2896470  -.0000000
                              3.3396906   2.3560274  -.0000000
                              3.3778231   2.4108029  -.0000000
                              3.4217060   2.4775731  -.0000000
                              3.4711196   2.5317758  -.0000000
                              3.5255913   2.5808927  -.0000000
                              3.5846002   2.6244541  -.0000000
                              3.6475618   2.6620433  -.0000000
                              3.7139339   2.6933009  -.0000000
                              3.7830219   2.7179279  -.0000000
                              3.8541850   2.7356888  -.0000000
                              3.9267427   2.7464136  -.0000000
                              4.0000000   2.7500000   .0000000

 19           ERROR DETECTED REF. NO.  110
```

Figure 9.18 Output from cutter-offset section.

Point-to-point facilities

The NELNC processor has been further developed to meet the requirements of point-to-point work and a 2P,L subset aimed mainly at programming for drilling machines has been produced. It can either be used on its own or combined with 2C,L to form the NELNC processor.

The geometry of 2P,L

A part programmer planning for drilling work does not encounter the large number of methods of constructing geometry which are met in contour-milling type of work; therefore, in a drilling subset used on its own, the number of allowable ways of defining geometry can be limited. In the 2P,L subset, there are:

6 ways of defining a POINT;
4 ways of defining a LINE;
5 ways of defining a CIRCLE.

In addition, the PLANE and MATRIX definitions of 2C,L are available, but all VECTORS and TABCYLS have been omitted.

In drilling work, the points at which machining has to be done frequently form groups which can conveniently be defined as a pattern of points. Within 2P,L, there are 9 ways of defining a pattern. The types of patterns which can be described are shown in Fig. 9.9.

The points in a pattern are given a value of zero in Z unless the part programmer includes a ZSURF statement.

The statement is of the form ZSURF/ and applies to all subsequent pattern definitions unless it is restated or exceptionally is cancelled when a TRASYS/ statement is encountered. Any defined point which is later incorporated in a pattern loses any Z value allocated in its original definition.

The order in which the points are stored in a grid pattern is shown by the point numbering for PAT3 in Fig. 9.9. At any time during pattern definition, the part programmer may make use of the OMIT and RETAIN facility where there is no need to preserve a complete pattern. For example, if in Fig. 9.9 the point in the second row, third column, is not required, PAT3 could have been defined as

PAT3 = PATERN/GRID, PAT4, PAT5, OMIT, 8

Similarly, in cases where it would be more convenient to list the points to be retained rather than omitted, the word RETAIN followed by the appropriate list of point numbers can be written in place of OMIT.

Another feature, INVERS, when tagged onto a pattern definition, causes the order in which the points in the pattern are stored to be reversed so that any subsequent operation on the pattern is automatically in reverse of the normal order.

In the 2P,L processor, a facility exists which allows the transformation of points to be programmed for. This allows the part programmer to define the geometry or a section of the part geometry in one co-ordinate system and have the results, when computed, stored in another. To use the facility, those geometric definitions which have to be transformed are put between the statements

TRASYS/   symbol for a previously defined matrix

and   TRASYS/NOMORE

Within a TRASYS bracket, Z values can only be allocated to points in a pattern from a ZSURF statement given within the TRASYS, since any previously set ZSURF will have been cancelled when the TRASYS is encountered.

Motion in 2P,L

In 2P,L motion to command the tool is restricted to the type

                    GOTO/ symbol for a point
or   GOTO/ symbol for a pattern.

When, at the point, the part programmer can program for making moves in the Z direction in the following ways.

(a) The motion instruction

            GODLTA/ distance to be moved in Z,

can be used to drill a hole or perform any similar operation at a point.

(b) The statements of the form

                    CYCLE/DRILL, list of parameters
or   CYCLE/TAP, list of parameters

allow the description of a series of movements of the tool to be made. For example, in DRILL, the tool will be sunk to a specified depth and retracted according to the values listed. In the processor, the only action which is taken is to check the validity of the statement. The generation of the precise information required to complete the operation is left to the post processor where depending on the machine tool a series of moves in Z may be generated or, alternatively, where canned cycles are available the appropriate function code is generated.

(c) The ACT/ statement can be used to cause a previously defined drilling sequence to be applied to previously defined points. The part programming statements used are as follows

                    ACT/DRL1
                    GOTO/PAT2
                    ACT/NOMORE

where DRL1 and PAT2 are a previously defined drilling sequence and a previously defined pattern of points.

The form of the drilling sequence definition is

    Symbolic name = DRIL/SO, string of parameter words and values

The following parameters are allowed

DEPTH, d   depth of hole at full diameter
TOOLID, a, b   where 'a' is tool identification number and 'b' is position
                of tool in turret.

SPIRET, a   describes the spindle motion on the return stroke

PLGFED, a   the feed in in./min at which the tool advances into the metal.

SPEED, a   the spindle speed in rev/min.

In any drill statement, the following are mandatory:

SO, DEPTH, TOOLID, PLGFED, and SPEED

At the moment, SO has no meaning within the language, but is there so that extensions can be made to allow for automatic selection of tools. The parameters in the above statement can be arranged so that the statement will describe other types of operation like tapping or reaming.

Turning facilities—2C

The processor, a subset of NELNC, has been developed to meet the special requirements of part programmers planning work for NC lathes.

In this type of work, the geometric elements encountered and the ways in which they are defined are similar to those in two-dimensional contour milling work; therefore, in the 2C subset all the POINT, LINE, CIRCLE, and TABCYL facilities of the 2C,L subset are provided. PATERN, PLANE, VECTOR, and MATRIX definitions have been included.

In lathe work, description of the tools encountered is normally much more complex than that required for tools used in point-to-point or contour milling work. To cater for this, a comprehensive tool-description procedure has been evolved so that details of the tool, such as a description of shape, cutting edge information, tool material, and information on the tool-clamping device can be stored in a tool file and referenced easily by the part programmer in a technology statement. The tool description card uses the format described for EXAPT 2 (chapter 8).

The process of removing metal in turning work, to arrive at a finished contour, is analagous to the process of area clearance in contour milling. Therefore, in 2C, use is made of the 2C,L language facilities for the definition of contours with the facilities BEVEL, ROUND, DIA, and PLAN added and as described in chapter 8. Again, to meet the special requirements of turning, a new type of contour definition has been added. This is of the form:

CON2 = CONTUR
OVCONT/CON1, .375
TERCON

which allows the contour CON2 which is uniformly larger than CON1 to be specified.

As in other types of machining, the machining conditions to be used during a turning operation have to be specified. The types of definition allowed are DRILL/, TURN/, GROOV/, and CONT/. These have the following functions:

DRILL/ allows machining cuts like drilling which take place on the axis of rotation to be defined. The statement includes details of the depth of cut, feedrate, tool number, and the sense of spindle rotation.

TURN/ allows the definition of machining conditions to be used during automatic removal of metal in a series of straight line cuts. The statement must include details of the depth of cut, feedrate, tool number, speed, direction of cut, and the tool setting angle.

GROOV/ allows the part programmer to specify information similar to that described for TURN. The difference is that in GROOV the tool retracts along the cutting path.

CONT/ used to describe the cutting conditions during contour turning. In this definition the direction of cut may be specified.

10. Multi-axis machining with APT

James A. Baughman

James Baughman is Senior Staff Manufacturing Engineer of the Corporate Manufacturing Services Department, TRW Inc., Cleveland, Ohio, USA. Majoring in mathematics at Washington University in St Louis, Missouri, and Georgia State College in Atlanta, Georgia, he has since spent 17 years in the aerospace industry, working in production processing and planning; special and standard tooling; tool designing; NC programming, engineering, and management; company management; computer graphics development; and computer-aided manufacturing techniques. He is the author and presenter of numerous papers to such organizations as the APT Long Range Program, the Aerospace Industries Association, the Numerical Control Society, the American Society of Tool and Manufacturing Engineers, and the Machinery and Allied Products Institute.

APT is a generalized system designed for programming simple to complex parts. It is basically oriented toward producing output for those parts containing three-dimensional solid geometric configurations requiring continuous-path tool motion. These are the troublesome parts requiring many fixtures, tools, and machine operations when produced by conventional equipment. These are the long time-span and high cost parts. APT helps produce the solution by harnessing the knowledge of the experienced programmer with the digital computer to produce parts quickly and economically.

APT 3 (Automatically Programmed Tools, Version 3) is now in effect with APT 4 in the 'field try' stage. APT, developed through the joint efforts of the Aerospace Industries Association members with support by the Massachusetts Institute of Technology and the United States Air Force, is continuously expanding. Refinements, new developments, and management are being provided by the APT Project at the IIT Research Institute in Chicago, Illinois. The APT 3 system has been working in a production environment since 1961. Many man-hours and computer hours have been expended in a thorough testing of its capabilities.

Scope

In many facilities, APT is used to produce engineering data, tooling for manufacturing, and production parts. APT mathematical calculation routines and geometry construction routines are valuable in the preliminary design and product engineering areas. The FORTRAN-like mathematics section, the vector analysis section, and the ability to describe and manipulate various geometric configurations provide a powerful engineering tool. After design, tooling is produced with unprecedented accuracy and efficiency. Master gauges, master models, jigs and fixtures, templates, dies, form blocks, etc., are prime candidates for APT programming. APT can handle all types of tooling, from the simple template configuration to the complex, double-contoured surface of dies. Finally, all of APT's powers come into effect with the production of complex part configurations. Control of the machine tool is embodied in APT through the use of its motion statements and post processors. Control of part production is the final product, allowing management to guide and monitor the production parts to the engineering specifications in the most exacting fashion.

The following photographs provide illustrative examples of parts produced by APT on multi-axis machine tools.

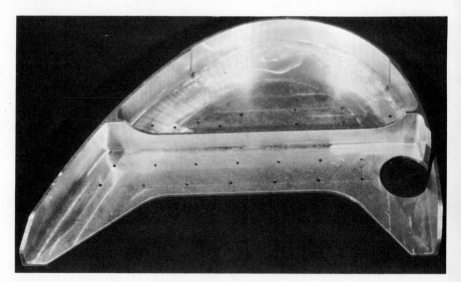

Figure 10.1 Machined fitting showing twisted outer contour, canted planes, external and internal radii. *Courtesy Lockheed-Georgia Company, Marietta, Georgia*

Figure 10.2 Machined housing showing hole patterns on several inclined planes. *Courtesy TRW Inc., Cleveland, Ohio*

Figure 10.3 Impeller blades, twisted contours. *Courtesy TRW Inc.*

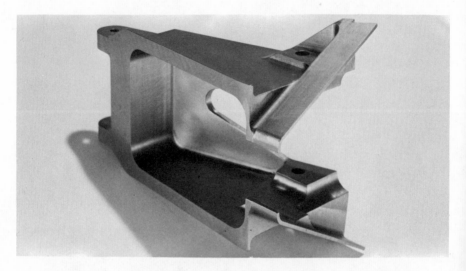

Figure 10.4   Inclined planes, steps, and contours shown on this bracket. *Courtesy TRW Inc.*

Computer requirements

The APT 3 system is currently implemented on IBM, CDC, UNIVAC, and GE computers. System software is written primarily in FORTRAN 2 or FORTRAN 4. System updates are prepared by the APT Project, usually on a quarterly basis, which requires an APT-trained computer programmer for maintenance. At least six man-months per year should be allocated for the task of maintaining the standard system and providing troubleshooting. If modifications or additions are required to any extent, computer programming time allocations should be revised accordingly. For instance, an active part programming organization with a staff of 60, with extensive developmental activities, would probably require five full-time computer programmers. For the more modest part programming organization with a staff of 10, one computer programmer may suffice. It should be stressed that the computer programming personnel must be trained in the nomenclature and structure of the APT computer system. This training will normally require a one year on-the-job span before the individual will become competent. Another important aspect is to provide at least one person who can bridge the gap between part programming and computer programming. Since two different languages are spoken, this person should be able to communicate the part-programming needs to the computer programmer in terms of computer programming requirements, and vice versa. This communication ability cannot be over-emphasized in terms of quick answers to computer and programming problems and providing a smooth functioning arrangement between the two disciplines.

Part-programming management

In general, the functional part-programming organization will consist of a pyramid structure as shown in Fig. 10.5. The part-programming groups will normally consist of three to seven programmers working with one NC engineer. Each supervisor usually handles two to four engineer-programmer groups. The programmer's job is to prepare the NC job package—consisting of NC drawing, APT input/output, operator's instructions, and tooling set-up requirements—then to try out the tape to produce the part to the specifications. This total function is under the cognizance of the NC engineer, assisting and trouble-shooting as required. Each group engineer reports to a supervisor whose function is about evenly divided between normal supervisional duties and providing

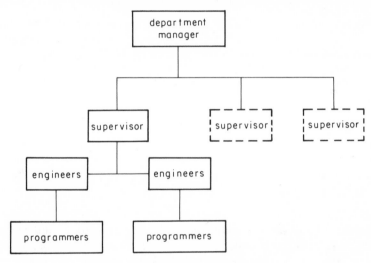

Figure 10.5   Part programming organization.

technical assistance as needed. The department manager should have some background and training in programming, besides the normal managerial requirements, in order to be aware of the technical depth and breadth embodied in computer-assisted part programming.

Probably one of the most important functions of the NC manager is to assure that proper co-ordination is established and maintained with related departments. For instance, co-ordination with the Engineering Department during the design stage will apprise engineering personnel with NC machine capabilities, allowing the designer to produce configurations that were not feasible by previous conventional machining practices. Also, blueprints can be dimensioned at this stage to facilitate NC manufacture. Co-ordination with Tool Design is essential to assure that NC tape and fixture are aligned correctly and that cutter collision will not occur. Co-ordination with Process Planning can provide the proper balance between conventional and NC operations. Since downtime on an NC machine is critical in economic terms, a close relationship with Maintenance and the NC shop is essential in order to pinpoint the problem and achieve quick correction. As mentioned above, co-ordination with Computer Operations and Programming is essential for troubleshooting activity and for development needs. The manager who provides for close co-ordination removes functional barriers and facilitates the work flow for a smoother, integrated system.

Part-programming methods

The proper method for part programming multi-axis machines follows a particular pattern. Each step should be scrupulously followed to assure minimum computer runs, minimum tape try-outs, and adherence to production schedules. These steps are as follows:

Print familiarization   The part programmer must 'know' the print to the extent that he can visualize the stock and part configuration in three dimensions. Each of the major views, auxiliary views, section cuts, and related dimensions and notes must be examined and correlated. Marking, on the print, the dimensions and notes which affect NC operations is a good practice to follow at this time.

NC drawing   A full-size scale-drawing of the part and stock should be made showing those areas to be machined by NC. Pertinent fixture details should also be included, such as clamps and bolts. This drawing will be used for several purposes. First, it provides a check on print dimensioning, such as locating double-dimensioned surfaces. Second, it provides a valuable 'feel' for the part programmer to sense the amount of metal removal and locate possible troublesome machining areas. Third, the drawing can serve for layout of intricate cutter paths and for measuring centre-line co-ordinates for such items as rough machining and machining the top surfaces of flanges, bosses, etc. This will expedite the coding sequence in many cases. Fourth, symbolic labels will be placed on each surface conforming to the symbolic names used in the part program. This serves to correlate drawing and program and will remain in file with the input deck or listing. The importance of a clean, neat NC drawing cannot be overemphasized since it serves as an invaluable aid to the part programmer and as a lasting record of what is contained in the symbolic part program. This particular picture is worth 10,000 words, producing accurate programming and machining.

Determination of processing sequence   With the print requirements well under-stood, selection of the 'best' machining sequence should be made and com-mitted on paper for reference during manuscript coding. Cutting-tool selection, speed and feed data, and pertinent comments about special requirements, such as clamp movements, should be noted beside each operation. This information is then coded in REMARK statements into the part program, for ease in locat-ing each cut sequence and its associated APT programming. These REMARK cards are mandatory when making subsequent program alterations.

APT manuscript coding   Coding normally follows a sequence of parameter statements, geometry definitions, and motion statements. The parameter state-ments consist of such data as post-processor requirement, machining tolerance, and centre-line printout requirement. The geometry statements define the draw-ing configuration and provide symbolic names for each surface. During the coding sequence, these symbols are placed beside the appropriate surface on the NC drawing. It is best to standardize symbolic names, such as L1, L2, L3, . . . for line definitions; C1, C2, C3, . . . for circle definitions, etc., in order to eliminate confusion and reduce errors. For instance, unnecessary complica-tions are introduced into a part program if the three lines (normally symbolized as L1, L2, L3) are given as MARTY, WQYZ17, and SWINGR. Motion state-ments provide the means for guiding the machine tool and cutter path. Auxiliary functions, such as turning the coolant on and off, are intermixed with motion commands. The motion statements should be sectioned into a series of cut sequences with a REMARK card preceding each. A sequence will consist of all the motion required for a particular tool from spindle insertion to extraction. The part programmer should thoroughly check his coding at intervals, such as after each cut sequence, rather than after the coding is completed for the whole part program. Coding must be precise and highly legible for key-punching. A systematic approach to coding is essential for accuracy. When the part program is run on a £400 per hour computer, an accuracy score of 99 per cent must count as a failure, costing time and money.

Key-punch listing checkout   After key-punching, the program should be listed and checked by the part programmer. The listing, providing additional clarity to the part program statements compared to the hand-printed manuscript, allows a final opportunity to review it as a whole, detect errors, and add coding if required.

Troubleshooting   If the computer run is unsuccessful in terms of tape output, APT diagnostics provide the means of locating the error. Section 1 diagnostics provide error comments for statements that do not conform to correct format, spelling, punctuation, and the like. Normally, these are simple to correct. Section 2 and 3 diagnostic comments are indicators of incorrect motion command statements. For these, it is usually more difficult to determine the reason for failure, especially in five-axis part programs. APT data will provide the surface equations, cutter location and orientation, last cutter motions, and a comment suggesting the probable reason for failure. A knowledge of analytic geometry, simple calculus, matrix algebra, and vector analysis may be required to mathematically pinpoint the problem. Section 4 diagnostics provide post-processor comments locating errors such as movement outside of machine limits or incorrect spindle speed. The part programmer should carefully consider each diagnostic and provide correction if he understands the reason for failure. If there is some doubt, consultation with someone of greater experience is necessary to provide an objective viewpoint and a valid correction.

Output checkout   After output of listing and tape is received, a thorough check is essential before shop release. The best check of three-axis centre-line data is to plot the cutter path on tracing paper on an NC drafting machine and use this as an overlay on the NC drawing. Cutter radius offsets can be checked along the cutter surface coverage. Small to large errors in centre-line deviation can be ascertained, depending on the NC drawing accuracy. Probably those errors of 0·02 in. or less will not be found until after tape try-out. Questionable output points can be checked mathematically as required by using the APT-furnished surface equations. The checkout of four- and five-axis output is more difficult. Since the tool is in a skewed position, post-processed output is relative to the machine slide positions (linear and rotary) and a plot is not easily related to the NC drawing. Some special processors have been developed to expedite checkout, such as outputting a separate tape containing cutter end (X, Y, Z) co-ordinates only; however, the ultimate solution of plotting three-dimensional configurations on two-dimensional paper has not been solved. The part programmer can, nevertheless, make a reasonable check by layout of appropriate cutter centre-line co-ordinates and tool geometry on the NC drawing, projecting as needed onto the auxiliary views containing the true view of the skewed surfaces. This selective, hand-plotting method should be used for all part programs if an NC drafting machine is not available. During the dimensional checkout, post-processor data such as feeds, speeds, and spindle rotation can also be checked.

Tape try-out   The major consideration is to have the part programmer who wrote the program direct the tape try-out. This is necessary for several reasons. First, he is the only person who has an intimate knowledge of the part program's details, and as such, he can make corrections quite rapidly. Second, he can direct the machine operator in making manual moves to by-pass a defective area or correct a cut sequence in order to produce parts as needed before tape correction. Third, the experience derived from observing the result of his programming handiwork will make his next part program a better one. Tape try-out is obviously the critical point in the part-programming sequence, for if a part is not produced which corresponds to the print specifications, all previous work is in vain.

APT features

Before progressing into the APT features and programming details, it should be noted that the information contained here is necessarily condensed. A complete explanation of all details is contained in the official APT Encyclopedia and APT Dictionary.

Chapters 7, 8, and 9 have dealt with 2C,L type geometry and with motion routines involved in EXAPT and 2C,L. APT provides similar facilities plus those involved in 3C and 5C routines. A description of these facilities is as follows:

Geometry
The additional geometric features offered by APT are shown in Fig. 10.6.

Cone: Defined by its vertex point, axis vector, and half angle measurement.

Cylinder: Defined by a point on its axis, axis vector, and radius.

Quadric: Defined by equation form for real ellipsoid, hyperboloid of one sheet, hyperboloid of two sheets, elliptic cone, elliptic paraboloid, hyperbolic paraboloid, and elliptic, hyperbolic, and parabolic cylinder.

Ruled surface: Defined by two space curves, their end points, and a vector or point in the plane in which each curve lies.

Sphere: Defined by its centre co-ordinates and radius (given or calculated).

Polyconic: Defined by a family of conic sections in parallel planes.

GEMESH: Defined by a set of points in parallel or radially located planes.

APTLFT-FMILL: Defined by a matrix of points, or points and slopes.

Tool motion
Tool motion is controlled by the programmer by using APT words that indicate the desired direction of movement relative to the previous motion or given vector. These words and their meaning are GOLFT (go left), GORGT (go right), GOFWD (go forward), GOBACK (go backward), GOUP (go upward), GODOWN (go downward), and can be illustrated as shown in Fig. 10.7.

During a machining sequence, the tool is positioned relative to three surfaces which are called the Drive Surface (DS), the Part Surface (PS), and the Check Surface (CS). An illustration of these surfaces and the tool is shown in Fig. 10.8. The tool is positioned tangent to DS and PS and is approaching the stopping point at CS.

Tool positioning can be specified to maintain the tool to the left of the DS (TLLFT); to the right of the DS (TLRGT); or on the DS (TLON), with the tool end tangential to the PS (TLOFPS). These positions are illustrated in Fig. 10.9 with the assumed tool motion being into the paper. Figure 10.10 shows similar conditions except that the tool end is on the PS (TLONPS).

The tool axis command (TLAXIS) is used to orient the axis of the cutter in a variety of ways. The command TLAXIS/I, J, K (where I, J, K are vector components directed upwards from the tool end) will orient the tool in a fixed position relative to the basic co-ordinate system as shown in Fig. 10.11. The command TLAXIS/NORMPS will align the tool axis perpendicular to the surface of the PS as shown in Fig. 10.12.

The use of TLAXIS/A, B, R, H, Θ, I, J, K, ∅ allows positioning the cutter through the use of a disc control method. The parameters R and H specify the

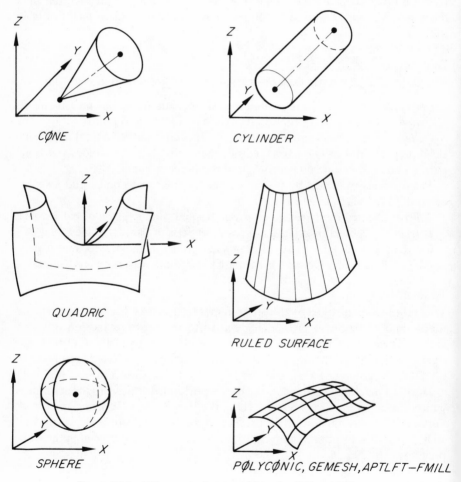

Figure 10.6   APT geometry features, additional to 2P,L and 2C,L types.

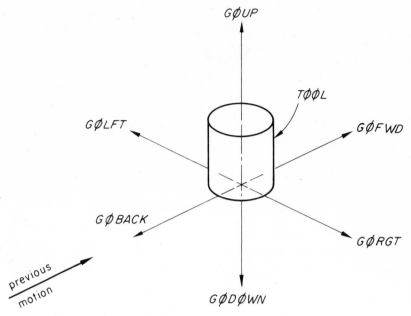

Figure 10.7   Tool motion, direction relative to previous move.

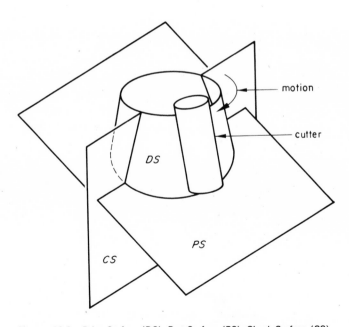

Figure 10.8   Drive Surface (DS), Part Surface (PS), Check Surface (CS).

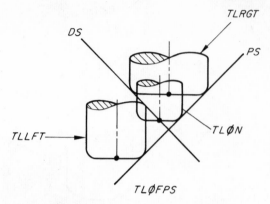

Figure 10.9   Tool tangent to part surface (TLOFPS).

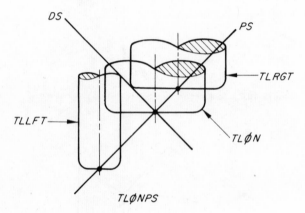

Figure 10.10   Tool on part surface (TLONPS).

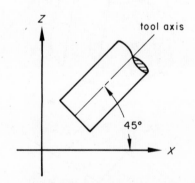

$$TLAXIS /.707107, 0, .707107.$$

Figure 10.11   Inclined tool axis.

tool axis

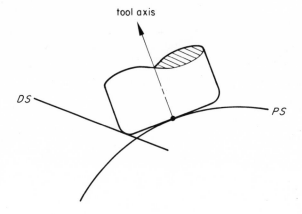

TLAXIS / NØRMPS

Figure 10.12   Tool axis normal to part surface.

location of a control disc on the cutter geometry; A and B set the cutter for three-, four-, or five-axis positioning and for tool positioning relative to DS, PS, or Ruled Surface Control. The Θ angle is used to tilt the tool axis relative to the DS or PS surface. The vector components I, J, K are used for four-axis control and ∅ is a lead or lag angle to tilt the tool forward or backward relative to the direction of motion. An illustration of the disc location and tilt angle relative to the DS is shown in Fig. 10.13. Lead or lag tilt is shown in Fig. 10.14.

It is apparent that the use of the tool axis commands coupled with motion and positioning statements will allow an extensive tool control for producing complex shapes, and this is a developing area of APT.

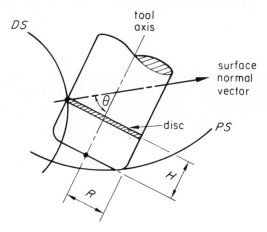

Figure 10.13   Tool axis controlled by disc.

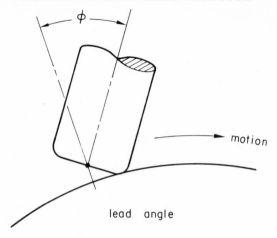

motion

lead angle

Figure 10.14 Leading angle tilt of tool axis.

Three- and five-axis part programming

Canted plane part surfaces

Multi-axis programming can be reduced to a set of simple elements producing the desired motion on canted planes, cylinders, quadrics, ruled surfaces, and double-contoured surfaces. The simplest type, and most common, is producing motion on canted planes. Figure 10.15 shows a set of holes to be drilled on an inclined plane. The hole dimensions are defined in the auxiliary view using a secondary co-ordinate system (X_1, Y_1). Part programming is simplified by enclosing the motion commands inside a TRACUT sequence which uses a matrix to transform the tool and co-ordinates and the tool axis to the desired location.

The part programmer must visualize the X_1, Y_1 co-ordinate system as first coincidental with the X, Y system and that matrix manipulation will occur from that point. Since the part program is first written with respect to the X, Y co-ordinate system, the matrix MAT1 is defined to first rotate the co-ordinates and tool axis in the YZ plane by 30° (YZROT, 30) and then translate 2 in X, 3 in Y, and 0 in Z (TRANSL, 2, 3, 0). When brought into effect by the TRACUT command, this will move the cutter centre points and their associated tool axes into the X_1, Y_1 working position. The CALL commands initiate a previously defined macro. Programming consists of the following.

TLAXIS/0, 0, 1 $$ Tool axis parallel to Z axis

MAT1 = MATRIX/YZROT, 30, TRANSL, 2, 3, 0
        TRACUT/MAT1 $$ Initiate matrix

GOTO/1, 1, 1 $$ Motion to a point
                                                above first hole
CALL/DRILLM $$ Drilling sequence
GOTO/2, 3, 1
CALL/DRILLM
GOTO/3, 4, 1
CALL/DRILLM
TRACUT/NOMORE $$ End matrix usage

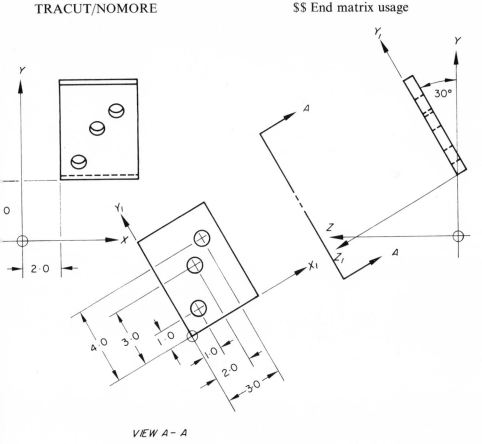

VIEW A- A

Figure 10.15 Drilling on an inclined plane part surface.

Many canted planes occur as chamfers on flanges, web surfaces, etc., which are machined either on three- or five-axis machine tools. If a three-axis machine is used, generation of the canted surface will be made by making machining passes close together in an up and down motion, producing a scalloped (or wavy) surface (Fig. 10.16). This type of surface generation is acceptable in most shops.

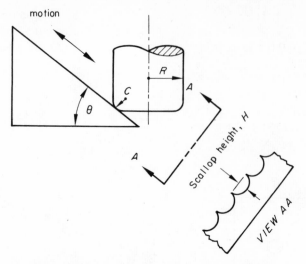

Figure 10.16   3C milling on inclined plane part surface.

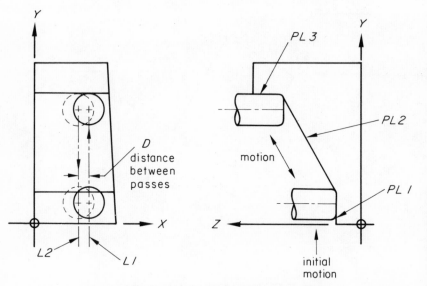

Figure 10.17   Example of inclined plane PS, 3C milling.

An example of three-axis canted plane machining is shown in Fig. 10.17. The initial motion is on the L1 DS in the positive Y direction using PL1 as the PS. The distance D between passes is calculated by using a formula to maintain a desired scallop height. The formula is approximate (using a standard ellipse definition) and is defined by the following method:

R = cutter radius (see Fig. 10.16)
C = corner radius on cutter
O = true angle of canted plane from XY plane
H = scallop height desired
Compute: E = R − C
          a = E + C sin Θ
          b = a sin Θ
          Y = b − H

Then $D = \dfrac{2a\sqrt{b^2 - Y^2}}{b}$ = distance between passes to hold scallop height to H.

The programming requirement to make two passes on the canted surface is shown as follows. If an extensive number of passes are required, an appropriate loop or macro would be used.

PSIS/PL1
TLON, GO____/L1, TO, PL2 $$ Movement to PL2. In the underlined space UP, DOWN, BACK, LEFT, etc. may appear, depending on the previous move.

PSIS/PL2
GOUP/L1 $$ Movement up PL2
TLLFT, GOLFT/PL3 $$ Movement along edge of PL3
TLON, GODOWN/L2, TO, PL1 $$ Movement down PL2 to PL1

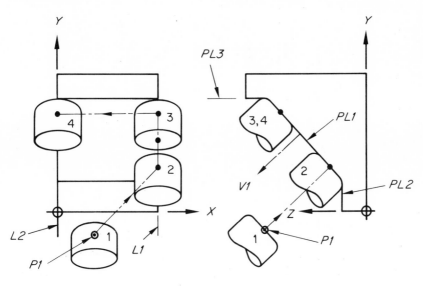

Figure 10.18   Example of inclined plane PS, 5C milling.

Generation of a canted plane using a five-axis machine tool is accomplished by positioning the cutter normal to the PS plane and commanding motion as desired. Figure 10.18 shows a canted plane with cutter positions 1 to 4 inclusive. The cutter is usually positioned normal to the PS in position 1 before plunging into the workpiece by obtaining the PS normal vector components. (Tool axis direction is from tool end upward through tool.)

V1 = VECTOR/PERPTO, PL1, POSZ	\$\$ Vector normal to PL1 plane
TLAXIS/V1	\$\$ Sets tool axis normal to PS
GOTO/P1	\$\$ Move to position No. 1
GO/ON, L1, PL1, PL2	\$\$ Move to position No. 2
TLON, GOLFT/L1	\$\$ Move to position No. 3
TLLFT, GOLFT/PL3, ON, L2	\$\$ Move to position No. 4

Outside cylindrical part surfaces

Cylindrical surfaces occur quite often such as on flange ends and around bored holes. If the cylinder is to be machined on its external surface, three-axis programming is similar to the following pattern as shown in Fig. 10.19

PSIS/PL1	
GO____/L2, ON, L1	\$\$ Position No. 1
PSIS/CYL1	
GOFWD/L2, TO, PL2	\$\$ Position No. 2

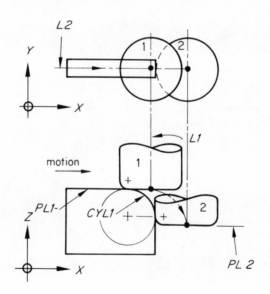

Figure 10.19   3C milling outside and on a cylindrical PS.

In this example, the cutter is positioned on L1 to provide a positive location, aligning cutter corner radius and cylinder centre-lines, before moving down the cylinder to the next positive location at PL2; again aligning centre-lines.

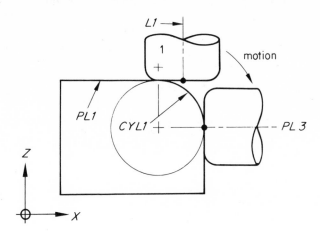

Figure 10.20 5C milling outside and on a cylindrical PS.

The five-axis method is shown in Fig. 10.20. The cutter is moved to position No. 1 (as in the previous example) and then the tool axis is maintained normal to the PS as the cylinder is generated. Usually, a linearization routine is also used with this type of surface generation to ensure a correctly processed cutter path. (See five-axis post processors in chapter 11.)

```
PSIS/PL1
GO    /L2, ON, L1      Position No. 1
PSIS/CYL1
TLAXIS/NORMPS         Tool axis normal to PS
GOFWD/L2, ON, PL3     Position No. 2
```

Other surfaces, such as cones, spheres, and quadrics can be programmed similarly to the cylinder examples.

Inside cylindrical part surfaces

Machining the inside of a closed surface, such as a cylinder, normally requires a restriction of cutter height (as in the following part program) in order to keep the top of the cutter from gouging (mathematically) into the upper closed surface area. Special attention is usually given to the initial position of the cutter prior to the start of the cut sequence for the same reason. An example of this type of program is shown in Fig. 10.21.

CUTTER/1, .5, 0, .5, 0, 0, .5 $$ 1·0 dia. ballnose with 0·5 height declared
TLAXIS/0, 0, 1 $$ Three-axis cut specification
FROM/PT1 $$ Initial starting point 1
INDIRV/− 1, 1, − 1 $$ Rough vector direction of start-up
                                     motion
GO/L1, CYL1, ON, L2 $$ Position 2
TLLFT, GOLFT/L1, PAST, PL1 $$ Position 3

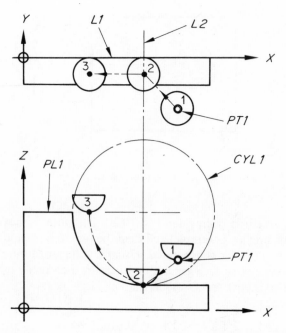

Figure 10.21   3C milling inside a cylindrical PS with very short notional cutter.

Another often-used method for making a three-axis cylinder cut is to program the cylinder as if it were a CIRCLE drive surface in the XY plane (circles in APT are defined only in the XY plane) and then to transform the cutter centre points into the required position by the use of the TRACUT routine. As this method allows flexibility, careful thinking is required, but it can be used to overcome system problems.

For example, part programs confined to CIRCLE and LINE definitions in APT use simpler two-dimensional calculations, different from the normal three-dimensional calculations. These two-dimensional calculations may produce a result when a problem in the three-dimensional calculation prevents a solution.

To illustrate the use of the two-dimensional calculation, consider the machining operation shown in Fig. 10.22 with the 'real' cutter shown in full lines.

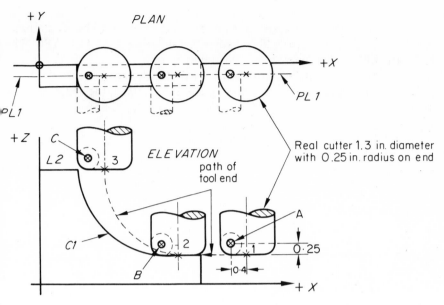

Figure 10.22   Desired position of real cutter and component. Real cutter axis in Z direction, end point at x offset by 0.4 in X and −0.25 in Z from the programmed cutter, axis in −Y direction with end point at ⊗.

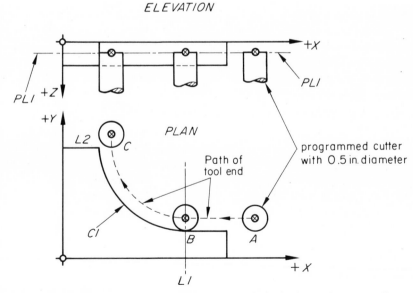

Figure 10.23   Programmed cutter and component for Fig. 10.22. Profile in XY plane to allow two-dimensional APT description.

A circle or LINE definition in APT cannot be used unless defined relative to the XY plane. Thus the imaginary task shown in Fig. 10.23 has to be programmed instead, taking the smaller 0·25 in. radius end mill round the path ABC. This tool path ABC is then moved into its desired plane by YZROT, 90 instruction which produces the ABC tool path shown with the dotted cutter in Fig. 10.22.

The required tool path is shown as 1, 2, 3 in Fig. 10.22, the end of the tool axis for the real tool. 1, 2, and 3 are offset by 0·4 in. in X and −0·25 in. in Z from A, B, and C respectively.

Thus, the complete translation from path ABC in Fig. 10.23 to path 1, 2, 3 in Fig. 10.22 is carried out by the second and third lines of the following part program:

CUTTER/.5	$$ Pseudo-cutter
MAT 1 = MATRIX/YZROT, 90, TRANSL,	
.4, 0, −.25	
TRACUT/MAT 1	$$ Use matrix
GOTO/PTA	$$ Location of 'A'
INDIRV/−1, 0, 0	$$ Approx. start direction
GO/C1, PL1, ON, L1	$$ Location of 'B'
INDIRV/0, −1, 0	$$ A reference direction
TLRGT, GORGT/C1, PAST, L2	$$ To location of 'C'
TRACUT/NOMORE	$$ End use of matrix

Canted drive surfaces

Quite often, part geometry will consist of canted intersecting planes that require the cutter axis to be tilted at a particular five-axis angle to cut both surfaces accurately at their intersection. This is easily achieved by using the vector definition of plane intersections and using this symbolic vector as the tool axis as shown in Fig. 10.24.

V1 = VECTOR/INTOF, PL2, PL3, POSZ	$$ Intersection vector
TLAXIS/V1	$$ Five-axis tool axis
GOTO/PT1	$$ Move to PT1 point
INDIRV/−1, 1, −1	$$ Rough vector direction
GO/PL2, PL1, ON, L1	$$ Location 1
TLRGT, GORGT/PL2	$$ Location 2
GORGT/PL3, ON, L2	$$ Location 3

Ruled surfaces

Twisted surface configurations are handled by APT's ruled-surface routines. Two space curves can be defined and straight-line rulings are generated between the curves to define the twisted surface. The resultant surface can be used similar

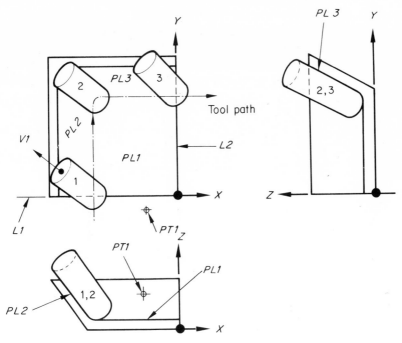

Figure 10.24 Tilted drive and check surfaces.

to other APT surfaces for motion commands. The RLDSRF definition consists of two space planes, two curves, and curve end points as follows:

$$RS1 = RLDSRF/S1, PT1, PT2, PT3, S2, PT4, PT5, PT6$$

where RS1 is a symbolic label of the surface

S1 is the first space curve, such as a circle, line, conic, etc.

PT1 and PT2 are end points on S1

PT3 is a point on the plane passing through curve S1 (PT1, PT2, and PT3 define the space plane)

S2 is the second space curve

PT4 and PT5 are end points on S2

PT6 is a point on the plane passing through curve S2

Fig. 10.25 shows an example of the above defined RLDSRF. Modifications can be made to the definition (if desired) to substitute a vector for the third point in each plane or to reduce one space curve to a point.

After geometry definition, the ruled surface is used to generate cutter paths. For example, Fig. 10.26 shows a ruled surface being used as a DS, requiring five-axis cutter path computations. The TLAXIS command requires the tool

11

to position its axis parallel to the RLDSRF rulings as it moves across the surface. The cutter height is restricted in order to reduce excessive arithmetic computations concerned with surface-to-cutter alignment.

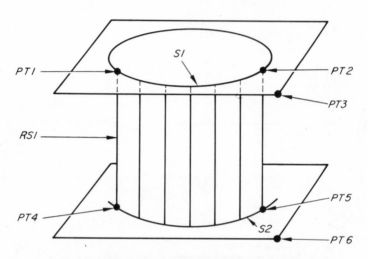

Figure 10.25   Ruled surface between curves S1 and S2.

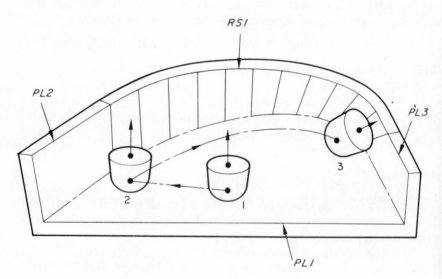

Figure 10.26   Cutter paths on a ruled surface.

CUTTER/1, .5, 0, .5, 0, 0, .5 $$ 1·0 dia. ballnose end mill with 0·5 height

TLAXIS/3, 2, .5, .5 $$ Relates tool to DS and locates 'disc' control

INDIRV/____, ____, ____ $$ Appropriate vector direction from Location 1

GO/RS1, PL1, PL2 $$ Position 2

TLRGT, GORGT/RS1, TO, PL3 $$ Position 3

Double-contoured surfaces

Double-contoured surfaces, such as those found on automobile body dies and aircraft skin form blocks, can be defined and machined by any of several three-axis APT routines. The polyconic surface routine (POLCON) allows a continuous family of conic sections to be defined in parallel planes to produce a smooth surface. After definition, the surface is used with the normal APT motion commands to produce CLDATA. The GEMESH routine uses a set of co-ordinate points (input by the programmer) to produce a smooth surface and also to generate an automatic cutter path. The features of GEMESH allow the surface to be machined longitudinally or transversally, allow the surface to be bounded by four 'fences', and allow a five-axis machine to be used if desired.

The APTLFT-FMILL

The APTLFT-FMILL routine is somewhat similar to GEMESH since it uses a set of co-ordinate points to produce a double-contoured surface along with an automatic cutter path. The routine is used in two steps, the first using the FMILL portion to define a complete set of co-ordinate points and normal vectors, and the second using the APTLFT portion to generate the cutter path. Input to the FMILL routine is a set of points (and tangent or normal vectors

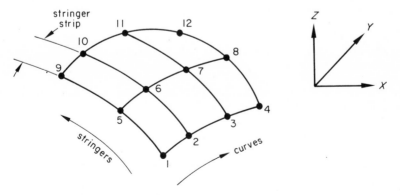

Figure 10.27 An F MILL surface.

if desired) plus tolerance and other parameters. This program uses the sparse array of input data points to blend a smooth surface through them, outputting an enriched array of points and normals necessary for cutter-path motion. If sharp breaks are required on the surface, provisions in FMILL will also allow this configuration. An example of FMILL input is shown in Fig. 10.27.

The point co-ordinates are designated to produce curves and stringers and a set of stringer strips. Output from FMILL is put on magnetic tape for later use by the APTLFT program.

CALL/FMILL			$$ Calls FMILL routine
.	3	4	$$ Designation for 3 curves and 4 stringers
6.5	6.1	3.8	$$ XYZ co-ordinate at point 1
.	.	.	$$ Co-ordinates of points 2 through 11
.	.	.	
.	.	.	
2.3	8.7	2.5	$$ Point 12 co-ordinates
.001 1	1 12	10	$$ Tolerance, divisions in stringer strips, chord height steps, etc.
FMEND			$$ End of FMILL routine requirement

The APTLFT program generates an automatic rough-machining tool path and/or finish tool path, provides feedrate control, allows several variations in direction of cutter path, and provides for matrix transformations for such purposes as machining male and female dies from the same input data. The cutter-size designation is left to the discretion of the programmer, allowing square end, corner radius end, or ballnose end mills. No check is made in this routine for gouge into adjacent surfaces—thus if the concave radius of curvature on the part is less than that of the designated end mill, the cutter will violate the adjacent area. The cutter path will follow the stringer direction in an alternating motion, or machine in the same direction across the surface, as desired by the programmer. The area machined is the total defined surface; no segmentation of the surface is allowed. The following is an example of the statements that are necessary to provide the APTLFT cutter motion.

CALL/APTLFT	$$ Calls APTLFT routine
GO/(LOFT/1, 30, 1, 10, 35, 2, 5, 0)	$$ Provides parameters necessary for cutter-path motion

As APTLFT-FMILL operates independently from APT it is not possible to bound FMILL surfaces by other APT surfaces.

Variable twist drive surfaces

Sometimes twisted part surfaces are defined on engineering blueprints by a series of section cuts that show angle of inclination at each location. The upper or lower controlling surface is a geometric element, such as a line, arc, splined curve, etc., and a smooth twist is desired along these controlling surfaces.

Figure 10.28 shows the controlling surfaces L1, C1, and L2 with section cuts at AA, BB, and CC locating angles a, b, and c respectively.

The VTLAXS routine provides the means for machining the surfaces shown in Fig. 10.28. The usual three-axis motion commands are used to move along the controlling surfaces and the output from these commands are modified by VTLAXS to provide five-axis motion. The commands for machining the surfaces are as follows.

GOTO/PT1	$$ Initial point at location 1
VTLAXS/ON, LEFT, a, 1, 0, RIGHT	$$ Sets VTLAXS routine to tilt cutter to 'a' inclination
TLAXIS/0, 0, 1	$$ Routine requirement
GO/L1, PS, CS	$$ Motion to location 2
TLLFT, GOLFT/L1, TANTO, C1	$$ Location 3
VTLAXS/OFF, RIGHT, b	$$ Sets cutter tilt to 'b' at end of cut
VTLAXS/ON, LEFT, b, 1, 0, RIGHT	$$ Turns routine on to 'b' tilt
GOFWD/C1, TANTO, L2	$$ Location 4
VTLAXS/OFF, LEFT, c	$$ Sets cutter tilt to 'c'

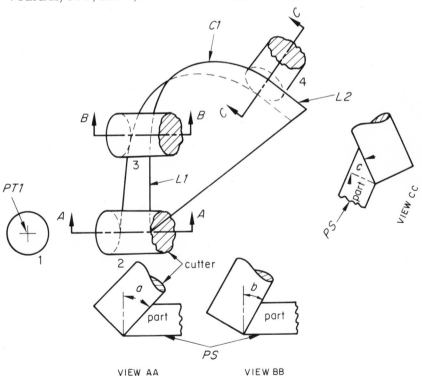

Figure 10.28 Twist milling.

The VTLAXS/ON statement tells the routine on which side of the surface cutter lies (left), angle of inclination (a), cutter diameter and corner radius (1, 0) and tilt with respect to cutter forward motion (right). The VTLAXS/OFF statement specifies tilt direction (right) and angle (b) at end of each section.

The foregoing examples constitute only a small portion of APT's capabilities. Combinations of the various routines, such as macro's, copy logic, looping, computing, multiple-check surfaces, reference-system transformation, etc., provide a multi-axis capability unequalled in any other present-day system. This computational capability and power exhibits itself in producing NC tapes that accomplish the near-impossible in comparison to conventional means. The impractical, conventional machining method becomes the natural, easily obtainable NC method. The tool to produce these complex parts, at minimum cost, is APT.

11. Post processors

R. M. Sim

R. M. Sim is a Senior Scientific Officer in the Numerical Control Division of the National Engineering Laboratory, East Kilbride, Scotland, responsible for Post Processor Development and Standardization, and a recognized authority on the subject. He entered industry as an apprentice with Glenfield and Kennedy (hydraulic engineers) in Kilmarnock, and left them eleven years later as design engineer 'collecting during that time an HNC in mechanical engineering'. Joining Ferranti Ltd in 1961, he worked on the design of precision potentiometers and other NC equipment, before moving to his present position at the NEL.

Any computer program designed to prepare control tapes for NC machine tools must consider both the geometry of the component to be machined and the dynamics and other particular requirements of the machine tool system. In present practice, this is achieved, for any one system, either by using a single special-purpose program written to suit the particular machine tool, or by sharing the generation of control tape between two distinct programs—a general-purpose processor and a special purpose post processor.

In this chapter, the role of the post processor will be discussed as an accessory to a general-purpose processor. It should, however, be emphasized that both the problems and solutions discussed apply equally within the context of the special purpose 'single package' program.

The concept of the processor has been dealt with at length in earlier chapters and it should suffice to remind readers that it provides a general solution, in terms of cutter-location data, to the problem of generating the cutter paths required by varying part-program specifications of component and cutter geometry. The resulting cutter-location data (or CLDATA) as generated by the processor, is usually independent of any properties of the machine tool or control system. This is not strictly true of the EXAPT and NELNC processors which use machine tool-dependent parameters within their inbuilt technology sections (see chapters 7 to 9).

In essence, the CLDATA consists of a series of sets of XYZ co-ordinate values for the tip of the cutter in successive positions as it moves around the defined geometry. Auxiliary part-programming instructions (e.g., to set spindle speed, feedrate, tool changing, etc.) are inserted by the processor in the appropriate sequence. However, these instructions are not considered computationally or logically by the processor.

As can be seen from chapter 6, the control tape requirements of NC systems can vary widely.

Individual control systems require that both the information describing the cutter path and the organization of that information into blocks be arranged in an exact and often unique fashion.

One control system may require that the positions supplied to it be calculated so that the maximum permissible accelerations of the associated machine-tool slides are not exceeded, whereas another system may incorporate electronic circuits to deal with accelerations of the slides in the hardware. Again, one control system may require absolute positions of every point on the control tape, another may require incremental distances from each previous point.

It is necessary, therefore, to carry out particular calculations to suit each combination of machine tool and control system. With the general-purpose NC programming systems—APT, ADAPT, NELNC, EXAPT, etc.—these machine-tool-dependent calculations are normally performed by a further computer program known as the post processor. A check list of the type of functions normally handled by post processors is given in Fig. 11.1.

In processing the part program, the two programs—processor and the required post processor—are run sequentially and to the user appear as one.

1. Read in data from CLDATA prepared by processor
2. Convert to machine tool co-ordinate system
3. Convert to absolute or incremental form
4. Check machine-tool limitations; e.g.,

    (a) travel of machine slides
    (b) interference between tools, workpiece and machine tool
    (c) allowable feedrates and spindle speeds
    (d) types of machine functions controlled (coolant, tool changing, etc.)

5. Develop feedrates and spindle speeds
6. Develop motion commands allowing for requirements of machine tool/controller; e.g.,

    (a) safe acceleration/deceleration of slides
    (b) acceptable overshoot at corners
    (c) tape reader time
    (d) servo settling time
    (e) pulse weight of control system

7. Allow for type of interpolation—linear, circular, etc.
8. Output control tape to suit control system requirements; e.g.,

    (a) correct media (paper tape, magnetic tape, etc.)
    (b) correct block format
    (c) correct code-set

9. Output printed listing as an aid to programmer
10. Produce diagnostics in the event of errors.

Figure 11.1 Check list of typical post-processor functions

Post processors are either unique to particular combinations of machine tool and control system or are generalized to suit families of machine tools. They are usually written in FORTRAN, are anything from 1,000 to 5,000 statements long, and can be considered to be mainly data processing (the proportion of arithmetic computation is usually quite small). The effort taken to write them can vary from three to six man-months for a positioning-type machine tool, to three to four man-years for a multi-axis continuous-path machine.

Post processors are an essential and extremely important part of the overall NC system. Although they are small in comparison with their associated processors (an average APT 3C post processor being only a tenth of the size of the APT processor), the total cost of their development for existing general-purpose programs is probably greater than the cost of the processors themselves.

This is not so remarkable when it is considered that each unique combination of machine tool, control system, and computer has usually resulted in a different post processor or at best an amended post processor.

With the large number of post processors which have been developed, there has undoubtedly been a duplication of effort and this can be accredited to a number of causes:

the diversity of application of machine tools;
lack of standardization between control systems;
lack of standardization in the design of post processors.

A measure of standardization can be achieved, both in control systems and in post-processor writing, and this will be considered later in the chapter. The diversity of NC machine tools, however, is wide and inevitably leads either to cumbersome generalized post processors or to a large number of smaller, single-purpose, or less generalized post processors.

Typical post processor applications

Turret drill with positioning control system

This class of machine has been produced in relatively large numbers and is usually designed to give rapid positioning of one of a number of tools to successive hole positions on the component. The work operations (drilling, tapping, boring, etc.) on these positions are often carried out in predetermined cycles which may be mechanically actuated by cams on the machine tool or built into the electronics of the control system. These predetermined or 'canned' cycles are usually called for from the control tape by the appropriate preparatory functions (e.g., G81 for a DRILL cycle, G84 for a TAP cycle, etc.).*

The control system is frequently a shared axis system, the Z axis, if controlled, sharing the same control register as the X axis. For effective drilling, tapping,

* See chapter 6.

11*

etc., the post processor must split up XYZ moves into two blocks, a move first in XY and then a move in Z if the component is being approached, or a move in Z followed by a move in XY if the tool is being retracted. A typical control system for such a machine might require absolute co-ordinates as input, and allow the datum to be set anywhere within the range of the machine. Alternatively, the datum may be fixed, or incremental dimensions required as input. The functions of turret indexing, spindle speed selection, feedrate, and coolant setting are usually commanded by tape, this normally being 1 in. wide paper tape, conforming in code and format to one of the EIA or ISO standards (see chapter 6).

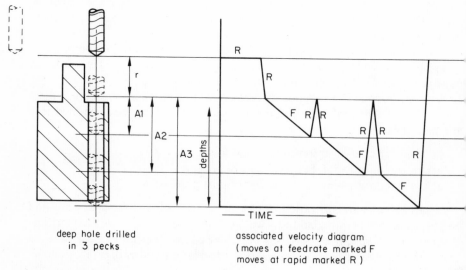

deep hole drilled
in 3 pecks

associated velocity diagram
(moves at feedrate marked F
moves at rapid marked R)

Figure 11.2   Deep-hole drilling. Ten moves are generated by the post processor from the single part-programming statement: CYCLE/DEEP, A1, A2, A3, F, IPM, R.

With this type of machine tool it is important that the combination of processor and post processor allows simple brief specifications in the part program of both the locations of the geometric patterns of holes, and of the machining operations to be performed at these locations. This can be achieved by the combination of PATERN instructions expanded in the processor and a variety of CYCLE instructions expanded in the post processors. An example of this is shown in Fig. 11.2, where, by the use of a single CYCLE/ statement in the part program, a number of blocks can be generated to give the complete machining operation. Of course, where the machine tool has built-in canned cycles, the post processor would only require to output a single block for each location, giving the relevant 'G' code and the required depths and feeds.

A further feature of many positional drills is their ability to mill at controlled feedrates along straight lines parallel to the X or Y axes of the machine. Using this feature, planners can part program XY contours, using continuous-path programming statements, and the drilling post processor will generate line-milling commands on the machine tool in a combination of moves parallel to the X and Y axes or at 45°, such that the contour machined will approximate to that defined.

Depending on the features incorporated, a post processor for this class of machine tool would be in the order of 1,500–3,000 FORTRAN statements in length and would take some three to six man-months to prepare.

NC lathe with 2C contouring control

Most NC lathes are fitted with an inexpensive 2C contouring control system. Such a system gives contouring control in the longitudinal and transverse axes of the lathes but, since it has no buffer storage of input blocks, machining is not continuous between blocks on the control tape. In other words, the cutter dwells at the end of each block until the succeeding block is read into the system from the control tape.

This means that, although these lathes have contouring control in two axes, the machine dynamics portion of their post processors need be no more complicated than for a simple positioning machine tool. This is examined in greater detail later in the chapter.

Because of their geometry configuration, however, lathes present the post processor with problems not encountered on other machine tools. On a milling machine, for example, a constant peripheral cutting speed is maintained around the component provided that the diameter of the cutter is not changed. However, on the lathe, with cutting conditions achieved by the combination of a non-rotating tool and the rotation of the workpiece about an axis, cutting conditions alter as the radius of the workpiece is altered. In consequence, for example in facing cuts, to maintain an optimum cutting speed on a lathe, the workpiece revolutions must be frequently adjusted to suit the effective radius of the component. This adjustment can be made by the part programmer, but it is preferable to leave the selection of spindle speed to the post processor to eliminate manual calculations. This is frequently done, the planner simply specifying the required cutting condition once, to suit the tool and material.

Most NC lathes employ multiple tools and unfortunately these tools do not normally index to a common tool-tip position. All movements on the control tape must refer to the same reference point on the tool turret though points on the CLDATA refer to the current tool tip. To simplify the part programmer's job, lathe post processors are usually designed to accept a part program statement which gives the distance from the tool tip to the reference point (in X and Y co-ordinates) for each tool. The part programmer and processor need then only be concerned with movements of the tool tips relative to the geometry of the component and can leave the post processor to translate from part co-ordinates to machine movements of the turret reference point.

Other peculiarities of lathe post processors are their requirement to handle threading operations and—in a few cases, where possible collision between tools, chuck, and workpiece is a problem—the development of an effective protective envelope.

The average lathe post processor will differ little in size from a post processor for a positional drilling machine, unless it provides a comprehensive tool-collision check, but, because of the lathe's peculiarities, it will differ widely in detailed design.

3C milling machines

Milling machines were the first type of machine tools to be widely fitted with NC. To a large extent because of the different economic conditions pertaining in the two countries, Britain and the USA have tended to develop milling machine control systems along different lines.

In the USA, development has favoured placing an interpolator with each machine tool on the shop floor and feeding dimensional information plus feed-rates to it on punched paper tape. The major differences among American control systems today are in the types of interpolation available—linear, circular, or parabolic—and in the handling of the machine tool dynamics. The problem of catering for machine tool dynamics is considered later in the chapter.

In most British control systems, the interpolator has been placed remote from the shop floor and has been designed as a device capable of supplying control tapes for a number of machine tools. In this way each control system on the shop floor is simpler and theoretically cheaper than the equivalent American system. However, as there is no interpolator on the control system, it requires much more input information, which is more conveniently handled by the use of magnetic tape. This magnetic tape normally carries 'time-based' information usually of a phase-modulated analogue form. A typical example would be the Ferranti continuous-path system.

Irrespective of whether the control system is designed for digital punched tape or analogue magnetic tape, the heart of the milling machine post processor is the section satisfying the requirements of the dynamics of the machine tool and control system, and this can be designed with similar input parameters and restraints. Basically these restraints are that, in moving the tool from one block of information to the next, the step changes in velocity for each axis of the machine tool should not exceed some acceptable figure.

The main difference between the two systems is in scale. The paper-tape system typically will allow a move in one block of up to 10 in. and a feedrate change to the next block of 20 in./min. The length of move contained in a block in the 'time-based' system will depend on the feedrate, but would typically be the distance moved in 50 ms. The change in velocity of the machine slides between one block and the next would be limited to that resulting from the maximum allowable change in distance moved between adjacent 50 ms blocks, typically 0·0004 in.

A drawback in designing APT post processors for time-based systems is that the linear-cut vectors output by the APT processor vary in length depending on surface curvature, tolerance conditions, etc., while the blocks required by the control system are linear moves to be covered in constant time intervals. This dictates that the post processor split up each APT cut vector into some whole number of control system blocks dependent on the required velocity, while at the same time restricting the resultant changes in velocity to some acceptable figure for the machine tool.

To overcome this drawback, when time-based post processors are specified, the NELNC processor has been designed to generate the canonical forms of the cutter path for each type of drive surface used in the part program and write these on to the CLDATA file. At the same time, it does not calculate linear-cut vector approximations to these drive surfaces. In this way, the processing time in 2C,L is shortened, while the post processor is supplied with information from which it can more easily interpolate the necessary time-based increments. This type of information is also suited to control systems which have circular interpolators.

Most post processors for milling machines also provide control of spindle speeds, coolant setting, block sequence number display, tool changing, and occasionally cutter radius compensation.

The multi-axis machine tool imposes additional problems on the post processor. This machine has usually a combination of the characteristics of milling and drilling machines, but has the additional facility of angling the tool axis relative to the workpiece. The difficulties caused by this are more fully dealt with later in the chapter.

The CLDATA interface

The CLDATA file forms the interface between the general-purpose NC processors and their post processors. On this file is held the information concerning the computed tool path and the auxiliary instructions for each part program: information which will be used by the post processor to produce a control tape. The file can be considered to exist on two levels: firstly, as a succession of logical records of differing word lengths each containing discrete information concerning one item, and secondly, as a grouping of these logical records in physical blocks on the magnetic tape, disc, or drum used on the particular computer implementation.

As can be readily appreciated, the layout of information on these logical records is of vital importance to the post processor. The APT 3 format is virtually a defacto standard for the CLDATA of general-purpose processors; for example, the CLDATA of the 2C,L program is fully compatible with APT 3. For this reason, it is worth considering the APT 3 format in detail.

APT 3 logical record format

There are nine classifications of information differentiated by codes from 1000 to 32000

Classification	Code
Original part program statement number	1000
Post-processor auxiliary commands	2000
Canonical form data	3000
Cutter location data	5000
Post-processor flags	6000
Axis mode	9000
Internal error flags	13000
Fini code	14000
User proprietary records	28000–32000

The ordering of words within each of these logical records is similar, but the contents vary. Each logical record is a separate block of data and would normally be called for by the post processor intact by a single FORTRAN call.

The first word of each record (W1) is the record sequence number. The second word (W2) is the classification code, as above. The third word (W3) and all subsequent words (Wn) up to a maximum of 245 words are based on the particular classification code of the record.

1000 type records   This record carries the card sequence numbers of the original part-programming statements.

W1	(integer)	= record sequence number
W2	(integer)	= 1000
W3	(BCD)	= identifier label (if used in looping)
W4	(BCD)	= columns 73–78 of input card
W5	(BCD)	= columns 79–80 of input card.

2000 type records   These records carry post-processor instructions.

W1	(integer)	= record sequence number
W2	(integer)	= 2000 (record type)
W3	(integer)	= n (record sub-type)

W4 onwards contain information dependent on the record sub-type 'n'.

There are several hundred possible post-processor instructions. Because of space limitations only a few will be listed here.

Post-processor instruction—END
W3 (integer) = 1 (sub-type code for END)

Post-processor instruction—STOP
W3 (integer) = 2 (sub-type code for STOP)

Post-processor instruction—DELAY
W3 (integer) = 1010 (sub-type code for DELAY)
W4 (floating point) = t (time in seconds)

Post-processor instruction—COOLNT
W3 (integer) = 1030 (sub-type code for COOLNT)
W4 (integer) = 71 (code for ON)
or = 72 (code for OFF)
or = 89 (code for FLOOD)
or = 90 (code for MIST)

Post-processor instruction—TRANS
W3 (integer) = 1037 (sub-type code for TRANS)
W4 (floating point) = x (value of x dimension)
W5 (floating point) = y (value of y dimension)
W6 (floating point) = z (value of z dimension)

Post-processor instruction—PPRINT
W3 (integer) = 1044 (sub-type code for PPRINT)
W4 (BCD) = first of n 6-character words $(1 \leqslant n \leqslant 11)$
W5 (BCD) = second of n 6-character words, etc.

3000 type records This record carries the canonical form of a circle, when that circle is used as a drive surface parallel to X–Y plane.

W1 (integer) = record sequence number
W2 (integer) = 3000 (record type)
W3 (integer) = 2 (drive surface)
W4 (integer) = 1 for TO, 2 for PAST, 4 for TANTO (check surface type)
W5 (integer) = 4 (circle)
W6 (integer) = 9 (number of words in canonical form)
W7 (BCD) = symbolic name of circle
W8 (integer) = subscript (if used)
W9 (floating point) = X value of centre of circle
W10 (floating point) = Y value of centre of circle
W11 (floating point) = Z value of centre of circle
W12 (floating point) = I component of circle axis vector
W13 (floating point) = J component of circle axis vector
W14 (floating point) = K component of circle axis vector
W15 (floating point) = R value of circle radius.

5000 type records This record carries the co-ordinates of the ends of the cut vectors generated by the processor. These positions represent the positions of the centre of the end of the tool.

W1	(integer)	= record sequence number
W2	(integer)	= 5000 (record type)
W3	(integer)	= 3 for FROM, 4 for GODLTA, 5 for GOTO and 6 for a continued GOTO (i.e., if there are more than 80 cut vectors in one drive surface)
W4	(BCD)	= BCD name of drive surface
W5	(integer)	= subscript (if any)
W6	(floating point)	= X co-ordinate of first cut vector ⎫
W7	(floating point)	= Y co-ordinate of first cut vector ⎬ triplet
W8	(floating point)	= Z co-ordinate of first cut vector ⎭
W9	(floating point)	= X co-ordinate of second cut vector ⎫
W10	(floating point)	= Y co-ordinate of second cut vector ⎬ triplet
W11	(floating point)	= Z co-ordinate of second cut vector ⎭

The sets XYZ co-ordinates will be continued until the cut sequence either is complete or the maximum size of the record is reached—245 words (80 sets of triplets).

Where a multi-axis tool has been specified (by the use of MULTAX), the APT processor will output IJK direction cosines of the tool axis with each set of XYZ co-ordinates. In this case, the maximum size of the record will be reached with 40 sets of co-ordinates.

6000 type records Tolerance and cutter records.

W1	(integer)	= record sequence number
W2	(integer)	= 6000 (record type)
W3	(integer)	= 1 for CUT-DNTCUT records, then

                          W4 (integer) = for DNTCUT
                                     = 0 for CUT

        or = 4 for INTOL
        or = 5 for OUTOL, then

                          W4 (floating point) = tolerance for part surface
                          W5 (floating point) = tolerance for drive surface
                          W6 (floating point) = tolerance for check surface

        or = 6 for CUTTER, then

                          W4 (floating point),
                          to
                          W10 (floating point) = seven parameters of cutter

9000 type records   Axis mode record.

W1	(integer)	= record sequence number
W2	(integer)	= 9000 (record type)
W3	(integer)	= 2 (MULTAX flag).

13000 type records   Error flag record.

W1	(integer)	= record sequence number
W2	(integer)	= 13000 (record type)
W3	(floating point)	= 1 through n (diagnostics 1 through n)

14000 type records   Termination record.

W1	(integer)	= record sequence number
W2	(integer)	= 14000 (record type).

NELNC logical record format

The NELNC processor is an expanded subset of APT 3. In other words, the NELNC and APT 3 processors are fully compatible wherever possible but NEL NC contains features not present in APT, just as APT contains features not present in NELNC. Because of these differences, the NELNC CLDATA format does not contain the 9000 and 13000 type records listed above, but has the additional record type 15000 described below. In all other records, the NELNC CLDATA is fully compatible with the APT 3 CLDATA.

The information that an interpolator requires for a circular drive surface is the centre of the circle, the radius of the offset circle, the end point of the circular arc, and the direction and amount of rotation around the circle. This can be conveniently handled by a single record type; in NELNC, the 15000 type.

W1	(integer)	= record sequence
W2	(integer)	= 15000
W3	(integer)	= 5 for offset drive surface
W4	(integer)	= 3 for ON
W5	(integer)	= 4 for circle
W6	(integer)	= number of words in canonical form
W7	(BCD)	= symbolic name of drive surface
W8	(integer)	= subscript
W9	(floating point)	= X value of centre of circle
W10	(floating point)	= Y value of centre of circle
W11	(floating point)	= Z value of centre of circle
W12	(floating point)	= I component of circle axis vector
W13	(floating point)	= J component of circle axis vector

W14 (floating point) = K component of circle axis vector
W15 (floating point) = R radius of offset circle
W16 (floating point) = $\pm$ Θ signed angle in degrees for arc of circle
W17 (floating point) = X value of end point
W18 (floating point) = Y value of end point
W19 (floating point) = Z value of end point

The record type 15000 also fulfils the requirements of other higher order drive surfaces. For example, each span of a cubic curve 'parallel' to and offset by the cutter radius from a TABCYL, can be defined as follows;

W1	(integer)	= record sequence
W2	(integer)	= 15000
W3	(integer)	= 5 for offset drive surface
W4	(integer)	= 3 for ON
W5	(integer)	= 50 for TABCYL
W6	(integer)	= number of words in canonical form
W7	(BCD)	= drive surface
W8	(integer)	= subscript
W9	(integer)	= span type indicator = 1 for beginning extension span;
		= 2 for intermediate;
		= 3 for final span.

W10 (floating point) = X co-ordinate for point at beginning of span
W11 (floating point) = Y co-ordinate for point at beginning of span
W12 (floating point) = Z co-ordinate for point at beginning of span
W13 (floating point) = co-efficient of U^3
W14 (floating point) = co-efficient of U^2
W15 (floating point) = length of interval
W16 (floating point) = maximum value of cubic
W17 (floating point) = minimum value of cubic
W18 (floating point) = X co-ordinate for end point
W19 (floating point) = Y co-ordinate for end point
W20 (floating point) = Z co-ordinate for end point

Circle and TABCYL information in NELNC processor

The NELNC processor has been designed to produce CLDATA efficiently for both dimension-based control systems and Ferranti-type time-based systems. In the first case, linear-cut vectors are required for both straight lines and circles. In the second case, the post processor is normally designed to interpolate time-based increments, frequently from the mathematical equations of the tool path geometry. Therefore, the generation of linear-cut vectors for circles and TABCYLS by the processor is usually superfluous, and is suppressed if the MACHIN statement indicates that it is not required.

Three courses of action can be taken by the processor at the CLDATA stage.

(a) Where APT type post-processors only are called in a part program, an output identical to APT output will be produced.

(b) Where time-based type post-processors only are called in a part program, the information for a circle or TABCYL will be given in the appropriate 15000 type record or records. No 5000 type linear-cut vectors are calculated or recorded on the CLDATA.

(c) Where APT and time-based type post processors are called in one program, then for circles the output will be given in a 15000 type record followed by a 3000 type record, and this will be followed by a 5000 type record. For TABCYLS the output will be a succession of 15000 type records followed by a 5000 type record.

EXAPT logical record format

The general structure of the CLDATA is the same for both EXAPT 1 and EXAPT 2. The EXAPT CLDATA format is based on APT 3 but differs from it in several areas.

Differences can be divided up into four categories:

structure;
coding;
levels of information;
new EXAPT records.

The EXAPT Association have stated that they will seek to remove the first two differences by 1970.

Differences in structure Each word of information in the EXAPT CLDATA is preceded by a word (integer) which classifies its type. The significance of the identification numbers is as follows:

Integer	Type classification
1	Text (alphanumeric)
2	Integer
3	Real

This means that each EXAPT record will contain twice the number of words contained in the equivalent APT 3 record.

Differences in coding Record classification codes in EXAPT run from 1 to 30 corresponding with APT 3 codes of 1000 to 30000.

At the time of writing, there are small divergences from APT in the integer codes used by EXAPT. This is an initial position which one hopes will be rectified soon.

Differences in levels of information Where there is the possibility of using the same post processors for both the APT and EXAPT processors within an organization, this will be where divergence from APT raises most problems. Differences in structure and coding can be easily handled by writing a small conversion routine for the input element to convert from the EXAPT to the standard APT format. Differences in the level of information are not so simply tackled and the logic of the program may require extensive modification. For example, in APT the co-ordinate information contained in 5000 type records refers to the tool end position. In EXAPT, the equivalent information refers to some control point on the tool mounting. For EXAPT 1, this control point differs from the tool end position by the length of the tool, while, for EXAPT 2, it differs from the tool end by an amount in both the X and Y directions, depending on the geometry of the lathe tool.

There is a basic difference in the manner in which tool and tool-loading information are handled in APT and EXAPT. In APT, each time a new tool is requested, the turret position is given explicitly by the part programmer and thus is part of the appropriate CLDATA record. With EXAPT, if the full automatic tool selection capability is used, each TOOLNO record will simply refer to the list of tools which the processor has selected for that part program and which appear at the beginning of the CLDATA, and it will be the responsibility of the post processor to specify which turret position should be used. However, by sacrificing the automatic tool-selection facility, turret positions can be specified by the part programmer enabling a simpler post processor to be used.

New EXAPT records Again, at the time of writing, there are numerous small additions to the APT 3 format. Of these, the most important are probably the TOOLST record, listing at the beginning of the CLTAPE all of the tools to be used in that run, and secondly, the records listing the contours of the raw material shape and the finished part.

Physical record format

Although, for each NC processor, the CLDATA logical record format is independent of different computer implementations, this is not generally so of the physical records (i.e., the groupings of these logical records in physical records on the magnetic tape, disc, or drum used to store the CLDATA). For example, in order to keep the CLDATA of the Univac 1108 version of the NELNC processor compatible with the Univac version of APT 3 both as regards logical and physical records, the physical record format of the CLDATA of the NELNC processor on the Univac 1108 was designed as shown in Fig. 11.3. This compatibility on the Univac 1108 was important, as it allowed interchange of CLDATA and standardization of tape reading and writing routines between APT and NELNC. However, as there was no equivalent version of APT 3 on the ICL 1900 series computer, the ICL version of NELNC simply uses

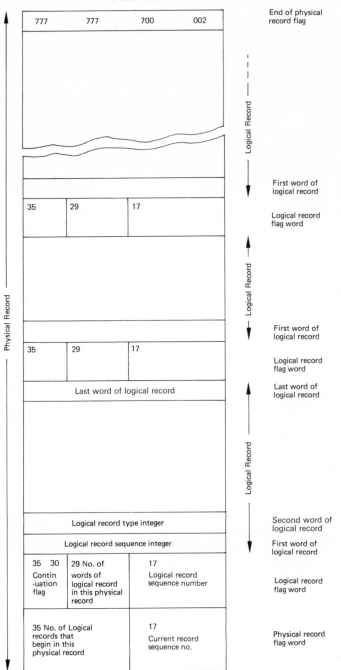

Figure 11.3 APT 3 and NEL/NC, physical record structure on the UNIVAC 1108 implementation.

FORTRAN read and write statements for its CLDATA, resulting in each logical record occupying one physical record. There is therefore no compatibility of physical CLDATA files between the various computer implementations of NELNC, and similar situations occur with the various computer implementations of the APT or EXAPT processors.

The fact that the physical CLDATA file can vary between implementations need not concern the post-processor writer, provided that the appropriate tape-reading routines are available to the post processor to match the method used to write the CLDATA, and provided that these different tape-reading routines use the same calling sequence from the input element of the post processor.

Two features therefore are important, if post processors are to be interchanged between processors with the minimum of alteration:

(1) standard CLDATA output from each processor;
(2) standard calling sequence for post processor utility routines.

Utility routines

Processors normally provide utility routines for the use of post processors to perform the following functions:

initialize and read the CLDATA;
punch operator-readable part identification on paper control tape;
convert information to binary-coded decimal characters for paper-tape output;
punch BCD characters on paper tape in ISO or EIA code set.

Proposals to have the following calls adopted as a standard are being discussed internationally.

TAPEOP

TAPEOP provides file-handling operations. Calling sequence

$$\text{CALL TAPEOP(FILE, CODE, ERR)}$$

where   FILE   is a label (normally CLTAPE)
        CODE   1, rewind
                  2, write end-of-file
                  3, backspace one record
        ERR   *negative*, normal return; zero, end-of-tape; *positive*, the file is not open.

BUFFTP

BUFFTP assigns files in read or write mode. Calling sequence

$$\text{CALL BUFFTP(FILE, CODE)}$$

where FILE   is a label (normally CLTAPE)
       CODE   1, establish write mode
                2, establish read mode
                3, release and rewind file.

TAPERD

TAPERD reads a single logical record from a previously written file. Calling sequence

CALL TAPERD(FILE, ERR, NOWRDS, 1, BLOCK, 0)

where FILE   is a label (normally CLTAPE)
       ERR   *negative*, normal return
                zero, end of file
                positive, error
       NOWRDS location to contain number of words in record
       BLOCK array to store logical record.

PARNOM

PARNOM is used to punch operator-readable identifier (usually PARTNO) on beginning of machine control tape. Calling sequence

CALL PARNOM(NX, A)

where NX   is the number of characters in array A
       A   beginning location of array A which contains the identifier.

OUTSET

OUTSET is used to initiate certain variables used by PUNCHA, PUNCHB, PUNCHC and PARNOM. Calling sequence

CALL OUTSET(ITYPE, IROW, IDS, NOM)

where ITYPE   type of output required
                  −2, no punched output
                  −1, punched cards, EIA tape image
                  zero, punched cards, Hollerith characters
                  +1, punched paper tape, EIA codeset
                  +2, punched paper tape, ISO codeset
       IROW   for card output, card row equivalent to tape channel 1
       IDS   BCD identification for cards, columns 73–77
       NOM   sequence number for cards, columns 78–80.

CONBCD

CONBCD converts a floating point number to its equivalent BCD format. Calling sequence

CALL CONBCD(A, B, IX, JX, KK, ERR)

where	A	floating point number to be converted
	B	first word of two-word array in which to store converted result
	IX	the number of characters to be left in B. If IX is negative, the sign character will be the first character in array B. IX will be increased by 1. Maximum length of IX is 10 if sign and decimal point are specified
	JX	the number of fractional digits to appear in the converted word. If JX is negative the decimal point will be included and IX be increased by 1. Maximum JX is 8
	KK	0, no zeros in B will be suppressed; *negative*, all leading zeros will be suppressed; $0 < n < 99$, all trailing zeros of the fractional part will be replaced with blanks $n = 99$, leading zeros will be suppressed and trailing zeros blanked
	ERR	error return flag.

PUNCHA

PUNCHA is used to provide EIA coded, punched paper-tape output. Calling sequence

CALL PUNCHA(NX, A, JS, KS, ERR)

where	NX	the number of BCD characters in the array A to be converted and punched
	A	beginning location of array A
	JS	0, normal, punch the contents of A 1, final, clear buffers
	KS	0, blank characters in array A will be ignored ± 1, blank characters will generate space codes
	ERR	error return flag.

PUNCHC

PUNCHC is used to provide ISO coded, punched paper-tape output. Calling sequence

CALL PUNCHC(NX, A, JS, KS, ERR)

where arguments are the same as in PUNCHA.

Standardization of the physical CLDATA records would also be necessary if it were desired to produce a CLTAPE by one processor on one computer and then to post process that CLTAPE on another computer.

Modularity

Concept

In order to simplify their design, large computer programs are built up of smaller logical sections. Post processors come into this category and they normally comprise five main sections. These sections are usually further divided into smaller units termed subroutines; in general each subroutine is designed to cope with one unique function.

The work of any post processor can be logically divided up as follows:

reading and sorting of the records on the CLDATA file;

developing preparatory and miscellaneous function commands;

computing suitable slide moves taking into account the dynamics of the machine tool;

producing printed listing and control tape;

monitoring control of the post processor during execution.

These five sections or elements are shown diagrammatically in Fig. 11.4 and will now be discussed in more detail.

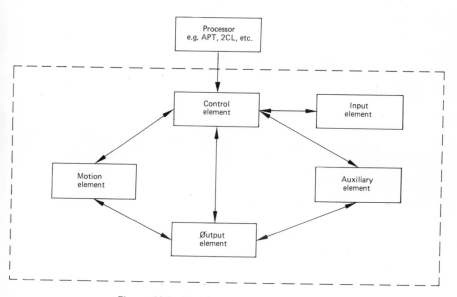

Figure 11.4 Modular structure of post processor.

Input element

In general, the input element of a post processor is designed to read logical records from the CLDATA, ensure that the records have been read successfully, and store the contents in some prescribed area for subsequent processing.

To do this, an APT or NELNC post processor makes use of special utility routines available as part of the standard processor system. The use of these utilities avoids the need to write assembler language routines within the post processor and makes the input element extremely simple.

In order to bring EXAPT into line with the other systems in this respect, utility routines will be provided to produce exactly the APT 3 CLDATA logical records and provision is being made for the post processor to use the necessary utility routines with the correct names and calls.

Auxiliary element

It is the function of this element to process all non-motion or machine environment instructions.

These instructions could include commands to the post processor, (e.g., to produce ISO or EIA coded punched tape); commands to the machine tool, (e.g., to turn on coolant or set spindle speed); and commands concerning both post processor and machine tool, (e.g., feedrates and tolerance conditions).

The auxiliary element either sets up parameters to be used by the other elements or selects preparatory or miscellaneous function codes to be punched out by the output element.

The following examples show the type of part programming statements handled by the auxiliary element and give some impression of its scope.

$$\text{COOLNT/}\begin{matrix}\text{FLOOD}\\\text{MIST}\\\text{OFF}\end{matrix}$$

This statement in the part program will specify the coolant condition required. The auxiliary element will select the appropriate function code to be placed on the control tape.

$$\text{SPINDL/f, }\begin{matrix}\text{RPM, CLW}\\\text{SFM, CCLW}\end{matrix}$$

The part programmer is frequently given the option of requesting cutting speed as either a spindle speed in revolutions per minute (f, RPM) or as a value in surface feet per minute (f, SFM). The auxiliary element will select the nearest machine spindle speed if the RPM modifier is quoted whereas with the SFM modifier it will first automatically compute a spindle speed depending on the surface feet per minute specified and the workpiece radius, and then select the nearest machine spindle speed.

LOADTL/tool number, LENGTH, tool length

This instruction will specify the tool to be loaded into the spindle of a tool-changing machine, at the same time unloading the current tool and replacing it in the tool store. The tool length will be made available to the motion element by the auxiliary element to compensate for differing tool lengths, the part-programmer only being concerned with the position of the tool end.

Motion element

The main responsibilities of the motion element fall into two categories—geometry transformations and consideration of the dynamics of the machine tool.

Translation of geometry is necessary since the information on the CLDATA is, firstly, in workpiece rather than machine tool co-ordinates, and secondly, concerns the position of the tool point and tool orientation rather than the required moves of the machine slide.

For normal three-axis machine tools with mutually perpendicular slides, the incremental movements of the tool end derived from the absolute CLDATA information would also be incremental slide moves. In consequence, it is a simple problem with these machines to translate from the CLDATA absolute position information to the absolute or incremental movements of the slides. However, when the machine tool has a further one or more rotational movements in the form of rotary tables, tilting heads, etc., the problem is more complex and the translation from the X Y Z I J K sets of CLDATA information (I J K being direction cosines of the tool axis orientation) to the slide movements is no longer trivial. This problem will be dealt with in more detail later in the chapter.

The motion element is also responsible for taking into account the dynamics of the machine tool slides. On a positional drilling machine, the motion element would be relatively simple, the accelerations and decelerations of the tool being dealt with by the machine control system.

With a contouring machine tool, this element could be more complex, often requiring both the calculations of safe speeds to limit overshoot at discontinuities, and permissible changes in speed while starting and stopping cutter movements.

For a contouring system, the motion element would be the most complex of the five elements which form the post processor. It is in the design of this element that the ultimate efficiency of machining times and accuracy of components depend.

Output element

The output element of a post processor collates the information prepared by the motion and auxiliary elements and outputs it in the specific coded form required for input to the control system of the machine tool.

In addition, the output element generates a printed listing of the information contained on the control tape together with estimates of machining times, tape lengths, and any comments or diagnostics which may have been developed by the other elements.

Control element

The control element is in effect the executive or management element of the post processor. It is responsible for making decisions about the work to be done and for delegating that work to the other elements of the post processor.

Normally, the control element accepts control from the processor program, initiates the post processor, and calls for input of information from the CLDATA. Depending on the type of records, it determines the branching to the appropriate elements for processing and, at the end of the CLDATA information, it finalizes the processing and returns control to the processor.

Typical post-processor operation

Figure 11.5 and the following description indicate how these elements are related. It should be emphasized that this description is only of a typical post processor; the details will vary considerably between post processors for specific machine tools and control systems.

The control element begins by calling on the input element to provide a CLDATA record. Depending on the record type, it then branches to the motion or auxiliary elements or does further processing within the control element.

If the branch is to the motion element, it will develop acceptable motion commands, taking into account co-ordinate translations, interpolation requirements, and system dynamics. These motion commands will form discrete blocks of output information and will ultimately include slide movements, feedrates, preparatory and miscellaneous instructions, etc. As each block is developed, its influence on its neighbouring blocks will be determined. If the influence is not acceptable, suitable correction will be made by reworking previous blocks retrieved from the output buffer. Once satisfactory, these blocks will be stored in the output buffer.

If the call is to the auxiliary element, non-motion records will be processed. These records may set up switches and parameters, or develop miscellaneous or preparatory commands for inclusion within the output blocks prepared by the motion element.

As the output blocks are prepared by the motion and auxiliary elements, they are passed on to the output element and stored in a buffer. This buffer allows blocks to be recalled to the motion element for rework. From the contents of this buffer the output element will create a printed listing and prepare a control tape in the correct code and format.

When processing of each logical record is complete, control will always be returned from the element concerned to the control element.

A further impression of the work done by a post processor can be gauged by comparing the CLPRINT shown in Fig. 11.6 with the printed listing produced

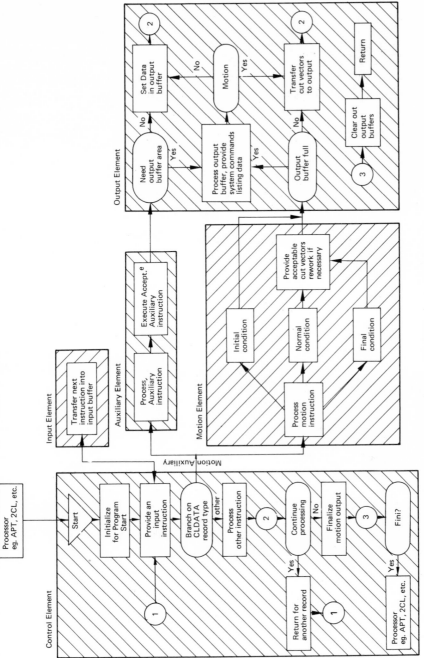

Figure 11.5   Typical post-processor system.

```
ENU COVER MOULD                                                      RECORD    2

MACHIN/CLEV01  1./  LINEAR                                           RECORD    4
        1.0u0000u     3.0u00n00     8.0u0000u    1.0u00n00    9.0000000
        1.0000000

UUTTOL/ .0050000 .0050000 .0050000                                  RECORD    6

INTUL/ .0u00000 .0000000 .0000000                                   RECORD    7

CUTTER/   .2640                                                      RECORD    9

FEDRAT/  60.0000                                                     RECORD   11

FROM /  PUFF                                                         RECORD   13
                      X             Y             Z
                  .0000000    -.5000000     1.5000000

SPINUL/ 1200.0000   RPR    CLW                                       RECORD   15

FEDRAT/  60.0000                                                     RECORD   17

DS IS/  C11                                                          RECORD   18
                      X             Y             Z
                  .9031459    -.0387993     .0000000

C11( U) = CIRCLE/  .0000    .0000    .0000   1.3750                  RECORD   20

DS IS/  C11                                                          RECORD   21
                      X             Y             Z
              1.0366364    -.0675881    -.0000000
              1.0322485    -.4881621    -.0000000
              1.1970758    -.2954634    -.0000000
              1.2293555    -.0947313    -.0000000
              1.2282102     .1085765    -.0000000
              1.1936708     .3089322    -.0000000
              1.1266766     .5008883    -.0000000
              1.029u490     .6792256    -.0000000
               .9034422     .6390955    -.0000000

DS IS/  L7                                                           RECORD   23
                      X             Y             Z
              1.2847534     1.2204067    -.0000000

C12( U) = CIRCLE/  .0000    .000    .0000   1.6250                   RECORD   25

DS IS/  L12                                                          RECORD   26
                      X             Y             Z
              1.4529860     1.0143056    -.0000000
              1.5778293     .8064978    -.0000000
              1.0731410     .5835951    -.0000000
              1.7371371     .3497695    -.0000000
              1.7686198     .1093973    -.0000000
              1.7670000    -.1330223    -.0000000
              1.7323079    -.3729523    -.0000000
              1.6651928    -.6059019    -.0000000
              1.5609100    -.8275110    -.000000u
              1.4393010    -1.0356318    -.0000000
              1.2847534    -1.2204068    -.0000000

DS IS/  L5                                                           RECORD   28
                      X             Y             Z
              1.3455837    -1.2812371    -.0000000

C13( U) = CIRCLE/  .0000    .0000    .0000   2.0000                  RECORD   30

DS IS/  L13                                                          RECORD   31
                      X             Y             Z
              1.5184775    -1.0706961    -.0000000
              1.0509077    -.8524481    -.0000000
              1.7521721    -.6181076    -.0000000
              1.0203590    -.3720985    -.0000000
              1.0541811    -.1190649    -.0000000
              1.6530001     .1362164    -.0000000
              1.0168381     .3889263    -.0000000
              1.7463778     .6342939    -.0000000
              1.6429495     .8676874    -.0000000
              1.5685055     1.0847007    -.0000000
              1.3455838     1.2812370    -.0000000

DS IS/  L7                                                           RECORD   33
                      X             Y             Z
              1.7268029     1.6624562    -.0000000

C14( U) = CIRCLE/  .0000    .000    .0000   2.2500                   RECORD   35

DS IS/  C14                                                          RECORD   36
                      X             Y             Z
              1.9266066     1.4261122    -.0000000
              2.0924249     1.1693445    -.0000000
              2.2242198     .8935629    -.0000000
              2.3198482     .6032518    -.0000000
              2.3777551     .3031315    -.0000000
              2.3969990    -.0019177    -.0000000
              2.3772670    -.3069358    -.0000000
              2.3188798    -.6069629    -.0000000
              2.2227670    -.6971207    -.0000000
              2.0945510    -1.1726910    -.0000000
              1.9243220    -1.4291930    -.0000000
              1.7268029    -1.6624562    -.0000000

DS IS/  L5                                                           RECORD   38
                      X             Y             Z
               .9034421    -.6390955    -.0000000

DS IS/TLAXIS                                                         RECORD   40
                      X             Y             Z
               .9034421    -.6390955     1.0000000

DS IS/  PUFF                                                         RECORD   42
                      X             Y             Z
               .6000000    -.5000000     1.5000000

SPINUL/   OFF                                                        RECORD   44

STOP                                                                RECORD   46

END                                                                 RECORD   48

FINI                                                                RECORD   50
```

Figure 11.6 CLPRINT of MACRO MAC 1.

by the Cincinnati Acromatic 4 post processor and shown in Fig. 11.7. The macro MAC1 defined within the part program given in chapter 9, Fig. 9.1, was used to produce the CLPRINT and post-processor output. The feedrate was programmed as 60 in./min.

CIMTROL / APT-III INC., POSTPROCESSOR

FOR CINCINNATI 30 INCH VERTICAL HYDRTEL

CIMTROL ELECTRONICS EUROPE ACRAMATIC IV WITH LINEAR INTERPOL TION

H/I.	ω	P	X	Q	Y	R	Z	F	M	CLTAPE	IP
END COVER MOULD											
LEADER/ 72.0											
H001* G01*	•	•	X+000000*	•	Y-004999*	•	Z+015000n*	F0600*	M055	0 13	00100.0
N002*	•	•	X+008527*	•	Y-008198*	•	Z+000830*	•	M035	0n18	0006n.0
N003*	•	•	X+009031*	•	Y-008387*	•	Z+000000*	F0596*	•	5 0 18	00050.6
N004*	•	•	X+010366*	•	Y-006675*	•	•	F0600*	•	5 0n21	00060.0
N005*	•	•	X+011322*	•	Y-004681*	•	•	•	•	5 0021	0006n.0
N006*	•	•	X+011971*	•	Y-002954*	•	•	•	•	5 0021	0006n.0
N007*	•	•	X+012294*	•	Y-000946*	•	•	•	•	5 0021	0006n.n
N008*	•	•	X+012282*	•	Y+001086*	•	•	•	•	5 0021	0006n.n
N009*	•	•	X+011937*	•	Y+003089*	•	•	•	•	5 0021	0006n.0
N010*	•	•	X+011267*	•	Y+005009*	•	•	•	•	5 0021	0006n.0
N011*	•	•	X+010290*	•	Y-006792*	•	•	•	•	5 0021	0006n.0
N012*	•	•	X+009636*	•	Y+007626*	•	•	•	•	5 0021	0006n.n
N013*	•	•	X+009034*	•	Y+008391*	•	•	F 568*	•	5 0021	0005n.8
N014*	•	•	X+012226*	•	Y+011583*	•	•	F0600*	•	5 0n23	00060.0
N015*	•	•	X+012849*	•	Y+012204*	•	•	F0500*	•	5 0n23	00050.0
N016*	•	•	X+014530*	•	Y+010143*	•	•	F0600*	•	5 0026	0006n.c
N017*	•	•	X+015778*	•	Y+008065*	•	•	•	•	5 0026	0006n.0
N018*	•	•	X+016731*	•	Y+005836*	•	•	•	•	5 0026	0006n.n
N019*	•	•	X+017371*	•	Y+003498*	•	•	•	•	5 0026	0006n.c
N020*	•	•	X+017666*	•	Y+001094*	•	•	•	•	5 0026	0006n.n
N021*	•	•	X+017670*	•	Y-001329*	•	•	•	•	5 0026	0006n.u
N022*	•	•	X+017323*	•	Y-003729*	•	•	•	•	5 0n26	0006n.0
N023*	•	•	X+016652*	•	Y-006058*	•	•	•	•	5 0026	0006n.0
N024*	•	•	X+015669*	•	Y-008274*	•	•	•	•	5 0026	0006n.0
N025*	•	•	X+014393*	•	Y-010335*	•	•	•	•	5 0026	0006n.0
N026*	•	•	X+013460*	•	Y-011463*	•	•	•	•	5 0026	0006n.c
N027*	•	•	X+012848*	•	Y-012203*	•	•	F 552*	•	5 0026	00055.2
N028*	•	•	X+013456*	•	Y-012811*	•	•	F 500*	•	5 0n28	00050.n
N029*	•	•	X+015165*	•	Y-010706*	•	•	F0600*	•	5 0031	0006n.0
N030*	•	•	X+016509*	•	Y-008523*	•	•	•	•	5 0031	0006n.0
N031*	•	•	X+017522*	•	Y-006180*	•	•	•	•	5 0031	0006n.c
N032*	•	•	X+018204*	•	Y-003720*	•	•	•	•	5 0031	0006n.n
N033*	•	•	X+018542*	•	Y-001190*	•	•	•	•	5 0031	0006n.n
N034*	•	•	X+018530*	•	Y+001362*	•	•	•	•	5 0031	0006n.n
N035*	•	•	X+018168*	•	Y+003889*	•	•	•	•	5 0031	0006n.n
N036*	•	•	X+017464*	•	Y+006343*	•	•	•	•	5 0n31	0006n.0
N037*	•	•	X+016429*	•	Y+008677*	•	•	•	•	5 0 31	0006n.0
N038*	•	•	X+015085*	•	Y+010847*	•	•	•	•	5 0n31	0006n.n
N039*	•	•	X+014068*	•	Y+012074*	•	•	•	•	5 0031	0006n.n
N040*	•	•	X+013456*	•	Y+012812*	•	•	F 551*	•	5 0 31	00055.1
N041*	•	•	X+016643*	•	Y+015999*	•	•	F0600*	•	5 0 33	0006n.n
N042*	•	•	X+017268*	•	Y+016625*	•	•	F0500*	•	5 0033	00050.n
N043*	•	•	X+019266*	•	Y+014261*	•	•	F0600*	•	5 0n36	0006n.c
N044*	•	•	X+020904*	•	Y+011693*	•	•	•	•	5 0n36	0006n.n
N045*	•	•	X+022242*	•	Y+008936*	•	•	•	•	5 0n36	0006n.c
N046*	•	•	X+023198*	•	Y+006033*	•	•	•	•	5 0n36	0006n.n
N047*	•	•	X+023778*	•	Y+003031*	•	•	•	•	5 0n36	0006n.n
N048*	•	•	X+023970*	•	Y-000018*	•	•	•	•	5 0036	0006n.n
N049*	•	•	X+023773*	•	Y-003068*	•	•	•	•	5 0n36	0006n.D
N050*	•	•	X+023189*	•	Y-006069*	•	•	•	•	5 0n36	0006n.D
N051*	•	•	X+022228*	•	Y-008970*	•	•	•	•	5 0n36	0006n.n
N052*	•	•	X+020906*	•	Y-011726*	•	•	•	•	5 0036	0006n.c
N053*	•	•	X+019243*	•	Y-014291*	•	•	•	•	5 0n36	0006n.0
N054*	•	•	X+017840*	•	Y-015948*	•	•	•	•	5 0036	0006n.0
N055*	•	•	X+017268*	•	Y-016624*	•	•	F0500*	•	5 0036	00050.n
N056*	•	•	X+009683*	•	Y-009038*	•	•	F0600*	•	5 0038	0006n.n
N057*	•	•	X+009034*	•	Y-008390*	•	•	F0500*	•	5 0038	00050.0
N058*	•	•	•	•	•	•	Z+01000n*	F0600*	•	5 0040	0006n.n
N059*	•	•	X+005589*	•	Y-005220*	•	Z+014674*	•	•	5 0042	0006n.n
N060*	•	•	X+000000*	•	Y-004999*	•	Z+015000n*	F0500*	M055	0042	0006n.n
N061*	•	•	•	•	•	•	•	•	M005	0046	00050.0
N062*	•	•	•	•	•	•	•	•	M025	0048	00050.0
LEADER/ 72.0											
END COVER MOULD											

MACHINING TIME   TAPE LENGTH
   .28 MINUTES   16.96 FEET

Figure 11.7 Post-processor listing of MACRO MAC 1.

Notice, in particular, the segmentation of cut vectors by the post processor (e.g., the last cut vectors of the CLDATA records 021, 023, 026, and so on in Fig. 11.7 have been split into two moves, one at the programmed feedrate and one slower) in order to minimize overshoot of machine slides. See also the estimates of machining times and control tape lengths.

The correspondence between the two figures can be studied by using the CLDATA (or CLTAPE) numbers listed in the last column of the CLPRINT of Fig. 11.6 and the penultimate column of the post processor listing of Fig. 11.7. For example, CLTAPE record 013 appears as FROM/POFF in Fig. 11.6 and leads to the first block in the post-processor listing of Fig. 11.7, and other record numbers can similarly be compared.

Typical problem areas

In the previous section, the general concept of a modular construction for post processors was discussed. This section will now consider some particular post processor problems in more detail.

Machine slide dynamics

NC control systems can be divided into two distinct classes:

(a) those which execute each control tape motion block as a discrete operation; and

(b) those which give true continuous path contouring by interpolating motion information smoothly and continuously from one block to the next.

In addition to positioning and straight-cut systems, many of the cheaper contouring systems can be included under (a). Under (b) can be catalogued those systems which are controlled by continuous digital or analogue information on the control tape, and those systems which are designed to buffer input information (i.e., to read and store the next block as the current block is being executed).

Class (a) control systems present few, if any, dynamics problems to the post processor. Their control-system hardware is generally designed to give both automatic acceleration to feedrate at the beginning of each motion block and deceleration to rest at the end. Further, as each motion block is not influenced by its adjacent blocks, there is no requirement for the post processor to check and allow for linear discontinuities between adjacent cut vectors.

For true continuous-path control systems [class (b)], the post processor has normally a far greater responsibility for the generation of smooth control information. Generally, the post processor must endeavour to ensure that the feedrate programmed by the planner be obtained as quickly as possible and then maintained over the path length of the tool, while at the same time restraining slide velocities and accelerations to within those allowed by the dynamics of the servo-system controlling the particular machine tool. Typically, this type of post processor would deal with the following:

tape reading time;
permissible corner velocities;
tool path velocity changes.

Tape reading time

In order to allow a smooth transition from one block of information to the next, class (b) paper-tape systems employ temporary buffer storage of one block. The post processor must ensure that the time taken to execute each block is sufficient to allow the following block to be read into the buffer storage. In the case of a series of short cut vectors being produced by, say, the APT processor, the velocity of the tool may have to be artificially reduced to increase the cutting time and thereby allow each succeeding block to be read in completely, before interpolation of each current block terminates.

Corner velocities

The servo-systems used in continuous-path contouring have widely differing characteristics. One of these characteristics, gain, and the resulting velocity-lag error, has particular relevance to the post processor. (The velocity lag error is a lag between the commanded and the actual tool position; this depends on the velocity of the tool.)

Systems range from those having a high velocity gain (typically 20 to 100 in./min for a lag of 0·001 in.) resulting in small velocity lag errors, to those with low gain (typically $\frac{1}{2}$ to 5 in./min for a lag of 0·001 in.) with resultant large velocity lags.

These velocity-lag errors are important to the post processor since they result in overshoot or undershoot of the tool at changes in tool velocity or direction (Fig. 11.8).

Note that the velocity-lag errors do not cause significant errors in the component machined provided that straight lines or large radius circles are being cut and that the responses of each servo on the machine tool are equal, and provided that the errors are proportional to the velocity. This is illustrated on Fig. 11.9, where the dotted circle represents the commanded position for the tool and the solid circle represents the actual tool position. The velocities of the X and Y slides will be $V \cos \alpha$ and $V \sin \alpha$ respectively.

Since the errors E_X and E_Y are proportional to $V \cos \alpha$ and $V \sin \alpha$ respectively, they are proportional to $\cos \alpha$ and $\sin \alpha$ and it follows that the tool, though lagging, will be exactly on the tool path.

Unfortunately, real engineering parts do not often consist of one long straight line or a large radius circular arc and thus when these special conditions do not apply (i.e., corners, etc.), the slide velocities may have to be reduced to keep the resultant overshoot or undershoot errors within acceptable limits. The acceptable limit can be specified by the part programmer in APT-like languages by the statement MCHTOL/t where t is the limit; or built into the post processor as a constant.

As the overshoot or undershoot error is dependent on the velocity change, the post processor can keep the errors within the acceptable limit by adjusting

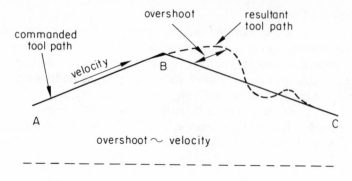

overshoot ∼ velocity

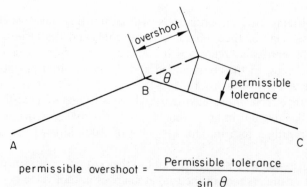

$$\text{permissible overshoot} = \frac{\text{Permissible tolerance}}{\sin \theta}$$

allowable feedrate at B is function of permissible overshoot

Figure 11.8 Overshoot at corners.

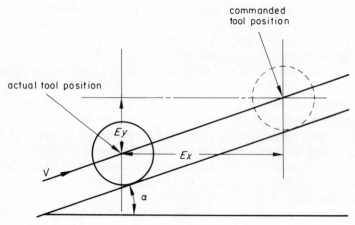

Figure 11.9 Dynamic lag.

the change in velocity. By using the specified tolerance and the angle of discontinuity, the post processor can compute, firstly, the allowable overshoot at each corner, and, secondly, from a table of empirical data relating errors to feedrates, determine the maximum feedrate permissible through the corner.

Tool path velocity changes

For most continuous-path control systems, each cutting operation is represented, within a control tape block, by the distance to be moved by each axis of the machine, and by a feedrate number. The feedrate number usually represents either the time in minutes for the execution of the block or the reciprocal of that time (inverse time feedrate code). The time is used subsequently by the interpolator, within the control system, to control the rate at which pulses are fed to the servos. The widely used Ferranti interpolator (mentioned earlier in the chapter) also complies with this description, its main difference being that, as the time for the execution of each block is constant, the time does not have to be specified. Within the execution of each block, therefore, both the post processor and the control system interpolator deal only in constant velocity. Acceleration or deceleration is handled by stepping from one velocity to another in separate blocks. As each such instantaneous change in velocity is commanded, the servo system must accelerate or decelerate the machine slides to the new velocity. In effecting this change, a velocity-lag error will be induced in the system dependent on the mass of the slides, the gain and frequency response of the servo system, and so on. In order to avoid exceeding the limits of the system, this velocity-lag error must be constrained and this is achieved within the post processor by restricting the change in velocity between any two blocks to some maximum [typically 10 in./min per block (approx 0·2 s) on a low-gain paper-tape system and 0·0004 in. per 50 ms per block (0·05 s) on the Ferranti system].

Typical problems met by the post processor in generating the step velocities are as follows: the cut vectors generated by the processor may be too short to achieve the requested feedrate within one cut vector; corner conditions and cut vector lengths may be such that the requested feedrate cannot be achieved; a cut vector may be too short to achieve a requested reduction in feedrate, say at a sharp corner or at a STOP condition.

Figure 11.10 illustrates these problems and their possible solution for a low-gain paper-tape system with a permissible 10 in./min per block velocity change. The base line A to H represents distance along the cutter paths as output by the processor—A to B, B to C, C to D, and D to E being cut vectors at various angles to each other and the programmed feedrate is 60 in./min; there is a programmed STOP command at E; E to F, F to G, and G to H are again cut vectors at 60 in./min and there is another STOP programmed at H. The angles at B, C, and D are such that corner feedrates of up to 80 in./min would be acceptable to the machine tool. Those at F and G are such that corner feedrates of only up to 30 in./min would be acceptable.

Starting at A, the post processor attempts to reach the lesser of the programmed feedrate and the maximum corner rate for B. As the desired feedrate is greater than the maximum allowable step of 10 in./min per block, the cut vector has to be divided into a number of steps; each step is selected from a previously compiled Table and is of sufficient time to allow the servos to settle and the next block to be read in. In this particular case, by the time the cut vector AB is transversed, the feedrate is up to 50 in./min. The programmed

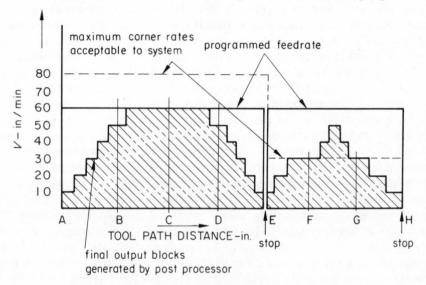

Figure 11.10   Stepped velocities generated by post processor.

rate of 60 in./min is reached in the next cut vector and can be then held over BC and CD. The stop condition at E will not be diagnosed by the post processor until the cut vector D to E is considered. Working back from E the post processor cannot reduce to zero the feedrate of 60 in./min previously computed for the CD vector within the length of the DE vector. To cover this problem, the post processor is designed to store the segmented output blocks in a temporary buffer as it generates them, and these are then available for recomputing or 'rework'. In this example, the block for C to D would be recalled and segmented into two blocks with feedrates of 60 and 50 in./min, while DE would be segmented into five blocks. As the corner feedrate at F is less than the programmed feedrate, the vector E to F is segmented from the STOP condition at E to the corner feedrate at F. The post processor then endeavours to reach the programmed feedrate during F to G by simultaneously segmenting from the corner rates at F and G, but has to be satisfied with 50 in./min at the centre of FG as the vector length is used up. Finally the GH vector is segmented to achieve the STOP condition.

It will be seen from this example that the final velocity of the slides during machining depends on many factors (cut vector length, angle of discontinuity, gain and frequency response of the system, acceptable machine finish, etc.) and can often be significantly different from the feedrate selected by the part programmer. Further, with so many variables to allow for, a carelessly written post processor can unnecessarily increase machining times.

Control tape generation for differing codesets

Unfortunately, at the time of writing, there is no 100 per cent use of one standard codeset for punched paper tape. Many users, having purchased NC machines

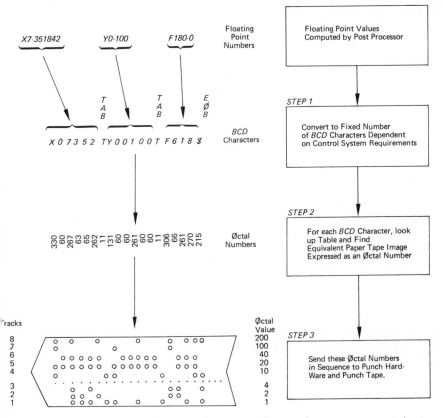

Figure 11.11 Sequence of operations to punch paper tape. The example presupposes a control system requiring ISO codeset, word address tab sequential format with five digits in X and Y fields, reading to 0·001 in., decimal point suppressed; no suppression of leading or trailing zeros. Note that the paper tape image is expressed as an octal number by counting 1 for the hole in the first track, 2 in the second, 4 in the third, 10 in the fourth, 20 in the fifth, 40 in the sixth, 100 in the seventh, 200 in the eighth and then adding the counts.

over a number of years, have a variety of control systems within their organizations which require not only differing codesets but paper tape of different widths. Where manual programming is used, this presents a problem in the variety of tape-punching equipment required. Fortunately, with computer programming, post-processors can be organized to produce paper tape punched in virtually any code without requiring additional hardware, although it may prove difficult to list or duplicate tapes if suitable tape handling equipment is not available for each type.

In addition to the consideration of codesets, post processors are usually organized to arrange the information to be contained within each output block into the layout or 'format' required by the particular control system. This may entail representing dimensions by a fixed number of digits, inserting or excluding leading or trailing zeros, etc. The current standards for control tapes have been discussed in chapter 6.

Because the computer hardware is involved (tape punch, etc.) this area of a post processor tends to be computer-dependent. However, the basic steps involved in producing punched paper tape in the correct code and format are similar for many computers (Fig. 11.11).

By suitable coding, step 1 can be organized to give the variable-length fields required by different control systems. Differing codesets can be simply catered for by inserting in step 2 appropriate tables of octal numbers against BCD characters for each codeset.

Constant cutting speed for lathes

In a lathe the workpieces are rotated while the tools, normally non-rotating, are limited to movement in the axial and radial planes of the machine. This machine tool geometry causes the cutting speed of the tool relative to the workpiece to be a function of the spindle speed and the radius of the workpiece at the tool point.

The surface finish is a function of the spindle speed and the axial or radial feedrate of the tool.

Current control systems for lathes require the spindle speed and tool feedrates to be input as rev/min (rpm) and in./min (ipm) respectively. If the diameter of the workpiece is changing, as when facing, constant rpm and ipm will give varying chip loads and metal-removal rates, and will produce poor surface finish. Better cutting conditions can then be obtained by dividing up the facing cut and changing the spindle speed and feedrate for each segment. This can be done by the part programmer, but obviously involves him in work which is time-consuming and error-prone. Most lathe post processors therefore allow the part programmer to specify cutting-tool velocity in surface feet per minute (sfm) and tool feed in inches per revolution (ipr). The post processor is then responsible for automatically segmenting the tool path and computing the spindle speeds and consequent feedrates.

The APT statements for part-programming cutting conditions would typically be as follows:

$$\text{SPINDL/s, } \begin{matrix} \text{SFM} & \text{CLW} \\ \text{RPM} & \text{CCLW} \end{matrix} \qquad \text{e.g., } \quad \text{SPINDL/250, SFM, CCLW}$$

$$\text{FEDRAT/f, } \begin{matrix} \text{IPR} \\ \text{IPM} \end{matrix} \qquad\qquad \text{e.g., FEDRAT/.01, IPR}$$

Where a constant depth of cut and good surface finish are required, the combination of SFM and IPR should ideally be chosen.

The implementation of this feature in post processors is complicated by the variety of spindle speed controls available. Some NC lathe headstocks, using hydraulic motors for instance, have infinitely variable spindle speeds over the full range of the machine and may permit the actual SFM to be given as a number on the control tape. Other headstocks have mechanical gearboxes giving discrete speeds, and still others have combinations of these two types, giving a number of overlapping ranges of infinitely variable speeds.

The infinitely variable headstock is the simplest to cater for, even if it requires to be programmed in rev/min. With it, and assuming that the control system being used has buffer storage, the post processor can perform an optimum segmentation of any radial tool path. The only restriction on the number of segments is that the time taken to feed the tool across each segment should be greater than the time taken to read each subsequent block on the control tape.

Mechanical gearboxes are not so simply handled. Shift positions can be computed across the radial tool path which will give changes from one discrete speed to the next and this will give a series of roughly similar stepped cutting conditions. However, each change in spindle speed will normally be accompanied by a dwell to allow the change. This will result in bad surface finish and, when machining certain materials, may also result in the blunting of the cutting edge of the tool. While this bad surface finish would not be acceptable for finish machining, it would be justified in roughing, in order to achieve maximum metal-removal rates. In this situation, two methods of designing the post processor are feasible. The first would allow the part programmer to select RPM, IPM when surface finish is important, and SFM, IPR for roughing cuts. The second method, and the more satisfactory, would be only to part program in the SFM, IPR mode, but to specify to the post processor when it would be safe to change spindle speed. The EXAPT 2 processor (see chapter 8) allows a percentage range in SFM to be programmed and outputs suitable blocks to the post processor.

Headstock controls having ranges of overlapping speeds cause the same type of problem as above. If the SFM mode is being used, changes from one range to another may be computed, and this again will probably necessitate a dwell. Again, a solution would be to only part program in the SFM, IPR mode but specify to the post processor when it would be safe to change from one range to the next.

Geometry transformation in multi-axis machine tools

Chapter 10 describes the use of the APT processor in multi-axis machining. Basically, this involves the processor in computing tool-axis orientations in addition to the normal tool-path generation. In this situation, the CLDATA produced by the APT processor contains XYZIJK information within each motion record, the XYZ values being successive positions of the tool end point in the part co-ordinate system and the IJK values being the direction cosines of the tool axis at each position.

Multi-axis machine tools have one or more rotational freedoms in addition to the normal three orthogonal slides. These machines can be divided into three main families.

Type A. Machines with a rotary table and a tilting spindle head, rotation of the table and head being each restricted to one plane: example—Sundstrand Omnimil (see Fig. 6.13).
Type B. Machines with a fixed spindle head and a table capable of rotation in two perpendicular planes, example—Kearney & Trecker Milwaukee-Matic.
Type C. Machines with a fixed table and a spindle head capable of rotation in two perpendicular planes; example—Cincinnati (see Fig. 6.12).

Many variations on these three basic types are possible. Fig. 6.12 shows a machine in which a quill motion, W, of the spindle is used to achieve a Z movement. As the spindle axis is allowed to rotate in two planes, the machine co-ordinates are non-orthogonal; the W slide is not always perpendicular to the X and Y slides of the machine, but the Z value is obtained from W cos A cos B, where A and B are the angles the quill makes with the Z axis of the workpiece.

A problem peculiar to multi-axis machines is that, under certain circumstances, there are several ways of achieving successive tool positions. A simple case of this would arise if successive positions of the tool were diametrically opposite on the outside of a cylinder mounted on a rotary table. The move could be achieved by rotating the table 180° either in a positive or negative direction or by simply moving the table unrotated by a distance equal to the sum of the diameters of the cylinder and tool. Currently the CLDATA does not carry information about surfaces being generated (other than circles) and the post processor would carry out a move of this nature in accordance with a fixed set of rules of which the part programmer must be aware.

With each of the types described above, if the Z or W slide is kept perpenpendicular to the XY plane and the tool axis remains unchanged, the slide movements will be the tool end movements, i.e., the XYZ values from the CLDATA. However, a change in the angle of the tool axis or a change in the direction of the W slide will mean that the slide increments will no longer correspond directly to the linear moves of the tool end. This is due to the fact that these machines normally have the centre of rotation of one or both of their rotational axes non-coincident with the cutting point of the tool. This in turn necessitates a translational move of the XYZ slides, and consequently the centres of rotation of the rotational axes, to compensate for any reorientation of the tool. In Fig. 11.12, the rotation of the tool axis has been restricted to one

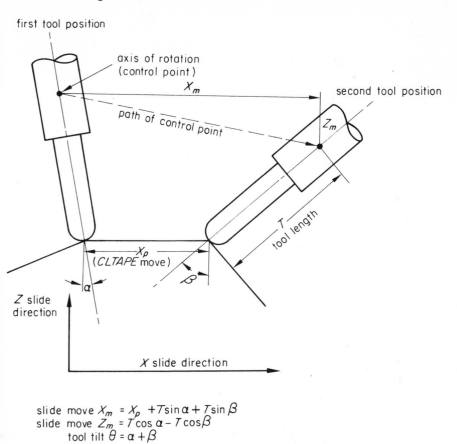

slide move $X_m = X_p + T\sin\alpha + T\sin\beta$
slide move $Z_m = T\cos\alpha - T\cos\beta$
tool tilt $\theta = \alpha + \beta$

Figure 11.12 Translational moves resulting from tool tilt.

plane and it can be seen that the translational move of the X slide of the machine, X_m, will differ from the requested CLDATA move, X_p, and will be a function of the angle of tilt, Θ, and the distance, T, from the tool tip to the axis of the rotation.

Post processors for multi-axis machines must therefore be capable of converting the XYZIJK values of the CLDATA to unique slide moves appropriate

12*

to the specific geometry of the machine. For example, a post processor written for the Sundstrand Omnimil OM 3 of Fig. 6.13 could use transformation equations as follows.

To convert from part co-ordinates to machine tool co-ordinates

$$X_m = (X_p - X_t)\frac{J}{\sqrt{1 - K^2}} - (Y_p - Y_t)\frac{I}{\sqrt{1 - K^2}} + H$$

$$Y_m = (X_p - X_t)\frac{I}{\sqrt{1 - K^2}} + (Y_p - Y_t)\frac{J}{\sqrt{1 - K^2}} + (R + T)\sqrt{1 - K^2}$$

$$Z_m = Z_p - Z_t + K(R + T)$$

$$A = \arcsin (K)$$

$$C = \arctan \left(\frac{I}{J}\right) - \Theta$$

To convert from machine tool co-ordinates to part co-ordinates

$$X_p = (X_m - H) \cos (C + \Theta) + (Y_m - (R + T) \cos A) \sin (C + \Theta) + X_t$$

$$Y_p = (Y_m - (R + T) \cos A) \cos (C + \Theta) - (X_m - H) \sin (C + \Theta) + Y_t$$

$$Z_p = Z_m - (R + T) \sin A + Z_t$$

$$I = \cos A \sin (C + \Theta)$$

$$J = \cos A \cos (C + \Theta)$$

$$K = \sin A$$

where X_m, Y_m, Z_m are machine slide moves

A and C are machine rotary head and table slides respectively

X_p, Y_p, Z_p, I, J, K, are part co-ordinate data

X_t, Y_t, Z_t are the co-ordinates of the rotary table centre in the part co-ordinate system

H and R are constants

T is the tool length, and

Θ is the angle between the X axis of the part and the machine co-ordinate systems.

The problem of non-linearity

As discussed in the previous section, in most existing multi-axis machine tools, the two axes of rotation are not coincident with the physical end of the tool. When a re-orientation of the tool axis is requested, it is necessary therefore to translate the centres of rotation while simultaneously rotating one or both of the rotary axes. If these translational and rotational slide moves are contained in one block on the machine tool control tape, equal proportions of the linear and rotary moves occur in equal time increments and result in the physical end of the tool moving in some curved path relative to the workpiece. This can be seen in Fig. 11.13a, the broken line representing the actual path traced by the tool end.

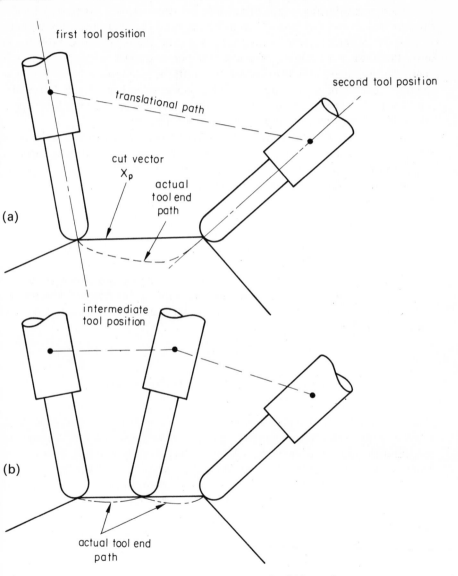

Figure 11.13   Reducing tool path error by dividing path.

It is the responsibility of the post processor to correct for this deviation from the linear path assumed by the APT processor. In practice, current post processors simply ensure that the tool end remains within a specifiable tolerance band of the desired linear path. They do this by simulating the geometry of the machine tool and determining where the tool end would be in part-program co-ordinates after the machine tool axes have moved some percentage of the total movement of each vector (usually 50 per cent). Any discrepancy between this computed tool end position and the required position on the linear path will be the estimated error. If this error is greater than some specifiable tolerance, the linear cut vector is halved and the error midway along the first half vector recomputed. This loop is repeated until the estimated error is less than the part-programmed tolerance. The effect is shown in Fig. 11.13b, where the cut vector X_p has been halved and the resultant physical tool end motion, shown chain dotted, may be within the linearization tolerance.

It does not follow that non-linear paths traced by the physical tool end always require linearization. In fact, when the curvature of the tool end path approximates to that of the required machining surface, it should be possible for the post processor to incorporate a number of linear cut vectors into a single output block and, by taking advantage of the curved physical tool path, improve surface finish and machinery efficiency. This situation would apply to any drive surface with its centre of curvature concentric with, say, the centre of the rotary table, e.g., a cylinder or cone mounted concentrically on the table. In APT, neat implementations of this optimization procedure are currently handicapped by the lack of information passed on to the post processor about the surfaces being machined; post processors attempting this technique require to re-synthesize the original surface from the information on discrete tool end points of the CLDATA.

Impact of cutting technology

The problems considered in the last section are typical of the type encountered in conventional APT-like post processors (i.e., where the processor is concerned solely with the component geometry, and where the specification of tooling and machining conditions is the responsibility of the part programmer). This section examines some further changes forced on the post-processor area by the addition of automatic computer selection of feeds, speeds, and tooling in the processor, as incorporated in the EXAPT and NELNC families of programs.

On the basis of the following examples and because of differing CLTAPE information, it would seem that post processors for the technological programs EXAPT or NELNC must necessarily differ from post processors for the geometry programs based on APT. However, post processors can be designed to run (without modification) with all three processors.

Tool loading

When using the geometric processors, the part programmer selects each tool as required and allocates it a turret or magazine position on the machine tool. As he develops his part program, he is conscious of the tools he has used and is able to optimize to some extent the loading and unloading of the turret or magazine. With the automatic selection of tools by the processor as in EXAPT, neither the part programmer nor the processor undertakes this tool loading, and it becomes the responsibility of the post processor to allocate the most efficient turret position for each selected tool. To do this it is necessary to design the post processor to read the CLTAPE twice: once to store away details of each tool used and the order of their appearance in the program; and once to load these in an efficient manner and complete all of the other tasks necessary to produce the control tape. This process may react on the processor: for example, if nine tools are selected for an eight-position turret drill, it is possible for the post processor to put out a command to manually change the last tool; however, this may not be the most efficient method of producing the component (depending on the component and the batch size) and it may be advisable to return either to the tool-selection area of the processor or to the original part program.

Spindle speed calculations

The problem in lathes of when to change spindle speed to achieve constant peripheral cutting speed, while at the same time ensuring that dwells do not mar the surface nor blunt the tool, was discussed in the previous section (constant cutting speed for lathes). In EXAPT 2, this work is done in the processor, which has access to machine information such as the range of speeds available. EXAPT post processors are therefore correspondingly simpler than the equivalent APT. Using EXAPT 2, the part programmer has the facility of requesting speed-change steps as a percentage of the preceding peripheral cutting speed and leaving the processor to select the appropriate rev/min.

The statement he would use is as follows:

$$CSRAT/tp$$

where   tp = cutting speed as the percentage $V_m/V \times 100$

in which V = specified or computed peripheral speed

and V_m = peripheral speed at which spindle speed change should be made.

In the event of the planner wishing to avoid dwells, a CSRAT of 0 per cent would of course force the processor to use only one spindle speed right across the face. Currently, as EXAPT 2 pays no attention to the exact speeds available on the machine tools, its instructions in this respect may still require modification by the post processor.

Collision avoidance

As long as the part programmer is responsible for the selection of tooling, it is possible to expect him to check that his choice of tools, both cutting and passive, will not foul the machine tool or the component during machining operations.

With the automatic computer selection of tools from a tool library, as in EXAPT, this is no longer possible except at the end of the computer run. By that time, of course, if the part programmer's check finds a possible collision course, it can only be eliminated by a further computer run with some change made to the part program input. The output of this second run would again need to be checked, however, since the selection and loading of the tools would again be made by the computer.

An attractive, though undoubtedly expensive, alternative to this manual intervention would be to include logic within either the processor or the post processor to check for collision course and, where found, to modify the tool loading or machining sequence automatically until the entire machining of the component could be successfully completed without collision. For the first half of this activity (i.e., the checking for collisions) to be effective, it would be necessary to take into account the geometry of the machine tool and any tool posts, turrets, and work-holding means, in addition to the geometry of the tools and the changing geometry of the component in machining from blank to finished shape. To allow a post processor to perform this checking, the EXAPT CLDATA contains the necessary information about the tool geometry and the current shape of the workpiece. However, this checking would be no mean task, requiring large amounts of computer core store and execution time, but would probably still be simpler to write than the second half of this activity, namely the automatic modification of tool-loading and machining sequences to avoid the collisions discovered during the first stage.

The correction of tool collisions by choosing new tool positions, even if dealt with by suitable logic in the post processor, would require a complete re-calculation for collisions for the whole part program, even if only one tool were repositioned.

It is unlikely that lathe post processing can be given the computing time, and the cost that this implies, until computing costs are reduced by yet another order.

Areas of possible standardization

The difficulties involved in achieving compatibility between post processors are considerable. The problem is complicated firstly by the lack of standardization in other related areas, and secondly by the fact that post processors are designed and written by a wide variety of users and control-system and machine-tool manufacturers.

The areas where lack of standardization affects post-processor compatibility are as follows.

(a) Part-programming vocabulary. Each post processor has associated with it a set of words which can be used by the part programmer to control it and to cause certain effects at the machine tool. These words should be selected from the pool of available words, but many existing post processors, particularly in the USA, have been allowed to use new words (when there were existing words available) with little attempt at standardization.

In some cases, two post processors use different words to indicate the same function, while others use the same word to indicate different functions. Attempts are now being made by various standards organizations to restrict the variety of words available while at the same time giving each word an explicit meaning.

(b) CLDATA formats, e.g., APT 3 and NELNC, EXAPT, etc.

(c) Different levels of information on CLDATA, e.g., different technological information on APT and EXAPT.

(d) Different control system requirements, e.g., 'hard' or 'soft' servos, absolute or incremental dimensions, punched paper or magnetic tapes, etc.

(e) Non-use of internationally agreed standards for NC, e.g., axis nomenclature or codeset for punched paper tape.

(f) Computer implementations.

While these are formidable problem areas, the concept of modularity in post processor design can simplify the task of introducing a measure of standardization between post processors.

An example of this is in the construction of input elements. Because of problem (b) above, different input elements have so far been necessary for APT and EXAPT. However, if they are each designed to store their CLDATA logical record information in similar arrays using the same internal format, then the auxiliary and motion elements of each post processor may be alike.

Similarly, if each miscellaneous and preparatory function is designed to be handled by a separate module within the auxiliary element, then differences between requirements of differing control systems can easily be handled.

The wide adoption of FORTRAN as the computer language in which post processors are written has simplified the problem of incompatibility between computers. By restricting the set of FORTRAN instructions to that covered by the USASI FORTRAN set, further standardization can be achieved and post processors can be implemented on differing computers with the minimum of modification.

Generalized post processors

Where an NC control system is used on a number of different types of machine tool, it has become the practice, especially in the USA, to write a single generalized post processor capable of accommodating each of these different machines. The structure of these post processors is a logical extension of the concept of modularity described earlier, each of the post processors being made up of a large number of subroutines or building blocks, each of which has been written to cover one unique function.

Because generalized post processors are normally written around one control system, many functions of machine tool and control system and post-processor procedures can be handled for different machine tools by the same basic building blocks. The generalized post processor, therefore, can eliminate much of the duplication which would be inevitable if separate post processors were written for each machine tool.

Each generalized post processor consists essentially of the basic routines, applicable to all the machine tools, plus a set of machine-tool-dependent routines defining the specific peculiarities of each machine. To add a new machine tool to the post processor, it is only necessary to add a new machine routine, perhaps a man-week of effort.

General Electric's post processor, GECENT 3, goes still further and divides the basic routines into separate packages, each for a specific type of machine tool—lathes, mills, drills, etc., and each capable of being loaded separately into core with its appropriate machine-dependent subroutine. This design was an attempt to reduce the core-storage requirements and execution time of the generalized program, for undoubtedly large core size and slow execution time are major drawbacks to the generalized approach. Far outweighing these drawbacks, however—at least for large and powerful computers—are the benefits of flexibility and speed of implementation of new machine tools.

Standardization via library of modules

An alternative approach to that of generality is to establish a library of subroutines or modules of sufficient range that newly required post processors can be built up by a suitable selection from the library. This approach has the advantage that each post processor built up in this way need only contain the modules sufficient to its requirements. Each post processor would therefore be smaller and probably faster than the equivalent generalized package. The main disadvantage lies in the sustained effort required over a long period of time to build up the library—effort which is currently used to produce diverse post processors for particular applications, often in minimum time.

The library approach to post-processor writing would require the adoption of much firmer disciplines than are in use today: the planning and use of standards of structure, coding, and nomenclature; the ability to resist the short-term advantage of the special one-off post processor; the pooling of individually written modules in a national computing centre.

There is little evidence that the NC industry is interested in this approach or yet appreciates the advantages of co-operation in this field. IITRI, the sponsors of the APT program in the USA, have endeavoured for years to have standards for post-processor design accepted—without success.

Compiler-compiler approach

A neat, but complex and probably remote, solution to the problem of post-processor standardization is the concept of a computer program, written in a high-level language, generating a post processor in machine language from the specification of the combination of machine tool and control system.

The problems involved in this solution are daunting—they are similar to the problems of the library approach with the additional complexity of having to write the compiler.

Post processor specification

It should now be apparent that the post processor is an essential link in computer-aided NC. Management concerned with planning and procurement of NC machines must be sufficiently familiar with the functions of post processors so that they can specify and obtain post processors which allow them to make full and efficient use of their machine tools. While post processors are occasionally written by computer bureaux, machine-tool manufacturers, or users, they are most frequently prepared by the control-system manufacturers, as they are in the best position to know the peculiar requirements of their system. However, irrespective of who actually prepares the post processor, its purchase should, if possible, be through the prime contractor for the machine tool and control system. Wherever possible, the procurement of machine tool, control system, and post processor should be treated as a single issue. Experience has shown that it is unwise to consider ordering the machine tool and post processor separately.

The single procurement has two advantages.

(a) The user is in a stronger bargaining position with the prime contractor to ensure he obtains his requirements.

(b) The discipline of specifying both post processor and machine tool simultaneously may indicate that certain features either on the machine or in the post processor are redundant; e.g., CYCLE operations or circular interpolation can be handled either on the machine tool or in the computer program. A decision to use computer-aided programming should consequently reflect in a correspondingly cheaper control system.

In fact, where a prospective purchaser of a machine tool has made a serious decision to use one of the processors—APT, ADAPT, NELNC, EXAPT, IFAPT, etc.—he should ideally go further and, before ordering the machine tool, insist on demonstration cuts being made on the proposed machine using control tapes produced by the processor and post processor implemented on the computer of his choice.

The machining times he would obtain from the post processor would be a valuable aid to accurate costing of his components on the new machine. Even more important, as the software would exist at the time of order, his lead time before the machine cuts metal on a production basis would be significantly reduced, since training of programming staff could proceed in parallel with the manufacture of the machine tool.

Ideally, any specification of a post processor should cover the following areas.

Vocabulary The user must ensure that full use can be made of all the facilities available on his combination of machine tool and control system. To do this, he must have available a full range of part-programming statements and the post processor must be designed to handle these correctly. The exact definition of each of these statements may vary from one machine to another and it is the user's responsibility to ensure that the definitions given in the specification adequately cover the full and efficient use of his machine tool.

(An actual incident, involving a user meeting with a machine-tool designer, control-system designer, and post-processor writer illustrates this point. A firm price had been quoted for the machine tool and control system. When the user asked how the cycle-drilling feature worked, the machine-tool designer said it was dealt with by circuits in the control system or in the post processor. The control designer said it was done by cams and micro-switches on the machine tool or by the post processor. The post-processor writer in turn asserted it was done by the control-system circuits or by the machine tool! The moral should be clear!)

Structure For ease of updating, it should be specified that the post processor be written where possible in USASI FORTRAN 4 and should use the modular concept detailed earlier in the chapter. The precise function and performance of each of the five elements comprising the post processor should also be defined.

Documentation The post processor should be so documented that it can be used correctly by the part programmer and maintained readily by the computer programmer. This implies that documentation should consist of three sections:

computer programs section;
part-programs section;
symbolic listing of computer program.

The computer programs section should contain prose and flow-diagram descriptions of the overall program and of each subroutine, details of errors diagnosed by the program, lists of variables and parameters used, and should give the analyses upon which the computations of, at least, each of the following elements are based:

increments of distance moved;
feedrates at discontinuities;
acceleration/deceleration of slides.

The part-programs section should list and describe in detail the meaning of all post-processor words capable of use by the part programmer and should give examples, where required for clarity, of the use of each word. Any special rules to be followed in using the post processor should be described.

All options available for use and all diagnostic error comments should be listed.

The symbolic listing of the computer program should include liberal comment cards to facilitate understanding of the coding.

If the post-processor documentation is to be helpful to the user it must be clear, neat, and concise.

Efficiency It is not always appreciated that the design of the post processor can have a significant effect on both the final machining efficiency and the program-execution time on the computer. Machining inefficiencies can be introduced both in effective floor-to-floor time, and in the final accuracy and finish of the component. Accuracy and finish can be affected for instance by the introduction of unnecessary dwells or abrupt changes in velocity by the post processor. These can be found by inspection provided, of course, that other errors introduced by the machine tool or control system can be eliminated.

In continuous-path post processors, the part programmer's feedrate may be artificially reduced to limit changes in velocity to steps acceptable to the machine tool and control system. The efficiency of floor-to-floor machining times is *not*, therefore, simply a measure of actual machining rates against part-programmed feedrates. The efficiency of the post processor can be measured only by comparing the time taken by a control tape produced by the post processor, with the time taken by a control tape prepared by taking into account the maximum changes in velocity acceptable to the machine tool and control system under conditions of angular discontinuities of the tool path. It is instructive to compare theoretical and actual times for a series of circular paths over a wide range of radii. If the velocities output by the post processor fall significantly below the theoretically possible, because of some lack of design (for example, lack of rework logic), the post processor is inefficient. However, to improve the machining efficiency may involve considerable expense in writing more sophisticated software and only a detailed knowledge of the range of components can allow an intelligent decision to be made on whether this would be worthwhile.

Machining tests Since the specification of some theoretical machining efficiency may be very involved and awkward to implement and test, probably the simplest performance check on a post processor involves acceptance tests on a range of typical components, using tapes produced by the post processor. Provided the components cover a good permutation of the facilities being tested, their successful machining, including checks on surface finish, accuracy, and floor-to-floor time, should constitute acceptance of the post processor. These selected components must, however, be suitable for accurate measurement. A portion of the surface of, say, a truncated cone, a paraboloid, or a propeller blade, should be avoided for this reason.

12. Making best use of a computer

T. M. R. Ellis

T. M. R. Ellis, AMInstCSc, MA, is a Product Marketing Officer with International Computing Services Ltd., engaged in the marketing of multi-access services. He obtained his MA in mathematics from Christ's College, Cambridge and then joined English Electric Computers, where he worked initially on writing programs for 'critical speed' calculations on turbo-alternators. In 1965 he joined the new team investigating APT, and visited IITRI in Chicago in this capacity, and again later, when he spent 4 months in Chicago assisting in the development of APT 4. He was subsequently in charge of the section at English Electric Computers which developed APT and EXAPT as well as other mechanical engineering programs. In March 1968 he transferred to Edinburgh to manage the company's medical computing bureau, and in September 1969 moved to his present position in Kidsgrove, Staffordshire.

In the previous chapters, the technical aspects of using a computer as an aid in the preparation of NC tapes have been discussed at some length. However, this discussion has inevitably assumed the existence of a suitable computer. In fact, this is rarely the case and the potential user of such computer assistance is, therefore, faced with the extremely important question: 'Which computer should I use; and what is the best way to use it efficiently?' There is no answer to this question; each case must be considered in the light of its own special circumstances. Indeed, as a user's involvement in NC grows, so these circumstances, and hence possibly the answer to the question, will change. In this chapter, the various methods of use of a computer will be discussed and their relative advantages and disadvantages compared, both in the light of present-day techniques and of the probable developments in the near future.

Methods of access to the computer

There are essentially four ways in which a potential user can have access to a computer, depending on:

(a) whether he will operate the computer himself ('workshop'), or whether it will be operated by one of the specialist staff whose services are provided with the computer ('bureau'),
and

(b) whether the computer is situated at the same site as the user ('in-house'), or at some other location ('remote').

While the services available from different organizations may vary quite widely, it will be found that within any one category they are very much of the same order. What, in general, can be expected from each of these four major types of service will now be discussed in more detail.

Remote bureaux

A remote bureau is an organization, situated somewhere other than at the user's premises, which consists of a computer (or computers) and a number of specialist staff, and which supplies a complete computational service to the user. It may be run by a computer manufacturer, by an independent company, or by a company for use by its various departments. In all cases the basic concept is the same, namely, the user supplies his data in some agreed format, the bureau processes it, and the bureau returns the results to the user.

The form in which the data is supplied will vary appreciably depending upon the application, the distance it has to be sent, the customer's preference and several other factors. The most common form is on preprinted data-sheets which the user will fill in by hand, but punched paper-tape or cards, or even direct 'telex-type' transmission from a form of typewriter, are often used. On receipt of the data, the bureau will, if necessary, put it into a form suitable for direct input to the computer, normally by punching it onto paper-tape or cards but possibly by use of optical or magnetic character recognition equipment. The correct program for the job will then be loaded into the computer and run using the data provided. The results so produced—normally printed, but often on punched paper-tape or drawn on a graph-plotter—will then be dispatched to the user by the fastest possible means.

Because everything between the initial collection of data and the interpretation of the results is taken care of by the bureau, this is very often the way in which a potential user of a computer first starts.

In-house bureaux

An in-house bureau offers the same type of service as a remote bureau but is situated at the same place as the user. It follows therefore that it is owned and run by the same organization as that to which the user belongs, and in all probability is in fact run by a special computer department in that organization. Its main advantage over a remote bureau lies in a faster turn-round.

In-house workshop

An in-house workshop, on the other hand, is a computer for which all the ancillary services such as program writing and testing, data preparation, etc., must be provided by the user. This means appreciably more work for the user and a greater understanding of the purely computational problems, which in a bureau environment he can ignore, but of course it leads to programs and turn-round times tailored to suit his needs. Because of the much greater involvement in the technicalities and the problems of computers, however, this method of access to the computer is only infrequently used in a production environment, and is mainly to be found in research, development, and educational areas.

Remote workshop

A remote workshop, which as its name implies is effectively a computer for hire in some remote location, was at one time a reasonably popular method of access to a computer, having most of the advantages of an in-house workshop without its capital cost. However, the use of remote workshops is dying out, and they are now mainly used by the smaller educational and research establishments which cannot justify the capital cost of their own computer.

Remote multi-access time-sharing bureaux

A new development in computing techniques which is leading to a further method of access to a computer is the concept of 'time-sharing' and 'multi-access'. This takes advantage of the fact that a modern computer works several million times faster than a human being to process sequentially a large number of programs which are controlled directly by the user, in the same way that a chess grand master may play several games of chess simultaneously while appearing to his opponents to give his full attention to each individual game. Thus, several users (sometimes even several hundred users) can be in communication with the computer at one time. The delay between action by the user and the corresponding action by the computer seldom exceeds a second or two—an insignificant delay to the human user. This method of access combines all the advantages of the bureau (standard or tailor-made programs and specialist service) with the faster turn-round and greater flexibility of an in-house workshop, but at a considerably lower capital cost than the latter.

Methods of use of the computer

Before discussing the relative merits of these methods of access, one must first consider the type, quantity, and frequency of the work for which the computer is going to be used. A further consideration is the possible requirement for computer assistance in fields other than numerical control; but, as this is clearly far too complex a question to be dealt with in general terms, it will not be considered here except in so far as it is relevant to the economics of the various approaches.

Once it has been decided to look purely at the NC side of the decision, the first question is: 'What type (or types) of NC work am I doing now, and what types am I expecting to do in the future?' and the second: 'How can a computer help me to prepare tapes for these types of work?' The answers to these will lead to the realization that there is, in all probability, a wide range of programs in existence which could be used. Some of the programs that are currently available from various sources are shown in Fig. 12.1.

Type of work \ Type of program	General-purpose		Special-purpose
	Free-format	Fixed-format	
Point-to-point and straight-line milling	EXAPT 1 2P,L AUTOSPOT (APT) (ADAPT) (2C,L)	KIPPS ROMANCE	Very many
2½ axis contour milling	2C,L ADAPT APT	MILMAP	PROFILEDATA SURF
Turning	EXAPT 2 2C (APT)	AUTOPOL	AUTOPIT
3, 4, or 5 axis contour milling	APT		

Figure 12.1

It is apparent from this Figure that, at the simpler end of the scale (in which some 75 to 80 per cent of all NC machine-tools fall), there is a large variety of programs available 'off-the-shelf', quite apart from the very large number of special-purpose programs. In this (relatively) simple area, the writing of a program to do a specific job is quite easy and, in fact, two of the programs shown as general-purpose (AUTOSPOT and KIPPS) started life as special-purpose programs, written for their own production departments by IBM and English Electric Computers respectively.

In Fig. 12.1, the selection of programs shown has been tabulated under the headings 'General-purpose' (by which is meant available for more than one type of machine tool) and 'Special-purpose' (for a specific machine tool). In addition, the general-purpose programs have been further divided into 'Free-format' and 'Fixed-format'. The choice of which program, or group of programs, is to be used us almost inextricably linked with the choice of the method of access to the computer. This distinction and the apparent multiplicity of programs will now be examined in more detail.

Free-format programs

When one talks about a free-format or a fixed-format program one is not, of course, really referring to the program but to the data for that program. In the case of the free-format programs, the form of the data, or part program, is normally called a 'language' and several such free-format languages have been discussed in detail in chapters 7 to 10. They all use an English-like method of describing both the geometry of the part and the manner in which the machining is to be carried out, and are based on the internationally accepted APT language.

The resulting part program is thus a sequence of pseudo-English statements which, when read together, describe accurately and unambiguously the desired mode of operation. Because of their form, it is quite simple to check or modify such a part program, particularly if meaningful names have been assigned to the various items.

In work where more complex geometry or workshop technology is required, the flexibility of this type of part-programming language becomes extremely important. Thus, for example, APT 4 contains the capability for defining and machining nearly 20 types of surface including circles, planes, cones, spheres, conics, and quadrics as well as surfaces defined by a set of discrete points or by their mathematical formulae in well over 100 different ways. This allows the designer a great deal of the freedom which is essential in some fields such as aerospace, but which is also highly desirable in most other fields, and which he has been denied in the past because of the constraints placed upon him by the limited capability of the man on the work-bench.

Owing to the similarity between the language used by these free-format programs, it is possible for a part programmer to change relatively easily from using one to using another. This is an important feature, both for an organization which is just starting in NC but anticipates increasing its investment while widening the scope of its activities in the future, and for an organization that already has a wide range of types of NC tool and does not wish its part programmers to be tied to one type.

Fixed-format programs

A fixed-format program, unlike a free-format one, generally accepts as its data a set of numbers laid out in a fixed format with no explanatory alphabetic words. Because of this, there is considerably less flexibility than with free-format programs. Fixed-format programs are, almost without exception, written only for machine tools with very similar capabilities. Because of this lack of flexibility, it is also rather more difficult to add new features to the program, and when new features are required a new program is frequently produced with an extended or altered data format.

However, despite these apparent disadvantages, the fixed-format program has several significant advantages over its free-format cousin. The first of these

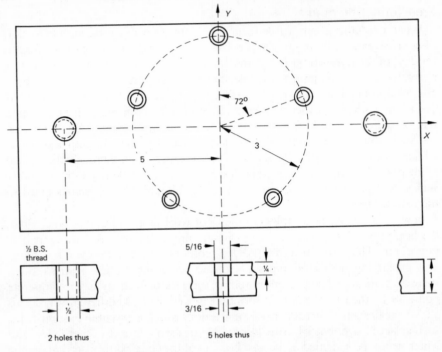

2 holes thus

5 holes thus

KIPPS   PART-PRØGRAM (including tool data, but without heading data)

Tool No. Type	Diameter	Description	Spindle speed (rev/min)	Feed rate (in/min)	Depth of cut
01 2	0.1875		2000	7.0	1.0
02 2	0.4466	TAP DR	1000	6.7	1.0
03 5	0.5	BSW	300	2.5	1.0
04 3	0.3125	SP FACE	375	2.0	0.25

(Group (No, Type	Pattern Code	X	Y	Z	Radius/ Pitch	Angle (degrees)	No. of holes	Tool Nos.))
1F	C				+ 3.0	+18	5	0104
2F	I	-5.0			+10.0		2	0203
3F	E							

EXAPT I   PART-PROGRAM (without heading data)

```
        ZSURF/1
P1    = PØINT/-5, 0
PAT1  = PATERN/LINEAR, P1, ATANGL, 0,INCR, 1, AT, 10
CIR1  = CIRCLE/0,0,3
PCIR1 = PATERN/ARC, CIR1, 18, CCLW, 5
        PART/MATERL, 20
D1    = DRILL/SØ, DIAMET, 0.1875, DEPTH, 1
D2    = DRILL/SØ, DIAMET, 0.4466, DEPTH, 1
T1    = TAP/SØ, DIAMET, 0.5, DEPTH, 1, TØØL, 100014
SP1   = SINK/SØ, DIAMET, 0.3125, DEPTH, 0.25, SPIRET, 1
        FROM/0,0,0
        WØRK/D1, SP1
        GØTØ/PCIR1
        WØRK/D2, T1
        GØTØ/PAT1
        FINI
```

Figure 12.2   Comparison of EXAPT 1 and KIPPS part programs. (All dimensions in inches)

is that, because writing a part-program becomes virtually filling in numbers in the appropriate columns of a preprinted sheet, the part programming can be done by less skilled staff than is required for free-format programs. This is largely psychological, since for example, writing an EXAPT 1 part program is intrinsically no more difficult than writing a KIPPS part program. It is, however, often longer (even if more quickly written), indubitably more verbose, and apparently, as a result, more complex (see Fig. 12.2); and, in any case, even the least skilled man or woman is used to filling in numbers on forms, whereas writing a computer program (and that is essentially what a free-format part program is) is a strange and unnerving experience! When this psychological barrier has been overcome, the advantage of 'familiarity' when using a fixed-format program will largely disappear, but it will, like all such barriers, be with us for an appreciable length of time. A more concrete advantage of fixed-format programs is that they are often cheaper to run, owing to the inherent simplicity of their data, than free-format programs. They are also easier to write and can thus be tailored more exactly to meet the user's needs; whereas with the general-purpose free-format languages attempting to be 'all things to all men'—and with a remarkable degree of success—any one user finds certain deficiencies which could be perhaps corrected in a smaller fixed-format program.

Special-purpose programs

In addition to the general-purpose programs so far discussed, there are a number of programs which have been written with a particular machine tool, or a particular job, or both, in mind. These special-purpose programs are often written by, or on behalf of, the machine tool or control builder, and are very nearly always the most efficient way of programming the particular machine tool or job for which they were written. Typical examples of this type of program are: PROFILEDATA, described in chapter 5, written by Ferranti for machine tools fitted with their Mark IV control system; AUTOPIT, written by Pittler for their Pinumat lathe; and SURF, written by Olivetti for their AUCTOR-CNZ machine tools. The last-mentioned program is also interesting in that, unlike the first two, it is not designed for general-purpose machining but for a specific type of 'faired' surface milling.

Perhaps the ultimate degree of specialization in this field has been reached (as it has in so many other fields) by the automobile manufacturing industry, especially in the area of component manufacture, where mass-produced specialization is the keynote. One of Britain's largest component manufacturers, in explaining their approach to NC programs, has said: 'Basically we make only six different components,' and explained that although there may be hundreds of different crankshafts (for example) being produced, they are all essentially the same, and, by using a special-purpose program, can be described by a handful of parameters (not necessarily all geometric).

Computer dependence

If the prospective user of computer assistance is not intending to sit down and write his own computer program, he must get it from someone else; and herein lies a further problem. All the general-purpose free-format programs in general use (APT 4, EXAPT, 2C,L, etc.) and their post processors are written in such a way that they can easily be run on almost any modern computer subject only to its being large enough, so that the size of computer available becomes the main criterion when deciding whether it is feasible to use one of these programs.

A fixed-format program, on the other hand, is by its very nature much more closely allied to the *modus operandi* of a particular computer. It is therefore usually very difficult to implement such programs on computers other than those for which they were originally written. Moreover, as this class of program tends to be written by an individual computer manufacturer for use by his customers, it is often not freely available in any case.

The final group of programs—special-purpose ones—are by definition very largely written for a specific computer; however, a minority of these programs have been written in a computer-independent fashion and can therefore be implemented on several computers.

Deciding how to use the computer

The potential user of computer-assisted NC programming can now therefore be considered to be familiar with the main concepts that he has to make a decision between. However, there still remain a number of very important questions to be answered before deciding how to make best use of a computer. All these questions can be summed up in the (apparently) obvious statement that he wants to be able to turn his designer's drawings into machine tool control tapes as quickly, as efficiently, as reliably, and as economically as possible. In the following four sections, each of these points will be discussed in the context of the more conventional methods of access to the computer; the way in which the advent of the data-grid and multi-access time-sharing computers affect the conclusions reached will then give a basis for the final decision to be made.

Turn-round

One of the main advantages claimed for computer assistance in NC is that it can cut out a great deal of the time-lag between the completion of the design drawing and the production of a working control tape. Figure 12.3 shows the ideal flow of information and/or physical data through the stages from design to production. The object of the exercise is to minimize the (unproductive) time spent in the second and third stages (planning and computing).

The first phase is the writing of a part program. In practice, the time spent at this stage is more likely to depend on the part programmer than on the type of part program he is writing. The shorter time required to write a free-format statement will, in general, be offset by the increased number of such statements.

The written part program must now be passed to the data-preparation department for punching onto paper tape or cards. Here the time varies appreciably

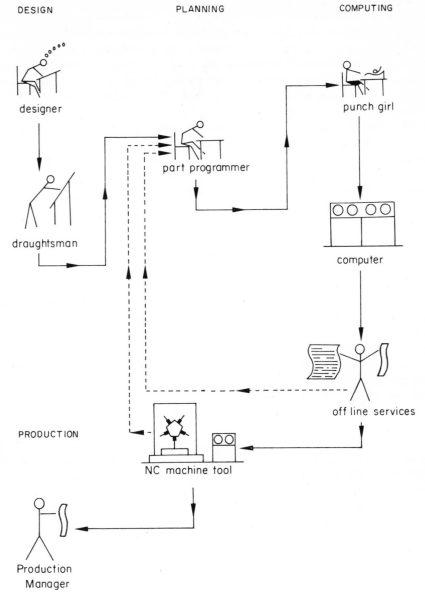

Figure 12.3   Information and data flow.

with the method of access adopted. With an in-house workshop, the data preparation is probably done by the part programmer himself and so the transmission time can be thought of as instantaneous. With a remote workshop, the data preparation is again done by the part programmer but he has first to get to the workshop and so the time could be taken as a few hours. In the case of a remote bureau, however, the part program will probably be sent by post or by data link unless the bureau is close enough for personal delivery. The transmission time could therefore vary between a few minutes and up to 2 to 3 days, depending upon the method chosen. An in-house bureau, on the other hand, will have a transmission time of however long it takes to get the written part program to the data-preparation department. This could also vary from a few minutes (personal delivery) to perhaps 24 hours (internal mail).

The time taken to process through from punch-girl, to the computer, and onto the off-line services for checking and dispatch of results will again depend upon the type of computer installation. In general, however, workshop times will be a few hours, and bureau times between 12 and 24 hours. It is at this stage that delays can start, however, as there may be errors in the part program which have caused failure and/or error diagnostics. In the case of workshop-computing, this can be corrected on the spot by the part programmer (who is also 'punch girl' and 'off-line services'—see Fig. 12.3) and only a few minutes need elapse before the part program is re-run. However, in a bureau environment, the off-line services department must contact the part programmer for his comments. Most in-house and many remote bureau will merely return the job to the part programmer to re-submit; the specialist computer bureaux, on the other hand, may endeavour to correct the error themselves if it is an obvious one, or to solve it by telephone or telex otherwise, only returning it to the part programmer as a last resort.

When an apparently satisfactory tape has been produced, it is returned to the part programmer by the reverse of the route that the written part program originally took (although occasionally a data-transmission link may be used to cut this return journey to a few minutes), and put on an NC machine tool for final test. This may show up further faults in the part program, necessitating its return to the part programmer and the repetition of the complete cycle before a tape is produced that can be used for production. The time taken for this final check will depend very greatly on the availability of the requisite machine tool. It is therefore very desirable to have enough checks at earlier stages to make this a final 'check' rather than a detailed 'test'. The total times that may be involved in the transition from drawing to valid control tape are shown in Fig. 12.4.

Efficiency

The remaining three features—efficiency, reliability, and economy—are all inter-related to a great degree and cannot therefore be discussed in quite such an isolated manner as can the turn-round. However, while there is an economic side attached to both the efficiency and the reliability of a service, this side will be temporarily ignored until the whole question of the economics of the various approaches is discussed.

Efficiency can therefore be taken as making the best use of the resources available (both material and human) in the transformation of the engineering drawing into a control tape. With this definition, it is clear that a bureau is more efficient for all but trivial work, as either form of workshop demands that the part programmer should devote a considerable proportion of his time to operating the computer; this is clearly inefficient as his time could be more efficiently employed writing further part programs. A solution sometimes suggested is that some form of clerical staff should be employed to look after the computer side, but this solution has rarely, if ever, been successful, for it

	In-house workshop	Remote workshop	In-house bureau	Remote bureau
Write part program	1–2 days	1–2 days	1–5 days	1–5 days
Send for data preparation	—	1–2 h	1–24 h	1 hour–3 days
Data preparation	$\frac{1}{2}$–4 h	$\frac{1}{2}$–4 h	1–10 h	1–10 h
Process to off-line services	$\frac{1}{2}$–3 h	$\frac{1}{2}$–3 h	12–24 h	12–24 h
Correct errors and repeat previous two steps	5–30 min	5–30 min	(non-specialist) 30–60 h	(specialist) 12–24 h
Return to part programmer	—	1–2 h	1–24 h	1 h–3 days
Verify tape on machine tool	several days	several days	several days	several days
Total time up to final verification (allowing for two false runs)	2–3 days	2–3 days	8–14 days	4–15 days

Figure 12.4   Typical processing times.

effectively sets up an in-house bureau without any of its advantages (skilled staff, etc.) while retaining many of its disadvantages. A genuine bureau, on the other hand, means that once a part programmer has written a part program he can forget about it until it returns, correct or otherwise, from the bureau. Experience in all forms of computing has shown that, at present, for all save the simplest jobs this is the most efficient way of running a computer or of using its services. The efficiency of the computer usage is also an important point if it is to be in-house; and it is virtually undeniable that computer time is most efficiently utilized when the computer is run by a specialist department which operates it and controls all work going onto it.

Reliability

The reliability of whichever method of computer assistance is used is clearly of paramount importance in NC and the level of reliability that can be expected must therefore be examined very carefully before any decisions are made.

In a workshop environment, the programs being used will almost inevitably be either general-purpose fixed-format, or special-purpose. The general-purpose free-format programs tend to be too large and complex for use, and especially for maintenance, in such an environment. Such fixed-format or special-purpose programs will probably, as mentioned earlier, be tied to a particular computer or range of computers. In all probability, therefore, a workshop user will obtain his program from the computer manufacturer, or from a machine-tool manufacturer, and will then run it himself. At intervals, the supplier will issue an updated version which will correct faults which have been found in the old program, or will add some extra feature to the program. If the user finds a fault therefore, he can either notify the supplier and then wait, possibly for several months, until an update embodying the correction is issued, or he can try to correct himself. In practice, the latter alternative is often the only way to get certain jobs processed within a reasonable time, and so it is essential to have at least one person who understands the program well enough to modify it where necessary. A further point to be borne in mind is that programs in this class have often been written by specialist bureaux for use by their customers (for example, KIPPS and PROFILEDATA), and these will not, in general, be made available to other computer users. Finally, there is the long-term problem of the maintenance of programs whose number will increase as the users' involvement in NC grows. An increasing amount of the NC user's time must, therefore, be spent in the maintenance of such programs, and the already low level of efficiency reduced still further.

When considering the reliability of a bureau service on the other hand, one must include not only the free-format general-purpose programs but also the off-line services. In this context, bureaux should be considered in a rather different manner than hitherto, namely as specialist and non-specialist bureaux —that is bureaux (either remote or in-house) which have specialist programming staff concerned with the maintenance of NC programs, and those that do not. In general, in-house bureaux tend to be of the non-specialist variety, while remote bureaux fall into both categories. The reason for this distinction is clear from the points made about workshops, especially when one considers the added problem of error correction by bureaux mentioned earlier. For special-purpose and fixed-format programs, the specialist bureau will have staff who are conversant with the programs (and often who will have written or modified them) and so the maintenance does not cause any problems. The free-format programs are rather different, however, as in general these programs have many man-years of effort in them and are very complex. As a result, they tend to be written by research institutes or universities under some form of multiple sponsorship scheme, and require a very considerable maintenance effort by the central organization, based upon a comprehensive error-reporting system. The results of this maintenance, and of any new development work, are issued in the form of an update at more or less regular intervals to all those organizations using the program. Thus, the bureau must have available staff who can recognize errors that are due to the program and must notify the appropriate maintenance body of these errors, in order that corrections may be made. However, because of the considerable time that is likely to elapse before such corrections are issued, it is clearly desirable that the bureau should be able either to modify the program temporarily or to suggest some way of getting round the problem by altering the part program. This is a service that can, by definition, only be provided by a specialist bureau.

This recognition and correction of errors is an important part of the reliability of the service provided by a bureau. As was stated when discussing the turn-round of the time of various approaches, the time taken in correcting errors in the part program is often drastically reduced if the bureau can either correct them at the bureau, or contact the part programmer by telephone or telex for his help. To be able to do either of these, however, demands that the off-line services department of the bureau understand the program and its part programming in order that they may correctly diagnose such errors. A bureau offering this type of service should request drawings with part programs.

Economy

One of the greatest problems faced by the prospective user of computer assistance is deciding whether it is economically justifiable, for in the final analysis it is probably this which is the overriding factor. The remote bureau is the easiest method of access to cost as there are virtually no hidden extras; a part program is dispatched and after one or more runs on the computer the control tape is returned. The cost of the bureau's services will normally be at a flat rate of £X per minute spent on the computer plus a small fixed set-up charge, or it may be charged according to some formula based on the quantity of input, output, or possibly both. There will, in general, even in the case of specialist bureaux, be no charge for the assistance of specialist staff in the eradication of any errors. Thus, the only problem is how many runs, on average, will be required to obtain a correct control tape. For the more straight-forward point-to-point work, the answer is that in 8 cases out of 10 one run will be enough, but for more complex milling several runs are often required. However, as it is in these areas that the greatest savings are normally to be found, a considerable saving over manual methods still results. A further economic advantage of the remote bureau is that all maintenance and development costs are absorbed by the bureau (and in the case of a specialist bureau these are considerable), the only occasion when these are directly passed on to a customer being when a program, or more probably a post processor, is being specially written for him.

13

Figure 12.5 shows how the average cost of using the KIPPS point-to-point program at the ICSL Kidsgrove Bureau compared with the manual cost for a major customer of the bureau in 1967. It shows that for small jobs (or rather small batches of jobs, since several part-programs, if run together, are charged as a whole) the cost of computer assistance was more than that of manual part programming (because of the £5 minimum charge for a run). However, batches of jobs with a total of 85 or more operations (into which category almost all their work fell) were more economically done with the aid of the computer. For all types of work, there is some such break-point, and as new techniques are used it becomes increasingly lower.

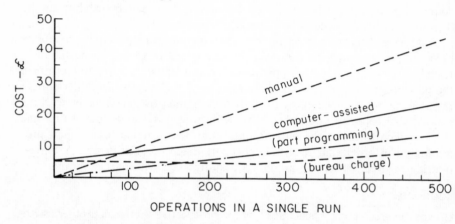

Figure 12.5 A major bureau user's NC costs.

The economics of an in-house bureau are far more difficult to assess. This is largely attributable to the problem of costing computer time. Very few in-house computers are fully utilized and therefore they have spare time available after processing the work for which they were installed and which (theoretically) justifies their very existence. Every organization has its own method of costing, usually slightly different from any other. However, almost without exception, all lead to a lower charge than that of a remote specialist bureau operating for the same period.

The in-house workshop has the same costing problem as the in-house bureau, although as it is often only used by one department it may not be quite so difficult as in the case of its bureau cousin. However, the computer time is often more expensive, even though the capital cost of the workshop computer may be appreciably less, as is shown in Fig. 12.6. The first three columns show how the cost to the customer for time on small and large computers in an in-house environment, and a large computer in a specialist bureau, is formed from its constituent components. The in-house figures are calculated on depreciation, plus 100 per cent overheads, plus maintenance (at £1 per hour per £100,000 capital cost), to produce basic cost; a further 100 per cent on this for program maintenance, administration, etc., gives the total charge to the user. The specialist bureau figures are calculated similarly but with an initial 125 per cent overhead reflecting the larger number of specialist staff, and a final 150 per cent extra charge reflecting the specialist programming, maintenance, off-line service, and profit margin. The final two sections show the approximate time and cost for a typical medium-size NC job using a high-level language such as APT or EXAPT. A large computer will be very much faster than a small one and approximate average times are shown reflecting this difference, and hence the cost of such a job.

Type of computer	Small in-house	Large in-house	Large specialist bureau	Large multi-access specialist bureau
Approx. capital cost	£65,000	£500,000	£500,000	£2,000,000
Cost per day of depreciation over 5 years	£52	£400	£400	£1,600
Overheads per day (100 per cent in-house, 125 per cent spec. bureau)	£52	£400	£500	—
Cost per day excluding maintenance	£104	£800	£900	—
Maintenance per hour	£1	£5	£5	—
Cost per hour for (a) single-shift (b) double-shift (c) treble-shift	£18·35 £9·65 (£6·80)	£138·35 £71·65 (£49·45)	(£155)* (£80) £55	—
Charge to customer per hour (cost + 100 per cent in-house cost + 150 per cent sp. bureau) (a) single-shift (b) double-shift (c) treble-shift	£37 £19·50 (£13·50)	£280 £145 (£100)	(£390) (£200) £140	£300 C.P.U. and £3 terminal, or off peak £200 C.P.U. £1·50 terminal
Time required for typical medium-size NC job	20 min	2 min	2 min	30 min terminal 40 s C.P.U.
Cost for above job (a) single-shift (b) double-shift (c) treble-shift	£12·35 £6·50 (£4·50)	£9·35 £4·85 (£3·35)	(£13) (£6·65) £4·65	£4·85 or off-peak £3·00

* Bracketed figures usually not applicable.

Figure 12.6

Comparison of these figures is not straightforward, since an in-house computer will not usually justify three-shift working, whereas a specialist bureau will rarely, if ever, operate with less than three-shift working. Thus one finds that a specialist bureau will charge about £4·50 for this job, while a large in-house computer bureau would charge about the same if they could justify two-shift working, but almost twice this if they were using a single-shift. When using a small computer in a bureau mode, the costs would be about 30 per cent higher than when using the larger computer, although if this were used in a workshop mode (single-shift without the final 100 per cent overhead) the cost would be of the order of £6. The specialist bureau will, however, have a minimum charge per run on the computer (which may be more than one job) to cover the basic set-up time on the computer. This is typically of the order of £10; thus two such jobs, or one larger one, would need to be submitted together before the computer time became cheaper in reality as well as in theory.

Thus far, economics have been considered purely in terms of £x for processing y tapes. However, as was mentioned earlier, the efficiency and reliability of a service also have an economic side. The prospective user must decide exactly what he can or cannot afford and what he does or does not need, and endeavour to balance the two. Thus, if he wishes to have a reliable service using APT, 2C,L or EXAPT, he must either use a specialist bureau or use his own (large) computer and pay both membership fees to the appropriate organization, if applicable, and the cost of employing the extra skilled staff. Similarly, a workshop enables a much faster turn-round to be obtained and 'time is money'; but this is at the cost of wasting the particular skills of the part programmers on other (computer) work and their time is, of course, also money.

In general, except for the small infrequent user, it is probably fair to say that a bureau (specialist or otherwise, depending upon the work) is the most economically justifiable approach when all things are taken into consideration, unless a turn-round of hours rather than days is essential. If, however, such a rapid turn-round is required, and the specialist services of a bureau are also required, arrangements can often be made for special delivery or transmission of data by means of a data-link over GPO lines.

Multi-access

The idea of sending data, and receiving results, by means of a data-link leads to a new concept of computing which has arrived in the last few years, and which promises to alter radically the methods of computer use. This is the concept of multi-access on-line computing mentioned earlier, using a terminal consisting of either a small typewriter directly linked to the distant computer, or a small input/output 'station' consisting of perhaps a card or paper-tape reader and punch together with a line printer or typewriter. Provided with this type of equipment, the user has all the advantages of a specialist bureau as regards the writing and maintenance of programs, without the inevitable delays in turn-round. He gets the rapid turn-round and error diagnosing of an in-house bureau or workshop, without the necessity for detailed computer knowledge. Such systems can either be used in a 'conversational' mode via the typewriter, when

computer will interrupt with error messages as soon as an error is made, or
a more conventional mode using either the typewriter or input/output station
input of data and output of results.
The economics of this new approach are also startling, as can be seen by
erence to the fourth column of Fig. 12.6. While it is not easy to give a detailed
eakdown, the charges shown are of the order to be expected from a specialist
reau having a £2 million computer at its disposal. Here one finds that to
ocess the 'typical job' considered earlier, during peak hours (probably 9 am
6 pm), would cost just under £5. Outside that period, it would cost just
er £3. Turn-round time would be 30 minutes! There is of course, in addition,
cost of hiring the terminal equipment and GPO modem (to connect it to a
ephone line) which will average, for a typewriter terminal with paper-tape
:achment, about £2 per day (assuming a 5-day week), and the cost of the tele-
one line. This latter could vary between £150 and £1,500 per annum depend-
g upon the method used. Even assuming the most expensive method and
lely peak-hour working, it would cost barely £1 per hour, thus raising the
st of the 'typical job' to £5·35, comparable with all except three-shift in-house
reau working.

Making best use of the computer

om the foregoing, it will be apparent that there is no one answer to the ques-
)n posed at the beginning of this chapter. However, the large number of
)parent possibilities should, for any particular NC user, have been reduced

	Programs available	Turn-round	Efficiency	Reliability	Economy
In-house workshop	G.P. fixed	1–4 h	poor	fair	poor
	S.P.	1–4 h	poor	fair	poor
Remote workshop	G.P. fixed	3–9 h	very poor	fair	very poor
	S.P.	3–9 h	very poor	fair	very poor
In-house bureau	G.P. free	3–24 h	good	fair	poor
	G.P. fixed	3–24 h	good	good	good
	S.P.	3–24 h	good	good	good
Remote non-spec. bureau	G.P. free	48–96 h	good	fair	poor
	G.P. fixed	48–96 h	good	good	good
	S.P.	48–96 h	good	good	good
Remote specialist bureau	G.P. free	24–96 h	very good	good	good
	G.P. fixed	24–96 h	very good	very good	good
	S.P.	24–96 h	very good	very good	good
Multi-access bureau	G.P. free	½–2 h	good	good	very good
	G.P. fixed	½–2 h	good	very good	very good
	S.P.	½–2 h	good	very good	very good

Figure 12.7   Summary of computing aids to NC.

to a much smaller number. In the conventional field, the decision between economy and turn-round must be taken with due regard to efficiency and reliability. In future, the growth of multi-access computing will tend to give the best of all worlds, particularly for the medium-scale user.

Someone starting in NC would be well advised, in the first instance, to go to a specialist bureau (either in-house or remote) as it will be able to help him most. He will also be well advised to standardize on APT-like part programming as it is towards this that the trend is moving. As he gets more experienced and wishes to spread his wings, he must decide whether to stick with the specialist bureau or to take on more responsibility himself via a non-specialist (presumably in-house) bureau or an in-house workshop. In the near future, he will probably attempt to justify installing a terminal to a specialist multi-access bureau which will give him a rapid turn-round as well as the specialist assistance he is used to.

Figure 12.7 shows, in summary, the advantages and disadvantages of each approach discussed. From it and the arguments mentioned, it is clear that the trend in other forms of computing is likely to be repeated in NC. In short, the best use of a computer is probably made by using it on a bureau basis, either in-house or remote, preferably via some form of multi-access link.

13. Less common NC machines

James A. Baughman

For career summary see chapter 10.

Numerical control equipment is usually thought to be primarily for chip-cutting requirements; however, there is a definite trend of expansion to include a variety of applications. Several of these different machines and applications are described below.

Electron-beam welder

Electron-beam (EB) welding equipment uses a contaminant-free vacuum chamber in which the workpiece is positioned relative to an EB head. The beam produces a deep, narrow weld with practically no heat conductivity to adjacent areas. The beam is focused to a point in a vacuum of 10^{-4} mm Hg, bombarding the surface metal to such an intensity that it turns into a near vapour. The resulting hole in the metal permits deep penetration and, as the beam is moved, the near-vapourized metal solidifies to produce an excellent weld.

The EB welder is capable of welding any material from 0·010 in. to 2 in. thick stainless steel or 3 in. thick aluminium or titanium. It can penetrate to a 6 in. depth in materials where 1 in. would be maximum for arc welders. The width of weld can be as narrow as 0·005 in. Welding speeds are 20, 30, and 50 in./min. Also, the welding process can be controlled so that the beam will penetrate an outer metal and weld an inner, normally inaccessible, area.

Figure 13.1 shows an electron beam welder. It features a Sciaky programming system for 2 axes contouring control, using the Bunker Ramo 3100 NC control. Additional features include automatic seam tracking, and adaptive control. The NC system is used for the usual variety of two-dimensional contour welding. Programming can be accomplished using APT or any other system with a suitable post processor. The seam tracker (which functions separately from the NC system) uses a probe, positioned directly in front of the beam, which traces a contoured surface. The tracing process uses a transducer to convert the lateral movements into electrical signals that drive a positioning servomotor to maintain alignment within 0·0025 in. Parameters such as accelerating voltage, focal distance, beam current, and welding speed are preset by the operator; however, the Sciaky adaptive control system provides surveillance of the welding parameters and adjusts as required to assure weld quality.

EB WELDER-68x68x78- WITH NC CONTROL

Figure 13.1   Electron beam welder. (*Courtesy Sciaky Bros., Inc.*)

A Sciaky EB Welder is currently being used at a US Naval Air Rework Facility. The primary use is to facilitate overhaul and repair of jet engines. The machine has a head and tailstock with 24 in. diameter faceplates inside a 68 in. long × 68 in. wide × 78 in. high vacuum chamber. Also included is a work carriage and rotary indexing table. Current applications are welding of jet engine reinforcement bands, engine fuel nozzles, and turbine exhaust cases.

Riveter

NC riveters are being produced for two-axis to multi-axis requirements. These machines position the workpiece under the head, clamp, drill the hole, inject wet sealant, insert the desired rivet or bolt, squeeze the insert (upset), and retract tool. Their use is quite practical since they can solve crucial problems involved in the manufacture of large sheet metal panels where strength-to-weight ratios are stringent, and where exotic metals are used.

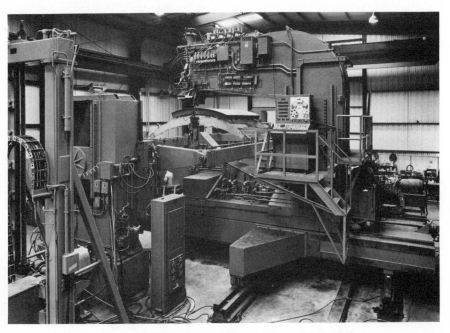

Figure 13.2 NC riveting machine. (*Courtesy Conrac Corporation*)

Figure 13.2 shows an NC automatic fastening machine produced by the CONRAC Corporation. These machines can accommodate workpieces of various size; for instance, CONRAC's largest can handle a 42 ft × 16 ft double-contoured panel with five-axis positioning requirements. It is capable of drilling titanium sheets and installing titanium rivets. The machine itself is constructed with high-strength alloys, ball-bearing leadscrews, and up-to-date hydraulic motors and actuators for fast positioning. Some features of the machine include a production rate of 12 cycles/min (including positioning), positioning accuracy of ± 0.005 in., pre-programmed automatic tool selection, and normality and tracing sensors to prepare control tapes using an integral record/playback device.

For double-contoured surfaces, programming is usually accomplished by use of the APT system and a suitable post processor.

Turret punch press

The turret punch press is designed for producing parts with varied hole patterns with openings of many shapes and sizes. Parts requiring notching, nibbling, and piercing can be made in one set-up. Other items such as knockout holes, extruded forms (countersinks, louvres), centre-punch locating points, and multiple piercing, can be easily accommodated with this equipment.

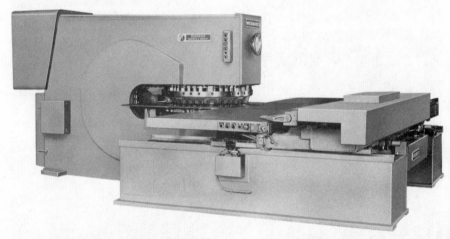

Figure 13.3 Wiedemann turret punching press. (*Courtesy Warner and Swasey Co.*)

Figure 13.3 shows a turret punch press produced by the Warner & Swasey Company, Wiedemann Division. Some typical parts produced are shown in Fig. 13.4. The rotary turret on the press contains 30 punches and dies. Each tool is located in an individual station mounted in a pre-aligned punch-holder and die-holder. The turret will index and lock within 4 s. Hole piercing is accomplished at 100 to 200 crankshaft strokes per minute. The workpiece is positioned at a rate of 1,000 in./min. Another feature of the equipment is self-adjusting pneumatic clamps to secure the workpiece. Press capacities run between 15 to 100 tons; maximum material thickness is $\frac{1}{2}$ in. in mild steel of 50,000 lbf/in.[2]

shear strength. Maximum hole diameter in $\frac{1}{2}$ in. steel is 5 in. Absolute hole positioning accuracy is ± 0.004 in. to ± 0.008 in., depending on press size.

Part programming is relatively simple and is accomplished either by hand programming or computer-assisted methods. All functions of the Wiedematic press are controlled by tape.

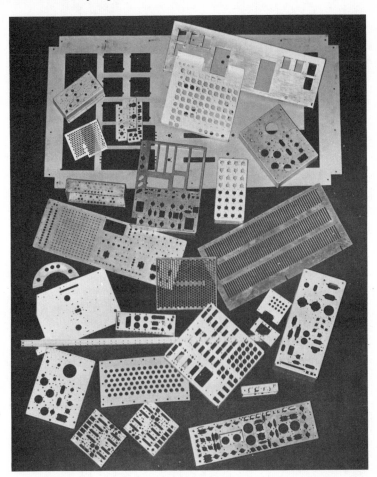

Figure 13.4 Typical punched parts. (*Courtesy Warner and Swasey Co.*)

Controlled inspection machine

One of the problems brought about by the increased productivity inherent in NC machine tools is the inability of conventional inspection methods to maintain the required pace. The use of gauge blocks, height gauges, micrometers, etc., by inspection personnel simply cannot match the NC productivity rate.

Several techniques have been introduced to overcome this problem. These include part checkout by the NC machine tool itself, inspection by an operator-controlled co-ordinate-measuring machine (see chapter 14), and inspection on a programmed NC inspection machine.

Figure 13.5   Four-axis NC inspection machine. (*Courtesy New Britain Machine Co.*)

An example of the latter is shown in Fig. 13.5. The Lucas four-axis NC measuring inspection machine, equipped with a Sperry UMAC5 Control, is designed to inspect and verify the part geometry or to digitize unknown geometric configurations. The inspection probe is shown in Fig. 13.6. The inspection machine has a longitudinal travel of 108 in., transverse of 72 in., and vertical probe travel of $33\frac{3}{4}$ in. with a programmed rotary table of 48 in. diameter. The traverse feedrates are 200 in./min for X and Y, 60 in./min for Z, and 2 rev/min for the rotary table. Accuracy is ±0·0005 in. between any two points within 12 in., and a repeatability is ±0·0001 in.

Figure 13.6 Lucas inspection probe. (*Courtesy New Britain Machine Co.*)

Inspection can be accomplished by manual or tape input. Inspection points and any deviations from desired co-ordinates are printed for permanent record. An example of part inspection with the operator checking the printed output is shown in Fig. 13.7. When complex contours, high accuracy, and large parts are involved, the use of this type of equipment gives inspection productivity improvements from 10 to 25 times over conventional inspection methods.

Figure 13.7   Lucas inspection machine in use. (*Courtesy New Britain Machine Co.*)

Tube bender

In the past, tube bending consisted of estimating the working tube length and amount of overbend to compensate for spring-back; then the tube was experimentally formed until the correct parameters were obtained. This tedious process has been replaced by a simpler, faster method available in NC tube-bending equipment. The NC equipment allows parts to be programmed by simply inputting workpiece dimensions either manually or by tape. Optional features allow automatic springback detection and correction, plus tape preparations by the machine tool.

Automatic precision bending is accomplished on equipment such as the Pines bending machine shown in Fig. 13.8. This equipment will handle 2 in. maximum diameter tubing with 0·125 in. maximum wall thickness. Typical bends per hour are 300 with a length accuracy of $\pm 0·005$ in. and bend accuracy of $\pm 0·1$ degree. The Pines 'Cybermat' control features a memory system which accepts input by tape or by manual means and then produces parts without reference to the original input. Memory data can be displayed and any desired corrections can

Figure 13.8 NC tube bending machine. (*Courtesy Pines Engineering Co. Inc.*)

Figure 13.9 Control equipment for Pines tube bender. (*Courtesy Pines Engineering Co. Inc.*)

be put in manually as needed. When a permanent tape is required, memory data is punched onto standard 1 in. tape.

The economics of tube bending are concerned primarily with the time required for set-up and the time required to make one bend and move to the next bend. There is a series of motions required in a standard bend operation such as closing the clamps and pressure dies, advancing and retracting the mandrel, advancing and rotating the tube, and rotating the bending arm. In the past, these motions were sequential. The Pines equipment features a full three-axis control unit (Fig. 13.9) which speeds production by commanding five different combinations of simultaneous movements. Tool set-up is simple: dies drop into place on precision die holders, and pressures are gauge set. Also of aid is the control logic that enables the operator to control precisely the operation such as dialling various limits for forward travel, inhibiting one movement before another can start, etc., all of which reduces set-up time by as much as 70 per cent, as compared to standard hydraulic bending machines.

Stretch-forming machines

The stretch-forming process is a method whereby the material is subjected to a pulling force (tension) before forming. The material is elongated to its yield point where its condition is semi-plastic. During the subsequent forming process, the material is held at or slightly above the yield point, producing a part with minimum residual stresses and a minimum of springback. This process produces parts with few internal stresses and is thus particularly suitable for applications where high temperatures are involved, such as in rocket engines and in the welding process for ship hull plates.

Figure 13.10   Stretch-forming machine. (*Courtesy Cyril Bath Co.*)

Figure 13.10 shows a hull-structure-forming machine by the Cyril Bath Company. Initial specifications are for a stretch tension of 400 tons, minimum bend radius of 10 ft, maximum bend angle of 45 degrees and 40 ft maximum part length at a production rate of 8 to 10 pieces per hour with mechanized handling. The machine basically consists of two tension-ram assemblies, each riding along a track at each end, and a series of die segments which can be positioned to produce the desired shape. The NC system is shown schematically in Fig. 13.11. The tape is used to position each of the die segments to form any radius or shape within the machine capacity.

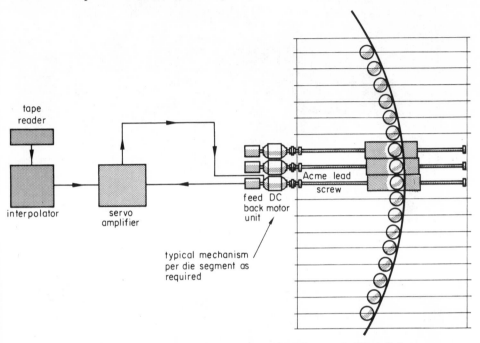

Figure 13.11 NC die segment positioning mechanism. (*Courtesy Cyril Bath Co.*)

It is estimated that programming would be relatively simple with tape output per part averaging only about 4 ft. Maximum time required for workpiece set-up would be 5 to 6 min with most cases averaging less than 1 min.

Ultrasonic tester

One of the most unusual NC machines developed is an ultrasonic inspection system which inspects parts for material flaws under water with sonar-like ultrasonic signals. A three-axis NC unit positions an ultrasonic transducer in correct relation to the part being inspected. The transducer sends high-frequency sound beams through the water into the material and, by alternately sending and listening for echoes from the material, scans the entire piece for discontinuities in the basic material. Since sound travels at a definite rate in water and in the

Figure 13.12   Ultrasonic testing equipment. (*Courtesy Bardel, Inc.*)

Figure 13.13   Ultrasonic tank. (*Courtesy Bardel, Inc.*)

material, the length of time for signal return can be calibrated and recorded. By monitoring the signal, flaws and metallurgical variations can be identified and located.

Figure 13.12 shows the ultrasonic tester developed by Bardel, Inc. A Bunker-Ramo 3100 NC continuous-path control unit directs the system. The stainless steel water tank measures 144 in. long × 53 in. wide × 27 in. deep. Figure 13.13 shows the two turntables in the tank and a special polar recorder which is slaved to the turntable and synchronized by the NC unit. The turntables carry parts up to 4 ft diameter and a maximum weight of 1,000 lb. Test speeds from less than 1 rev/min to 60 rev/min are possible. The two turntables allow one part to be loaded while the other is being inspected. Production rates are many times that of the conventional manual method which required highly trained technicians.

Programming is accomplished by means of the APT system. A Bunker-Ramo APT 3 Fortran post processor is used to match the normal metal cutting commands to the inspection requirements.

Wire processing

It has been estimated that some 5 billion solderless wrapped connections are made per year over a wide range of applications. For instance, some circuit boards require as many as 16,000 terminal connections. Also, the electronic and electrical manufacturers are continually faced with the requirement for reduced weight and size. This in turn places a responsibility on the wire processing

Figure 13.14 NC wire-connecting machine. (*Courtesy Gardner-Denver Co.*)

machine tool builder to produce high processing speeds, exceptional accuracy, and satisfy miniaturization aspects of automated manufacturing.

The Gardner-Denver Company has answered this requirement with its Automatic Numerically Controlled Wire-Wrap® Machine as shown in Fig. 13.14. The unit is designed to automatically attach inter-connecting wires with solderless wrapped connections upon electronic or electrical assemblies having modular terminal arrangements. The machine will produce a variety of wiring patterns, utilizing four dressing fingers which provide pivot points in a given pattern configuration. The ability to produce patterns allows great flexibility and also reduces wire density and cross-wire routings.

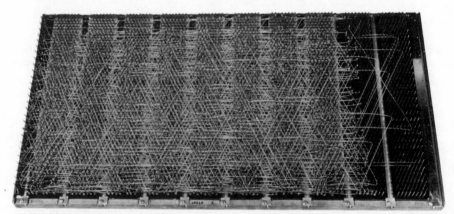

Figure 13.15   Typical panel wired by NC machine. (*Courtesy Gardner-Denver Co.*)

The wire, which is insulated, solid-conductor type, is fed into the machine from a barrel or bulk spool. The machine performs the desired pattern, strips the wire, and wraps the two ends simultaneously to the desired terminals. The average cycle time for the process is 5 s. An example of a typical workpiece is shown in Fig. 13.15. Customer data has shown that a panel can be wired 30 times faster with this method, compared to the normal hand soldering method, and without error.

Input to the machine is in punched card or tape form. Programming can be accomplished using straightforward manual programming techniques. Some companies, however, enhance the machine's capabilities by automating the engineering design process, using computer-aided design techniques in which the computer optimizes the wire routing and determines ideal sequencing. Output from this system produces cards or tape to operate the wire-processing tool. The computer program also outputs wiring lists, bills of material, connection lists, component counts, and other pertinent managerial data.

14. Applications of numerical measuring systems

H. Ogden

Harry Ogden, BSc(Eng), CEng, AFRAeS, is Manager, Measurement Department, Numerical Control Division, Ferranti Ltd, Dalkeith, Scotland. After graduating from the Halifax Municipal Technical College in 1944, he was employed by A. V. Roe and later by the English Electric Company on aircraft structural stress analysis. From 1948 until 1954, he was Scientific Officer with the National Gas Turbine Establishment, during which time he undertook extra-mural studies in electronic instrumentation and servo control theory. In 1954, he joined Ferranti Ltd as Chief Mechanical Engineer in the newly formed NC development team. He was responsible for establishing principles, techniques, and physical units for the successful application of NC electronic equipment to a wide variety of machine tools. The Ferranti Inspection machine, which he invented in 1956, is now used in many countries of the world to reduce inspection times by up to a factor of 20.

The increasing acceptance of numerical control in manufacturing industry has shown the need for NC techniques in less sophisticated form to be applied to established methods of manufacture and show considerable improvement in accuracy, repeatability, production economics, and operator convenience.

This is particularly true of the established use of numerical measuring systems for use in manufacture, inspection, and data preparation at the design drawing stage. By using the developed NC techniques of measurement feedback in association with electronic readout display, a wide range of applications has opened up which are economically justifiable and broaden the effective base of measurement and control technology.

Machine tool electronic readout

Conventional machine tools, until relatively recent years, have not been particularly well endowed with measurement facilities. The use of optical measuring scales for jig borers grew apace in the 'fifties but the extension of this form of measurement to larger machines on the one hand, and production-type machines on the other, has not been extensive because the measurement facilities do not match the functional requirements.

On the larger machines, facilities for datum shifting with centralized multi-axis display are now possible by the use of NC-derived measuring systems, of which possibly the Ferranti Moiré Fringe and Farrand Linear measuring systems have been most widely adapted in the Western world.

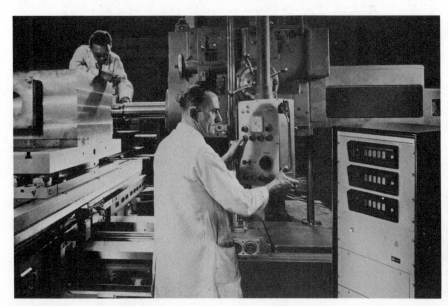

Figure 14.1   A large horizontal boring machine equipped with Ferranti 3-axis readout. (*Courtesy Ferranti NC Division, Dalkeith*)

The facilities of these electronic measuring systems include a fully floating datum facility, dial insertion for datum offset, and bi-directional electronic counters with display in illuminated figures, localized for operator convenience. Resolutions of 0·0001 in. (0·0025 mm) are available for production machines, with high resolutions for special applications, down to, say, one micron. Scale accuracies of better than 0·0001 in. per foot are achievable; the overall accuracy attained is dependent on the quality and orthogonality of the machine-tool slides.

Considerable development has gone into the packaging of linear measuring systems to facilitate mounting on existing machine tools without the need for re-design and indeed some of the more enlightened designs lend themselves to retrofitting on existing machine tools. A considerable retrofit market is being developed, both in Britain and the USA, particularly on large horizontal boring machines and milling machines such as that shown in Fig. 14.1.

Large lathes and vertical boring mills are also being fitted with numerical readout systems. It is rather surprising to note that the standard centre lathe has no ready means of measurement, particularly of diameter, and that the operator is dependent for quality control almost entirely on the use of a micrometer or other calipering device.

The Ferranti Digiturn, shown in Fig. 14.2, is a recently developed lathe electronic measuring system which, when applied to the conventional centre

Figure 14.2   A Digiturn system installed in a lathe showing the electronic counter display unit and the two measuring transducers for radial and longitudinal measurements. (*Courtesy Ferranti NC Division, Dalkeith*)

lathe, will achieve 20 per cent improvement in output and, in association with preset tooling on the lathe, can achieve a saving of up to 40 per cent on batch sizes of up to eight components.

The basic equipment comprises an electronic counter display unit and two transducers which measure respectively the longitudinal and cross slide movements of the lathe, these measurements being displayed on the decimal readout. The longitudinal movement may be zeroed at any point while the radial transducer movement is arranged to measure diameter. A multiple digit dial-in switch is used initially to align the diameter display to a reference diameter established on the lathe. Thereafter, by means of preset tooling techniques, no further diameter re-setting is necessary and all components can be turned by a semi-skilled operator dealing directly with printed instructions of length and diameter. The Digiturn system has become associated with a simple programming sheet, which instructs, line by line, an operator giving the appropriate turning tool, the lathe spindle speed, the feed per revolution, the direction of movement and the size in length or diameter, together with additional machining instructions.

Future developments in numerical measuring systems will be compact linear measuring transducers which will enable a wider range of application to other types of machines such as cylindrical grinders and toolroom milling machines. These dimensional facilities will enable operators on jobbing-shop work to increase their accuracy, consistency, and productivity.

Electronic radial measuring systems resolving in degrees, minutes, and seconds of arc, are available with both Moiré Fringe and Inductosyn radial scales. Commercially available radial measuring scales have resolutions of 1 to 3 seconds of arc and accuracies of 2 to 6 seconds of arc. Some large-diameter installations have used a novel measuring technique based on the use of thin stainless steel reflecting gratings stretched around a reference diameter with a precisely known number of lines. Such an installation has recently been applied to a Maag type gear-grinding machine by Ferranti in conjunction with the National Research Council in Ottawa, Canada, with an angular accuracy of better than 1 second of arc. The electronics associated with this measuring system leave the way clear to the machine's being put onto an automatic division mode of operation.

British industry has been encouraged to adopt metric units with the hope that by 1975 the greater part of the country's industry would have made the change. This impending change from Imperial to metric has led to the conclusion that in the interim there is a need for measuring equipment which can, at the turn of a switch, work in either the Imperial or the metric system. A number of British systems are now available which use either twin-track digitizers or linear measuring systems with twin pick-off arrangements or an electronic multiplier unit to enable this switch change-over from Imperial to metric to be achieved. It is considered that the flexibility of such electronic digital display systems will play an important part in British industries' metrication programme.

Inspection machines

In the pioneering of NC continuous-path systems, it became apparent at Ferranti, Edinburgh, that the proving of these NC products by conventional inspection was time-consuming, and parts that could be made in minutes took hours to inspect. It was thus conceived that a freely moving mechanical measuring machine with electronic numerical display would facilitate inspection of engineering components and, in effect, would change the whole economic aspect of conventional inspection methods.

The initial Ferranti development was an inspection machine with X and Y movements of 24 in. and 15 in. respectively and a Z movement of 10 in. designed for production inspection to an accuracy of ±0.001 in. with a measurement resolution of ±0.0005 in. This machine was equipped with an extendable vertical probe unit with interchangeable taper probe tips which established the constraint type of inspection process and revolutionized the approach to production engineering inspection.

Figure 14.3 Size 2 Ferranti inspection machine with workpiece, probes, etc. (*Courtesy Ferranti NC Division, Dalkeith*)

The machine was also fitted with lock and fine adjustment mechanisms on the X and Y axes so that measurements could be made by means of a projector microscope on components which were flexible or delicate. The locks also permit a marking-off function to be achieved. Figure 14.3 shows a number of these machines installed in a production inspection department.

This type of machine opened up a large market in the USA and led to the development of a number of similar machines with some increase in capacity, accuracy and resolution. On these machines, Z axis measurement has now been developed, time-shared on some and three-axis simultaneous on others. These NC inspection machines are now employed across the broad range of general engineering and not specifically in the new growth industries.

An increasing number of these inspection machines are being equipped with additional electronic facilities other than the numerical display of machine position, with devices such as printout and, more recently, as shown in Fig. 14.4, the use of the small computer. Printout provides a permanent record of inspected results and enables inspection departments to rationalize the inspection process and to provide records which may be used for a statistical analysis of the batches.

Figure 14.4 Cordax 3000 with PDP8S computer. (*Ferranti NC Division, Dalkeith*)

This analysis would lead to a review in detail of the manufacturing process and pin-point areas where inadequacies may be in the set-up or tooling or manufacturing plant. The use of semi-skilled personnel in the inspection function is dependent on this pattern of development. The teleprinter has been used for printout purposes in conjunction with a paper-tape input to print additionally the nominal dimensions in line-by-line juxtaposition with the actual measurements performed by the inspection machine, enabling direct comparison between nominal and actual.

Fundamentally, inspection is concerned only with manufacturing errors that lie outside the specified tolerance band. Indeed, the whole aspect of modern engineering is based on interchangeability of component parts and therefore inspection is concerned with the functional performance of these manufactured parts. Simply stated, a 'yes' or 'no' dimensional decision is required on these parts with further analysis potential to decide where the error lies in the manufacturing process.

By the association of a small computer in place of the simple teleprinter unit, automatic differencing between nominal and actual dimensions can be made to a pre-programmed inspection routine with automatic printout of the error and, if required, only those errors that lie outside the given tolerance band. By these electronic techniques, simplification of inspection routine and improvement in performance and productivity are key factors. Statistical evidence on batch sampling can be acquired as a by-product of this computer comparison process and by suitable programming a number of improved facilities become available. For instance, components being inspected by cartesian co-ordinates have normally to be aligned on the inspection machine precisely in accord with these co-ordinate axes. By means of an axis-transformation program, the inspection results can allow for mis-alignment and still print the appropriate true cartesian co-ordinates, the errors between actual and nominal dimensions, and indicate those errors which fall outside the agreed tolerance band.

Indeed, the whole aspect of inspection analysis is undergoing now a radical change, with development leading to 'sophisticated inspection centres' with NC inspection programming routines, and with automatic acceptance or rejection in accordance with a statistical quality-control analysis.

The associated development of in-process measurement is continuing apace and will enable stage-by-stage inspection to take place during the production manufacturing cycle and at the earliest possible stage in this process. From a logical point of view, the closer inspection is to the production process, the smaller the volume of possible rejected parts. A number of existing automatic transfer lines have automatic gauging and grading stations. By employing NC techniques, universality on a wide range of component sizes and shapes is possible and would give a similar degree of quality control to short-batch manufacture as is presently available to the volume manufacturer. The criteria that are used are usually concerned with maximizing the machine output from a profit point of view with restrictions on tool life, surface finish, and dimensional accuracy.

Drawing measuring machine

Numerical control display systems lend themselves to printout and, by further manual keyboard facilities on the printout unit, to preparing the paper-tape information for NC machines.

The drawing measuring machine is a development of this basic technique, such that the X and Y dimensions on a special draughting board may be measured by suitable digitizers and displayed on a digital readout with the printout facility. The Ferranti ADE Reader shown in Fig. 14.5 is an example of this development, which comprises a draughting board with diffraction gratings, measuring scales on X and Y axes, a graticule probe, and a sliding control panel.

Figure 14.5   ADE reader. (*Courtesy Ferranti, Crewe Toll, Edinburgh*)

It has an associated electronic logic cabinet with panels carrying numerical indicators displaying the X and Y movements of the graticule and the record number; it also has miscellaneous control switches and selectors together with a free-standing teletype.

A layout or detailed drawing to a choice of scale from 10–1 to 1–10 can be accommodated on the draughting board assembly.

The graticule probe is 'pointed' at a geometric element on the drawing such as a point, line, or circle, and, when the appropriate switch on the sliding control panel is pressed, the geometric definition corresponding to that element is simultaneously typed and punched onto 8-channel paper tape for subsequent transfer of this information to a computer. It is possible to digitize curves not mathematically defined and to indicate tangency constraints between successive elements. Additionally, digiswitches on the control panel specify depth and radial and angular dimensions.

Thus, the drawing reader provides the facility for defining elements geometrically and numerically via a general-purpose computer.

At the computer, a cutting sequence program uses the geometric information supplied to determine the contour required. This profile geometry information is then transmitted to a tool-offset program, either the Ferranti Profile Data System or 2C,L. The tool-offset program computes and transmits the tool centre path to a continuous-path post processor which produces a machine-tool control tape.

Within the ADE Software package, programs exist for point-to-point as well as continuous-path control machining and post processors for a number of NC machine tools.

The advantages of the drawing measuring machine in comparison with conventional procedures are numerous. The transfer of design data directly onto tape in one step instead of through the conventional detailed drawing, part programming, and manual teleprinting stage, saves drawing-office time and improves product-lead time. The one-step process must improve accuracy of information and minimize mistakes. Provision is made to enter information more accurately than can be achieved by probing a drawing.

Firstly, the probe can be used to get nearly the desired dimensions by positioning it to the drawing. Then the drawing can be ignored and the probe carefully moved until the desired X and Y values are indicated for the probe position. To ease this process, the 0·0001 in. units can be dialled in.

Secondly, the desired dimensions can be directly punched on the tape from the teletype keyboard.

The back-up software relieves the tedium of geometric calculations and further minimizes possible errors.

The standardization approach of dealing with all layouts for subsequent manufacture by NC fits well into an organized routine. Automated draughting equipment is becoming available which derives specific detailed drawings from the layout drawing using the drawing reader output tape as the medium in association with further computer software.

In essence, the developments taking place in the interface between the design draughtsman and NC manufacture have the capability of improving overall economics in a most remarkable manner, and will have an increasing influence in the organization of manufacturing industry.

Example of the use of the drawing measuring machine

This machine is used primarily to convert graphic information into taped data for the direct control of point-to-point drilling or straight-line milling machines

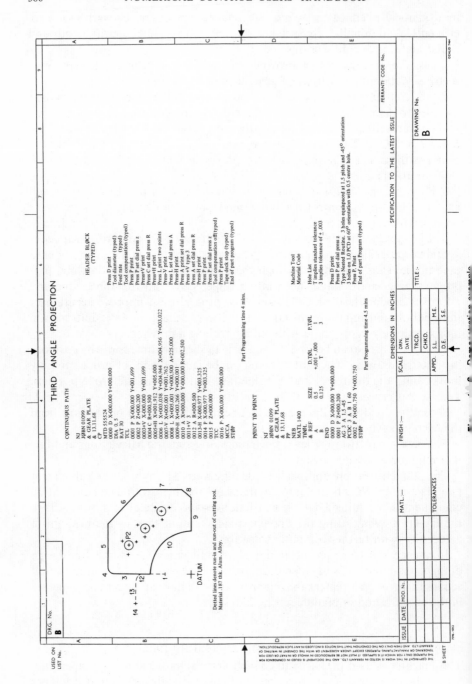

or, via a computer, for the control of point-to-point and continuous-path machines. It is the last case that the example illustrates.

Since a precision measuring machine generates the geometric information, the operator is only required to analyse the mechanical drawings and to describe them by using a set of push-buttons. In the computer process, the input information is accepted by special computer programs which expand the input information and produce tool-centre path data for a variety of NC machines, including a draughting machine.

In the drawing shown in Fig. 14.6, there is clearly a point-to-point section and a continuous-path section which may be defined and processed separately. The profile is defined by using the Simplified Cutting Sequence, and this is now described in some detail.

The facilities on the drawing measuring machine (or ADE Reader) can be used to assist the definition of lines and circles, to enable the changing of Z-levels, and also to specify the sequence of cutting.

When any printout button is pressed, the record number and code letter indicating the type of element to be defined are first punched onto tape and printed out,

e.g., 0001 P (where P indicates a 'point')

After the type of element has been established, the parameters defining the element exactly may be entered in several ways, depending on the requirements of the elements.

(a) *A point* only requires its co-ordinates

0014 PX − 000.977   Y + 003.325   (Fig. 14.6)

(b) *A line*

0006 L X + 003.038   Y + 004.944   X + 004.956   Y + 003.022

is defined and the parameters are determined by probing two points lying on the line and in a given direction. When the probe is aligned with each point, the printout button is pressed.

(c) *Depth, angles, and radii* may be entered by making use of a set of switches on the control box on which numbers may be entered (see Fig. 14.5). The pressing of an associated button will cause the number to be punched on tape preceded by a letter identifying it as a radius, angle, or Z-value, e.g., 0004 C R + 005.000 indicates a fillet radius of 005.000 units where the radius value has been set up on the control box.

(d) *A tangent* is defined by pressing a button indicating a tangency condition when a letter T will be punched automatically.

By using these facilities, the operator can enter on the tape a series of statements which, taken together, define the shape and size of the part and the sequence in which it is to be cut.

Directional elements

(a) A line has the direction given by its slope (or as previously implied). Thus, in Fig. 14.6, the line

$$0008 \text{ L X} + 005.001 \quad \text{Y} + 000.500 \quad \text{A} + 225.000$$

defines a line at an angle of 225° to the positive X axis, and

$$0003 + \text{V X} - 000.000 \quad \text{Y} + 001.699$$

indicates a vertical line in the positive or upwards direction with a zero X value. (The value of Y would be ignored by the computer program.)

(b) A circle or arc is defined as being clockwise or anticlockwise by the letter defining the type of element; thus,

$$0004 \text{ C R} + 000.500$$

indicates a clockwise circle and

$$0010 \text{ A X} + 000.000 \quad \text{Y} - 000.000 \quad \text{R} + 002.500$$

indicates an anticlockwise circle defined by its centre and radius.

Each element will be traversed until it meets the next element and the required intersection points are calculated exactly by the computer program. In addition, the operator can also enter any relevant machining information at appropriate positions on the tape. Examples of information for machining are shown in the drawing.

(a) Machine datum value 000 D X − 000.000   Y + 000.000
(b) Feed rates RAT 30
(c) Tool size DIA 0.5
(d) Tool compensation—indicating whether the tool centre lies to the left, to the right, or on the path as defined, e.g., TCL (Tool Centre lies to the Left of the path). The operator is also able to recall particular lines or groups of lines. Thus, 0011 + V 3 indicates a positive vertical line with the same parameters as the vertical line defined by the record number 0003. There are also various other facilities which are not illustrated in this example, such as curves defined using the interpolation, areas to be cleared to various depths, and construction lines and repeat sequences.

The computer program which accepts this simplified cutting sequence format carries out all the requisite calculations, using the geometric definitions punched on the tape, and produces an output which is fed through a post processor. The resulting control instructions are recorded on a tape which will activate the control system of the appropriate machine tool to produce the required part. The output from the simplified cutting sequence program can also be fed through a drawing program which produces a paper tape which in turn can be fed through a microplotter to produce a drawing on microfilm. This can also serve as a data check. Line printer output is also produced by the simplified cutting sequence program, mainly to record any errors which have been made by the operator defining the part but also to show the change points of the line and circle elements as they have been calculated.

The format for the point-to-point section is now described, again using Fig. 14.6. The main types of statement are:

(a) *Comment lines* preceded by an & (ampersand) and ignored by the program.
(b) *Program directives* contain information for the main program.
(c) *Post-processor directives* contain information for the post processor.

The heading block indicates the post processor required (e.g., NEB—indicates the Newall jig borer three-axis control), and a material directive (MATL) followed by a code number from a table ensures that spindle speeds and feed-rates appropriate to the material being cut will be allocated. Also in the heading block should be a list of holes with the required diametral and positional accuracy in each case. In this list, four parameters are required fully to define the hole required:

hole reference letter (e.g., A);
finished hole diameter required (e.g., 0·5);
tolerance which can either be selected from a standard table, in which case T is stated, or can be defined numerically (e.g., +0.001 − 0.000);
the bilateral positional tolerance expressed in thousandths of an inch (e.g., 1 = 0·001 in.).

Following the heading block, the machine datum is defined (0000 D X − 000.000 Y + 000.000) and the Z-level (0001 P Z + 000.200) defining the depth of hole required below the surface of the material. Following the Z-level, hole-defining information may be given. This will vary widely according to the hole distribution required but in this case the operator requires to position a hole-group routine (PCDC) on another hole-group routine (AG). He specifies first the routine for the principal hole pattern (AG).

AG 3 A 1.5 −45 indicates a row of three equispaced holes of the same diameter (hole reference: A). The distance between the holes is 1.5 and the orientation with respect to the X axis is −45°. He then specifies the subsidiary hole group routine which is to be superimposed on each of the holes of the principal hole pattern, i.e., PCDC 3 A B 1 60, indicating three equispaced holes on a PCD with a centre hole of different diameter. In this case, the centre-hole reference is A and the other hole reference is B. The PCD of holes is 1, and the orientation of the equispaced holes is 60°.

After the operator has defined this hole routine, he then probes the principal point at which the routine will be applied (e.g., 0002 P X + 001.750 Y + 003.750).

The special computer program then carries out all the necessary geometric

14

calculations and produces an output which is acceptable to any one of the point-to-point processors. The post processor then produces a tape to suit the machine on which the piece part is to be manufactured and adds standard planning information. In Fig. 14.7, the control tape information is on the left with the intermediate information on the right.

```
N001 G80 Z+000000 F999 M05 MACHINE LOADING DETAILS FOR MACHINING JOB NUMBER      1099 ON THE NEWALL 1520 JIG BORER.
N002 X+000000 Y+000000 F999 T31 M00
N003 X+020000 Y+041830 F999 S12 M13
N004 G81 Z+012000 W+008 F460

                                                    TOOL   CHANGE   REQUIRED

                                 FINISHED SIZE       TOOL SIZE          SPEED        FEED      TOOL NO.
                                      0.1250          0.1250            3000.0       6.0000      T 1

                                        X-CO-ORDINATE      Y-CO-ORDINATE      Z-CO-ORDINATE
N005 X+012500 Y+037500 F999                2.0000             4.1830             1.2000
N006 G81 Z+012000 W+008 F460

N007 X+020000 Y+033170 F999                1.2500             3.7500             1.2000
N008 G81 Z+012000 W+008 F460

N009 X+030607 Y+031224 F999                2.0000             3.3170             1.2000
N010 G81 Z+012000 W+008 F460

N011 X+023107 Y+026893 F999                3.0607             3.1224             1.2000
N012 G81 Z+012000 W+008 F460

N013 X+030607 Y+022563 F999                2.3107             2.6893             1.2000
N014 G81 Z+012000 W+008 F460

N015 X+041213 Y+020617 F999                3.0607             2.2563             1.2000
N016 G81 Z+012000 W+008 F460

N017 X+033713 Y+016287 F999                4.1213             2.0617             1.2000
N018 G81 Z+012000 W+008 F460

N019 X+041213 Y+011957 F999                3.3713             1.6287             1.2000
N020 G81 Z+012000 W+008 F460

N021 G80 Z+000000 F999 M05                 4.1213             1.1957             1.2000
N022 X+000000 Y+000000 F999 T32 M00
N023 X+017500 Y+037500 F999 S17 M13
N024 G81 Z+012000 W+008 F447

                                                    TOOL   CHANGE   REQUIRED

                                 FINISHED SIZE       TOOL SIZE          SPEED        FEED      TOOL NO.
                                      0.5000          0.4688            533.0        4.7853      T 2

                                        X-CO-ORDINATE      Y-CO-ORDINATE      Z-CO-ORDINATE
N025 X+028107 Y+026893 F999                1.7500             3.7500             1.2000
N026 G81 Z+012000 W+008 F447

N027 X+038713 Y+016287 F999                2.8107             2.6893             1.2000
N028 G81 Z+012000 W+008 F447

N029 G80 Z+000000 F999 M05                 3.8713             1.6287             1.2000
N030 X+000000 Y+000000 F999 T33 M00
N031 X+017500 Y+037500 F999 S17 M13
N032 G86 Z+012000 W+008 F405 M00

                                                    TOOL   CHANGE   REQUIRED

                                 FINISHED SIZE       TOOL SIZE          SPEED        FEED      TOOL NO.
                                      0.5000          0.5000            533.0        0.5330      T 3

                                        X-CO-ORDINATE      Y-CO-ORDINATE      Z-CO-ORDINATE
N033 X+028107 Y+026893 F999                1.7500             3.7500             1.2000
N034 G86 Z+012000 W+008 F405 M00

N035 X+038713 Y+016287 F999                2.8107             2.6893             1.2000
N036 G86 Z+012000 W+008 F405 M00

N037 X+000000 Y+000000 F999M31             3.8713             1.6287             1.2000

            TOOL  LIST  FOR  JOB.

            T 1           DRILL-SIZE    0.1250

            T 2           DRILL-SIZE    0.4688

            T 3           BORER-SIZE    0.5000

            JOB  COMPLETED.
```

Figure 14.7 Line printer output from example.

The above paragraphs by no means cover the scope of the ADE Reader but give some indication of the possibilities for mechanical parts requiring point-to-point and continuous-path machining.

Frame-bending machine

An interesting development involving in-process dimensional measurement of ships' frames during an automatic cold-bending operation has been carried out jointly by Hugh Smith of Glasgow and Ferranti Ltd, Dalkeith, with the sponsorship of the British Ship Research Association.

This machine is currently installed in the ship building yard of Swan Hunter of Wallsend and is illustrated in Fig. 14.8.

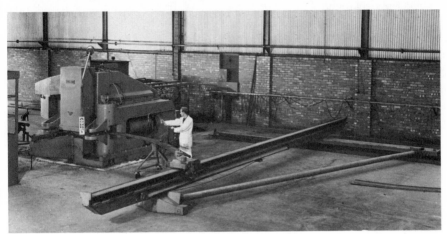

Figure 14.8   Ship-frame bending machine. (*Courtesy British Shipbuilding Research Association, Wallsend*)

The frame section is initially fed by means of a roller drive mechanism through the jaws of the bending machine and coupled onto the moving carriage of the X, Y co-ordinate measuring frame on the output side of the machine. The carriage position is then set to datum and the electronic control counters set to zero.

The paper-tape instructions to the electronic control unit of the machine carry co-ordinate offset dimensions as

a demanded dimension, X1, Y1;
anticipated springback;
accuracy required in coded number form;
frame-marking information.

The frame-bending machine is instructed to bend successive sections to the offset co-ordinates read in by the tape reader. Each section bend is confirmed against the measured deflection at the end of the frame relative to a datum in the jaws of the frame-bending machine. Should the offset co-ordinates differ from the taped instructions, the machine will carry out an iterative bending action until the co-ordinates lie within the given tolerance band. As the frame is stepped through the machine, the frame is also marked according to the instructions on the paper tape, so that successive operations (e.g., stringer-notching and frame setting-up) can be easily carried out.

A study of springback factors, which has been made on various frame sections and materials, permits the ready computation of the control tapes to achieve quickly and accurately the co-ordinate shape of ships' frames up to 21 in. in depth and 60 ft in length.

The force available in the Hugh Smith cold frame-bender is up to 500 tons and the average section takes 20 minutes to complete, but modifications to the machine and control system gained as a result of the prototype evaluation, will allow a considerably greater throughput per shift from the production machines. The accuracy of frame profile attained over a 60 ft length is better than $\pm 0{\cdot}125$ in.

15. The basis of cutting calculations in EXAPT

Herwart Opitz with W. Budde, W. H. Engelskirchen, and B. E. Hirsch

Professor Dr.-Ing., Dr.h.c. Herwart Opitz, DSc, is Director of the Institute for Machine Tools and Production Engineering at the Technical University, Aachen, Germany. His activities in the fields of industrial management and production techniques have brought him world-wide recognition, with honorary degrees from the Catholic University, Louvain, Belgium, Manchester College of Science and Technology, England, University of Strathclyde, Scotland, and University of Cincinatti, USA.

Dipl.-Ing. Wolfgang Budde, Dr.-Ing. B. E. Hirsch, and Dr.-Ing. W. H. Engelskirchen all studied at and graduated from the Technical University in Aachen, and were Scientific Assistants to Professor Dr.-Ing. H. Opitz.

The work reduction offered to the part programmer, and the economics of a programming system containing production technique information, depends particularly on the two main parts of the programming method—the input language and the processor. The input language should require as few part programming statements as possible, but the input should be easy to check.

For the processor, this implies that the thought processes of the planner will be replaced, as far as possible, by algorithms used in the computer to evaluate the instructions in the part program. This point of view is particularly important when it comes to the programming of technology where, with NC machining, the optimizing effect of the specialist team of planner and skilled machinist are missing.

When programming an NC machine, the complete work sequence must be predetermined in every detail by the part programmer. The machine operator has hardly any means of affecting the process. At the same time, the need to make full use of the expensive NC machine and its tools makes it necessary to use faster cutting conditions than the values obtained from experience on conventional machine tools. The part programmer will be overloaded if he has to calculate (from criteria of tool life and tool strength) the optimum cutting conditions for every tool path. If the processor does not free him from this task, then the determination of cutting values will be affected by his uncertainty in judging the various parameters which affect these cutting values (i.e., the material and cutting tool properties, the wear behaviour, the tool life, the tool stability, and the spindle power). Such uncertainties will reduce the economic advantages and productivity.

In the light of these remarks a special advantage of EXAPT[1] compared with most other programming systems, is the facility to program the technology of the workpiece and the machining conditions (for such processes as drilling, turning, and milling) with clearly constructed short instructions, leaving the computer to determine the optimum working conditions according to economic and machining criteria.

Any economic comparison with other programming methods, especially manual ones, must therefore not only compare programming time and cost, but also the saving in production time and cost which can be achieved by optimizing machining conditions.

The development of EXAPT had the aim of producing a programming system based on APT, for drilling, turning, and milling work which could deal not only with geometry but also with technology.

For the geometric descriptions for each different machining method, it is desirable to make a different selection from the permissible APT definitions, adapting the description of the workpiece to the task. The technology for drilling, turning, and milling, however, differs so much that the technological processor programs must be kept separate for each machining process.

This requirement led to the splitting up of the EXAPT system into three language parts, EXAPT 1 for drilling and simple milling processes, EXAPT 2 for straight-line and continuous-path turning work, and EXAPT 3 for 2C,L type milling work.

The efficiency requirements for the three language parts are basically similar and in the main are concerned with the determination of the:

tool path (avoiding collisions);
cutting conditions;
choice of tools;
machining sequence.

Efficiency of the EXAPT technology concept

The previous processor requirements determine the type and extent of the required input language and the technical solutions in the processor.

The input language determines the ease of learning, the correctness of hand-

(1) See references on p. 424.

ling, and the possibility of introducing errors into the language, as well as the cost of part programming. It should use the possibilities of the digital computer and suit the planner's normal way of thinking.

In conventional production, the planner chooses a suitable machine tool on the basis of the workpiece drawing, and then produces work schedules using the tools available for this machine. Factors such as clamping, work sequence, various machining operations, and cutting conditions, are determined. The latter are generally determined, more or less carefully and accurately, from the workpiece material and cutting tool material. Since the number of materials to be machined in most factories is limited, and since the availability of tools on most machines hardly changes, it is possible to divide information describing a particular machining task into two groups:

information which remains fixed (e.g., characteristic values of material, tool and machine);

information which varies with every new workpiece (e.g., clamping, work sequence, type of operation).

To reduce the work of the part programmer, the three EXAPT languages are constructed so that only the information required specifically for any particular case need be given in a part program. The data which remains constant is provided for in the EXAPT processor by values which, on introducing the system into a factory, must be put in once and for all in the form of tool, material, and machine cards. This splitting up of the input information appreciably reduces the cost of running the program on the one hand, while on the other hand the cutting conditions no longer depend on the varied backgrounds and the unreliable judgements of part programmers.

Since the cutting conditions are determined by the tool and material cards, these offer the possibility of quickly introducing into the factory, from one central point, new developments in machining techniques, or new values of previously used constants, merely by changing the corresponding cards. The data contained on the EXAPT cards can partly be determined, tested, and documented, by means of a service program run separately on a computer. Some processor routines which are used frequently in processing a part program are thereby made more reliable and effective.

Since the input information contained on these tool and material cards need not be processed afresh for every part program, it is possible to save computer analysis and translation time and to reduce the possibility of error.

Determination of work sequence, premachining and tools in EXAPT 1

Scope

In EXAPT 1 the following technological data can be automatically determined by using the processor[2].

(a) Work sequences.
(b) Tools.
(c) Reversals and feed changes in the case of deep holes.
(d) Feeds.
(e) Cutting speeds.
(f) Tool paths.

Requirements for determination of technological data include the previously mentioned material, tool, and machine cards, as well as the instructions for description of the workpiece and for the definition of the machining process.

Part material definition

The condition of the workpiece material is given by:

$$\text{PART/MATERL, w, } \begin{matrix} \text{UNMACH} \\ \text{SEMI} \\ \text{CORED} \end{matrix}, \begin{matrix} \text{ROUGH} \\ \text{SMOOTH} \end{matrix}, \text{CORREC, } c_1, c_2$$

The modifier MATERL, w must always be given; it causes the technological data of the material with group number w to be produced. By means of further modifiers it is possible to give the condition of the hole before machining (UNMACH = solid material, SEMI = premachined, CORED = precast), the surface condition (SMOOTH, ROUGH) as well as percentage figures for feed c_1 and cutting speed c_2, in case this is thought to be necessary because of weakness in the workpiece.

Machining definitions

The machining procedure is defined by the instruction

symbol = machining process/modifier list.

Tool card If the modifier list gives a tool identification number after the modifier TOOL, i (i = identification number) then the details of the tool with this identification number are chosen from the tool card, used, and also entered into the actual tool list. Figure 15.1 shows the tool card for a twist drill. The processor is only provided with the values given in the lower part of the card. This contains a type number for the rough classification of the tool, dimensional information for the accurate choice of tools, conditions of use such as feed factor, cutting speed factor, direction of rotation, and the tool identification number. If no tool identification number is given in the machining definition, then the tool is chosen according to the classification number and dimensional information. If, in addition, the modifier SO (single operation) is omitted, then the processor specifies premachining and the tools suitable for this operation.

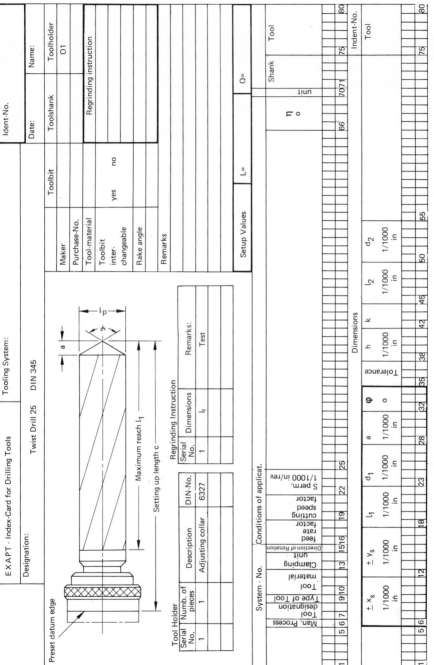

Figure 15.1 EXAPT 1 Drill card.

14*

Tapping The determination of the work sequence for threadcutting is represented on the left-hand side of Fig. 15.2 by means of a sequence diagram. On the right-hand side, the input information is given in the form of EXAPT 1 instructions, followed by the result in the form of a list of the sequence of machining operations. For this purpose, various formulae are used (see Fig. 15.2), which depend on the diameter and the depth of the required hole. The diameter for some premachining operations (e.g., predrilling) is given by a minimum and maximum diameter, while the final machining is described accurately by the diameter programmed into the machining definition.

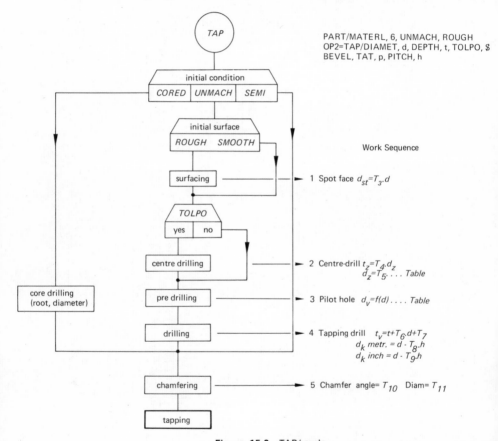

PART/MATERL, 6, UNMACH, ROUGH
OP2=TAP/DIAMET, d, DEPTH, t, TOLPO, $
BEVEL, TAT, p, PITCH, h

Work Sequence

1 Spot face $d_{st}=T_3 \cdot d$

2 Centre-drill $t_z=T_4 \cdot d_z$
  $d_z=T_5 \cdot \ldots$ Table

3 Pilot hole $d_v=f(d) \ldots$ Table

4 Tapping drill $t_v=t+T_6 \cdot d+T_7$
  d_k metr. $= d \cdot T_8 \cdot h$
  d_k inch $= d \cdot T_9 \cdot h$

5 Chamfer angle$= T_{10}$ Diam$= T_{11}$

Figure 15.2 TAP/ cycle.

In calculating the depth when premachining, account is taken of the fact that the premachining depth is sometimes different from the programmed depth of the final hole. This is so when calculating the excess depth for blind tapped holes and the depth for centre drilling and for countersinking. The constants T_n used in the formulae depend on experience within the firm, and on the final results required. They are therefore not programmed as constants into the EXAPT processor, but can be read into it from machine cards. In this way it is simple to adapt the processor to user requirements. After these methods have been used to determine suitable tools, diameters, and depths, other material-dependent tool requirements, such as tip angle and tool material, are taken from the tool card.

Tools are chosen for all the machining operations corresponding to the requirements given in this machining list. When doing this, tools previously chosen are taken into account, in order to keep the number of tools required as small as possible. If no suitable previous tool is found, then the tool cards are used to make a further choice.

Drilling Figure 15.3 shows the results of diameter determination for a drilling example. By altering the machine cards, this result can be easily changed and made suitable for the requirements of a particular firm. For this purpose, the

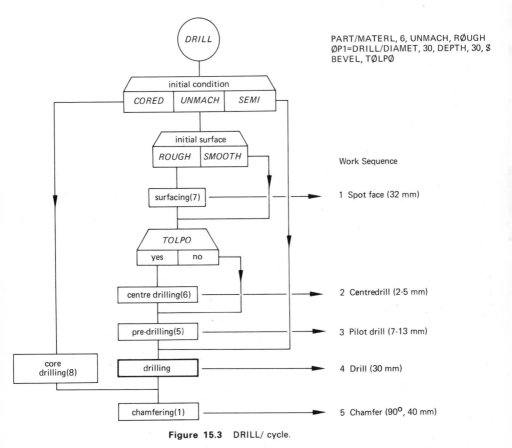

Figure 15.3 DRILL/ cycle.

machine cards contain approximately 80 values which represent the constants for the various required formulae. Among these values, there are, for example: centre-drill diameters for various final diameters; constants for the determination of the predrilling diameter in the case of large hole drilling, thread cutting, spiral drilling, counter-boring, and reaming: constants for the calculation of the excess depths required for reamed blind holes and tapped holes; and constants for the calculation of reversal depths in deep holes.

Drilling deep holes   The use of the constants contained on the machine cards can be demonstrated by means of an example which concerns the determination of reversals and feedrate reductions required in the case of deep holes.

The reversal depths are calculated by the processor using the following formulae:

$$1\text{st reversal at} \quad L_1 = T_{63} \cdot D + T_{64}$$
$$2\text{nd reversal at} \quad L_2 = L_1 + L$$
$$n\text{th reversal at} \quad L_n = L_{n-1} + L$$

where

$$L = T_{69} \cdot D + T_{70}.$$

The first feed reduction to 80 per cent of the feedrate is carried out at the depth.

$$L_{1F} = T_{65} \cdot D + T_{66} \cdot D^2$$

and the second to 50 per cent at

$$L_{2F} = T_{67} \cdot D + T_{68} \cdot D^2$$

where L = depth and D = hole diameter.

If the constants T_n used in these formulae are given the following values on the machine cards:

$$T_{63} = 2 \cdot 3 \qquad T_{64} = 14$$
$$T_{65} = 6 \cdot 0 \qquad T_{66} = 0 \cdot 05$$
$$T_{67} = 9 \cdot 0 \qquad T_{68} = 0 \cdot 05$$
$$T_{69} = 0 \cdot 7 \qquad T_{70} = 6 \cdot 0$$
$$T_{71} = 0 \cdot 8 \qquad T_{72} = 0 \cdot 5$$

then the diagram shown in Fig. 15.4 is obtained for reversals and feed reductions to 80 per cent and 50 per cent respectively.

Workpiece material cards for every drilling operation, cutting speeds and feeds are determined for the tool and the workpiece materials involved. The peripheral cutting speed is taken by the processor from the workpiece material cards (Fig. 15.5) and is recalculated into rotational speed. These material cards are divided into material groups of equal machinability and contain information

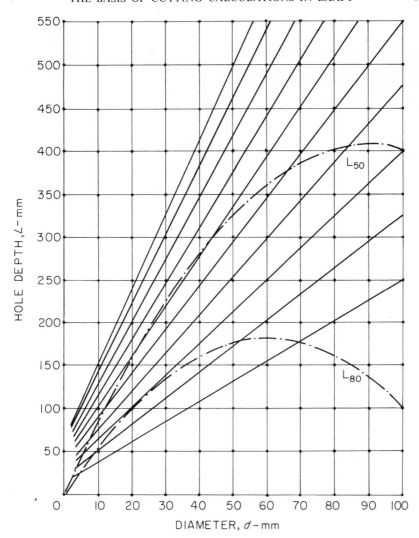

Figure 15.4 Deep hole drilling.

(for every machining process) on the type of coolant to be used, the tool material, the type of tool (i.e., in the case of spiral fluted tools a code for the pitch of the spiral flutes), the tip angle of spiral drills, the cutting speed, and the feed. The feed is determined by approximating the feed curve by three straight lines (Fig. 15.6), so that the feeds are derived from the formula

$$S = a_n D + b_n$$

Material Group No. ☐

Material characteristic data / Type of machining	Coolant	Cutter material	Tool type	Angle of tool tip	v (m/min)	a_1 ($\times 10^{-4}$)	b_1 ($\times 10^{-2}$)	a_2 ($\times 10^{-4}$)	b_2 ($\times 10^{-2}$)	a_3 ($\times 10^{-4}$)	b_3 ($\times 10^{-2}$)
Centre drilling					●						
Drilling					●						
Reaming					●						
Core drilling					●						
Spot facing					●						
Countersinking					●						
Tapping					●						
Boring					●						
Milling					●						
Recessing with special tool					●						
Other machining					●						
Other machining					●						

Figure 15.5   EXAPT 1 Material index card.

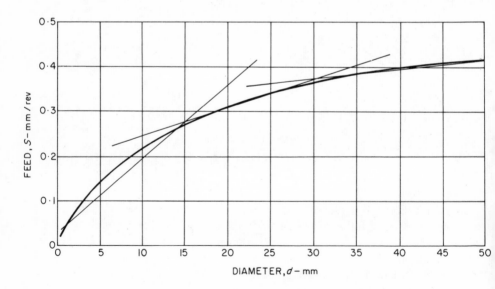

Figure 15.6   Feed versus hole diameter.

where the six factors a_n and b_n are taken from the material cards. This practical cutting-speed determination corresponds closely to the requirements of industrial production. If further limits have to be introduced from the point of view of the machine tool, or from considerations of tool life, then it is possible to introduce other constants into the card, and in this way to supplement the program as has now been done in EXAPT 2.

Determination of optimum cutting values in EXAPT 2

Scope
In the following, determination of the cutting values refers to the evaluation of the cutting depth, the feed, and the cutting speed, for all working movements during one pass of the turning tool. In this section of the chapter, the discussion is in general terms; a working example is given at the end of the chapter.

In explaining the economic and machining points of view on which the concept of the EXAPT 2 cutting value model is based, it is in the first instance not necessary to differentiate according to the feed direction and machining process. For all cases, the cutting values are calculated in the same processor subroutine. The influences which must be taken into account in superimposing longitudinal and transverse motions, or in the case of plunge cutting, will be given later.

Cutting-value model
The construction of the cutting-value model must be based on the requirements which it must fulfil within the context of a programming language.[3]

(a) The production of the values desired by the user even when these do not correspond to the optimum.

(b) The introduction of the initial data, the determination of the technically optimum cutting values using the workpiece material, tool, and machine to its best capabilities and taking into account the contact conditions for each element of the cutter path.

(c) Optimizing the cutting values, taking into account changing management aims and adapting to these with the least possible effort.

(d) Determination and output of data with acceptable computing times.

The problems which occur with the first two of these requirements must be solved by means of system analysis which, especially in the case of (b) must be based on a mathematical description of the machinability relationships. The requirement (c) presupposes that methods be developed which take into account the economic effects of management policy on the cutting values, and which ensure that the operational targets are achieved. The relationships found in this way must be arranged in an information sequence which, according to requirement (d) can be translated into a computer program of optimum effectiveness.

If, in a machining operation, one considers not only the many factors which influence it and which vary with changing contact conditions at the cutting edge, but also the inter-relationship of the cutting values, then one can see that it is not possible to reproduce all these values by storing them in a tabular manner in the computer, without exceeding an acceptable computer storage capacity.

An analysis of the system produces a set of mathematical equations which describe the actual relationships between the input and output information of the analogue model. Thanks to a great deal of investigation in the field of machinability, many of these relationships are available in the form of mathematical formulae which can be programmed, or at least made suitable for computer processing. In choosing and adapting these relationships, care must be taken that the accuracy is adequate for the requirements of workshop practice and that the numerical values for the input are available to the user. It is unreasonable to include theoretically based effects in the determination of cutting values if they have no noticeable effect on the actual setting values of the machine as this would only cause waste of computer time. The numerical evaluation of the equations used must not be carried out by means of program constants, but must use values which are input once and for all in accordance with the wishes of the user, and which need not be changed again in spite of very wide range of part programs. As already mentioned, the EXAPT 2 cards represent a very useful input medium for this. In this way, the technological and economic relationships can be taken into account in the determination of the cutting values.

The technological relationships which are taken into account can be divided into three groups:

relationships which describe the production technique limits set by the workpiece, tool, or machine;
relationships which take into account the chip form and chip flow;
tool life relationship which describes the inter-relationship of the cutting values.

Workpiece limitations As regards the first group, the limits which must be observed in relation to the workpiece must be given in each part program by the part programmer. In the context of a programming language this is the best solution, since it would cause considerable computing costs to produce as accurate a statement about workpiece strength and stability from the part program as the part programmer can produce by studying the drawing.

Apart from the possibility of giving cutting values in the part program, there are also instructions by means of which it is possible to take into account various factors in the cutting-life determination.[4] In this way it is possible, at particular machining stages, to give a torque limitation after the TORLIM/ instruction or to give cutting-value factors after the PART/—CORREC instruction and in this way influence the chip cross-section, or the feed and cutting speed, in order to take into account, for example, inadequate workpiece stability. A limit on cutting depth, in order to avoid chatter vibration, can be programmed by giving a maximum cutting depth, A, for the cutter calculations after the modifier DEPTH,. If the cutting depth is not programmed, then the processor first of all compares the departure of the actual unmachined contour from the final part contour in the region of the particular machining point in question, with the maximum permissible engagement length of the cutting tool edge. If the excess dimension can be removed in one pass, then this produces a cutting depth A, and chip width B. This normally happens at the last of a sequence of roughing cuts where, by reducing the cutting depth at the previous step, a minimum cutting depth according to the information on the material and tool card can be made available.

In the case of finishing and precision finishing, the oversize programmed in an OVSIZE/ instruction or after the modifier FIN, and/or FINE, can usually be removed in one pass. In these cases, the feed also comes out as being purely dependent on the workpiece. The contour elements designated in the part program as finished or precision finished areas, describe the maximum permissible roughness and therefore determine the feed, taking into account certain properties of the workpiece material.

If several passes are required, then the maximum permissible cutting edge engagement length, B_{ZUL}, determines the chip width and cutting depth.

Tool limitations   The boundary criteria set by the tool are described by the application data given on the tool card (Fig. 15.7). Only the most important can be explained here. The meaning of B_{ZUL} has already been described above. The card value should not correspond simply to a geometrical dimension of the cutting edge, but should be actually measured after the delivery of the tool.

Herein lies one of the possibilities of adapting the chip form, laid down in the calculations, to make it specifically suitable for certain types of machining processes and tool shapes, even if cutters of standard dimensions are used, such as tungsten carbide inserts. B_{ZUL} represents the one absolute limit to the chip cross-section. The other limit is laid down by the card value H_{ZUL}. With this maximum value of chip thickness, the effect of chip guides or chip breakers on the feed can be very easily taken into account. It is very easy to limit the chip cross-sectional area to correspond to the maximum permissible cutting force at the tool, if the workpiece material to be machined and the particular shape of the tool are taken into account. Reference will be made later to the tool card values for wear-mark width and tool life.

Machine-tool limitations   One boundary condition which must be observed with respect to the machine tool is the torque which, for a fixed chip width B, limits chip thickness H. Furthermore, maximum and minimum feeds available at the machine, and the available spindle power, are important and limit the maximum value of cutting speed. The rotational speeds calculated from the cutting speeds are tested against the maximum and minimum values available on the machine.

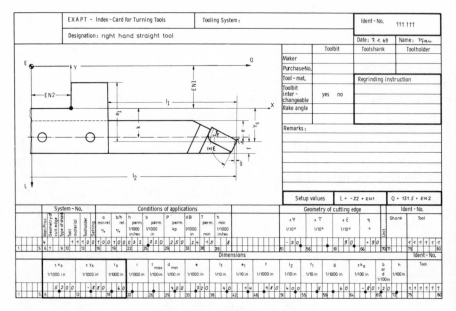

Figure 15.7 EXAPT 2 tool card.

Chip shape As regards the second group, the chip formation is taken into account by means of the chip-form equation. This describes the changes in chip cross-sectional form, depending on the chip width and the setting angle which are used in dividing up the cut. It makes it possible for the user to determine the variation of the desired aspect ratio of the chip over a large range of cutting depths, by giving a few characteristic values in the material cards. This function can be changed to correspond to the cutting edge geometry of each tool, by means of a factor in the tool cards. In this way, it is possible to control the critical chip cross-sectional form for each combination of workpiece material and tool material and, for example, to avoid over-square chip shapes in the region of the smaller roughing cutting depth.

Tool life relationship As regards the third group, the tool life relationship describes the effect of the three cutting values on the wear of the tool. The life of a tool is defined as the maximum time of use between two successive regrinds. The end of the tool life is reached when the wear limit (in this case a certain width of wear mark on the free surface) is reached.

The relationship between tool life and feed, or cutting depth, respectively, is laid down according to Taylor's relationship between tool life and cutting speed. The permissible width of wear mark, VB, is also included as a parameter in this formula. This seems desirable in the context of cutting-value determination, since the magnitude of the wear criterion will differ according to the type of use to which the tool is put. A numerical value for the permissible width of wear mark is introduced into the card for each tool and is then fixed. Compared with other wear criteria (for example, the crater wear) the use of the wear on the free surface (in a model to reproduce characteristic values) has the advantage of being easier to assess. This applies both to the assessment of the wear of the tool and also to the provision of numerical values in the tool card.

Based on Taylor's relationship between tool life and cutting velocity, the empirically determined dependence of tool wear on the other important factors in the tool life expression can be written

$$V = C.S^E.A^F.T^G.VB^H \tag{1}$$

where   V is cutting velocity (m/min)
          VB is land wear width (mm)
          S is feed per revolution (mm)
          A is cutting depth (mm)
          T is cutting time (min)

and C, E, F, G, H are constants and exponents whose values depend on the cutter and workpiece materials.

The equation has been written in terms of V since it is normally the cutting speed which has to be obtained in cutting-value determinations and since the results can be best assessed in terms of the cutting speed when compared with practical values.

Tool life constants

The constants and the exponents of the tool life equation are valid for a certain combination of workpiece material and tool material. The required characteristic values are stored on the material cards, for all workpiece material and tool material combinations used in a particular factory. Then, according to the material given in the part program, and the tool which is being called up, the corresponding factors are taken from the cards by the program and the cutting-value determination is based on these factors.

Since only very few users have available the numerical material in the form in which it must be entered into the material card, certain formulae have been developed into which the planner or machining specialist enters the values used in his factory. Auxiliary programs (not part of EXAPT) then evaluate these cutting-value recommendations, determine the required characteristic values, and produce the material cards for the user.

As far as the validity of the tool life equation is concerned, care must be taken that the region of cutting speeds within which the equations present sufficiently good approximation to the wear process, are not exceeded. The boundary values, V_{max} and V_{min}, are therefore included on the material cards and are checked by the program. The exponents and constants are chosen in such a way that they give the correct weighting to the wear effect of the three cutting values. It is then possible to determine the cutting speed which, taken together with the other two wear effects (already determined according to other criteria) fulfils the boundary conditions of the tool life given by the wear-mark width. The same boundary condition, by giving different tool life values, can produce an infinite number of equivalent cutting-value combinations with regard to the wear effect. Only one corresponds to the optimum with regard to cost and/or productivity.

Optimum cutting conditions

The product of the cutting values gives the volume which can be machined per unit time, or determines the machining time required for a certain task.

The concept of the NC machine tool makes possible an appreciable reduction in non-cutting times by increasing positioning speeds and speeding up tool changing, and allows the main part of the preparatory work (programming and tool preparation) to be carried out away from the machine during the actual machining time. Thus the proportion of actual machining time compared with conventional machine tools has risen considerably, so that any measures regarding rationalization must have the aim of reducing this machining time and hence must aim to increase the cutting values.

Every combination of cutting values corresponds to a certain intensity of use of the tool and the machine, which involves a different consumption of capital per unit time. Tool wear caused by a certain cutting condition is characterized by the resulting tool life. It is possible then to determine a tool life on the planned intensity of use which, as long as other factors do not play a determining role, is to be used for the cutting-value determination.

The intensity of use is determined by management policy. During a boom, a time optimum is appropriate (i.e., minimum time or maximum output, accepting the increased costs due to more rapid tool wear). When the market is steady, optimum cost conditions should apply, and the factory should be working at its designed load. In a recession, longer piecetimes are desirable to reduced tool-wear costs; this will partly offset the rising fixed costs.[5]

The last factor depends on the cost structure of the particular plant, and the possibility of changing this with time is not considered any further here. Neither will the industrial aims of maximizing profit and earning capacity, which are of particular interest when using capital-intensive NC machine tools, be taken into account within this context since the intensities of use corresponding to these lie somewhere between the time and cost optima.

Calculating optimum cutting values Practical observations show that normally the wear effect of the cutting speed is the greatest, that of the feed noticeable, and that of the cutting depth very small. The cutting depth should be chosen as a maximum, not only because of its small effect on wear but also in order to reduce the number of cuts, for this causes a reduction in the main machining time and also a reduction in the auxiliary time for tool retraction, return stroke, and resetting. Feed and cutting speed have equal effects on the machine time. The smaller wear effect and the advantage that, with increasing chip thickness, the cutting force, cutting moment, and cutting power, are relatively reduced, make it desirable to maximize the feed from other boundary criteria and only to vary the cutting speed in order to satisfy the tool life equation.

For both extreme cases of intensity of use, the optimum tool life criteria can be calculated from the following equations:

$$T_{opt.} \text{ time } = \left(\frac{1}{G} - 1\right) t_w \text{ minutes} \tag{2}$$

$$T_{opt.} \text{ cost } = \left(\frac{1}{G} - 1\right) t_w + \frac{W_k}{M_k} \text{ minutes} \tag{3}$$

where T is tool life index value in minutes
       G is exponent of tool life in equation 1
       T_w is tool-changing time in minutes
       W_k is tool cost per cutting edge
       M_k is machine tool cost per minute

If it is assumed that a certain tool is for use on a particular machine and will only be used for machining a limited group of materials, it is permissible to calculate the optimum tool life index value for each tool outside the processor, instead of doing this for each tool change during the processing of the part program. It is introduced into the cutting-value calculation from the application data on the tool card as a pre-calculated optimum time.

Operation of technological processor

In the EXAPT 2 cutting-value model, the above mentioned economic and technological relationships are combined into one information stream. Input information is contained in the part program and in the cards. The output is the cutting depth, feed, rotational speed, and co-ordinate values for the tool motion, which are given as intermediate results to the post processor. A determination of the intermediate points of the tool motion is not carried out in the cutting-value model itself, but takes place using its own optimizing criterion in the section of the processor which divides the material which has to be cut into its component parts by means of the information about tool geometry and the

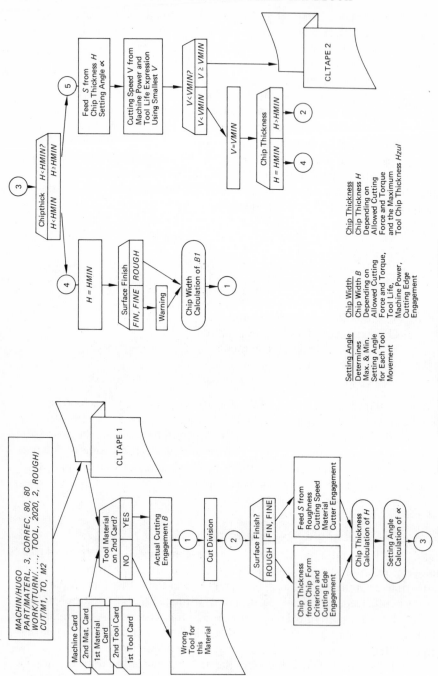

Figure 15.8 Determining cutting values in EXAPT 2.

geometry of the unmachined and finished component. The steps in the cutting-value determination can be seen in a simplified manner in Fig. 15.8.

By means of the machine name, the material number, and the particular tool number which is being called up, the data of the corresponding cards are made available in the core store. By means of the data on the material card it is easy to test whether the tool is suitable for machining the particular workpiece. If not, then a corresponding warning to the programmer is printed out.

With the aim of reducing the number of cuts as far as possible, a chip width B is chosen as large as possible; it is limited either by the permitted engagement length of the cutting edge (B_{ZUL}) or by the amount of material to be removed. The value obtained in this way makes it possible to divide into individual cuts the portion which has to be machined away. The calculation of the tool path for each of these cuts gives a table of co-ordinates which are transmitted to the cutting-value determination. For each portion, the calculation for feed is followed by that for cutting speed.

In the feed calculation, two different equations are used for roughing and finishing. In the case of finishing, the surface finish value to be achieved on the workpiece determines the feed. The values determined subsequently in the chip thickness sub-routine are used only for testing. In the case of roughing, a chip thickness value is calculated from every criterion which can provide an upper limit. The first value is obtained from the chip-form criterion; others are calculated in the chip thickness sub-routine, in relation to the maximum chip thickness H_{ZUL} and the permissible principal cutting force P_{ZUL} of the tool, as well as from the available spindle torque of the machine. The lowest of these values gives the chip thickness which, together with the chip width from the tool path calculation, is the maximum to be used. The setting angle is calculated in the setting-angle sub-routine for every element of a machining pass. A check is made to see whether, during a cut with a calculated chip thickness H, the feed has been less than the minimum feed for a machine. If this occurs, then in the program (portion 4, Fig. 15.8) the minimum required chip-thickness H_{min} is used in the calculation of a new chip width B_1, and this value is used to obtain a new distribution of cuts. The process then begins again, in this case, at 1 in Fig. 15.8.

If the values for chip thickness and feed lie within the limits of the feed available, then the maximum cutting speed can be calculated for every element of the tool path. It corresponds to the smallest of the values determined for B_{max} from the material card, from the machine power, or from the tool life, where only the latter case corresponds to the optimum. At the same time a check is made to see whether, for any of the elements of the tool path, the cutting speed is less than the minimum cutting speed V_{min}. If this is so, then, with the value of V_{min}, the possible chip thickness for the corresponding element is re-calculated and, if this results in a value less than H_{min}, then it is necessary in branch 4 to use the newly calculated chip width to arrive at a new distribution of cuts. If, during the pass, the cutting speed does not become less than V_{min}, then the output of the cutting values is given in the form of feed per revolution and rotational speed of spindle together with the corresponding motion instructions to the post processor.

For the purposes of facing, the part program can control the rotational speed in such a way that the processor calculates a sequence of switching points together with the corresponding rotational speeds, between which the speed remains constant.

The effect of the feed, and the reaction forces which change with the direction of feed, is taken into account by reducing the calculated longitudinal feed. The reduction in feed takes place steadily and progressively, the greater the departure of the feed direction from the longitudinal direction. The amount of feed reduction can be laid down for each material by giving a limiting value of S_{cross}/S_{along} in the material card. The unfavourable chip formation in the case of plunge cutting can also be taken into account by a feed-direction factor in the material card. The two correction factors are multiplied by one another in the case of plunge cutting.

An EXAPT 1 example

Details of the operation of the technology part of the EXAPT 1 are illustrated by showing the derivation of the various parameters for one line of a part program.

The part programmer sees an H 7 tolerance on the 20 mm diameter of the hole on the drawing and decides that it should be reamed. The hole is 20 mm deep in St 42 steel, and his factory has called this 'Material 1'. The required accuracy of positioning shows that centre drilling must precede drilling, so he adds the modifier TOLPO in his REAM/ statement.

The EXAPT 1 program contains the following instructions:

PARTNO/EXAMPLE
MACHIN/PP1
PART/MATERL, 1 $$ St 42
  ⋮
A = REAM/DIAMET, 20, DEPTH, 20, TOLPO
  ⋮
WORK/A
COOLNT/ON
GOTO/100, 100, 0
  ⋮
FINI

```
 ***************************************
 *                                     *
 *  TECHNOLOGISCHE KENNWERTE - EXAPT1  *
 *                                     *
 ***************************************


ARIABLE   WERT      BEDEUTUNG
 *****************************************************************************************************
   T1      1.0000  ZENTRIERBOHRERDURCHMESSER NACH DIN
   T2      1.6000  ZENTRIERBOHRERDURCHMESSER NACH DIN
   T3      2.5000  ZENTRIERBOHRERDURCHMESSER NACH DIN
   T4      4.0000  ZENTRIERBOHRERDURCHMESSER NACH DIN
   T5      6.3000  ZENTRIERBOHRERDURCHMESSER NACH DIN
   T5     10.0000  ZENTRIERBOHRERDURCHMESSER NACH DIN
   T7     16.0000  ZENTRIERBOHRERDURCHMESSER NACH DIN
   T8       .1100  FAKTOR FUER MIN. DURCHMESSER BEIM ZENTRIEREN
   T9      2.5000  FAKTOR FUER MAX. DURCHMESSER BEIM ZENTRIEREN
   T10     1.0E+20 KENNZEICHNUNG FUER UEBERTIEFE, DIE BEI MODIFIZIERUNG NICHT VERAENDER1 WERDEN DARF
   T11     0.0000  TIEFE BEIM FLACHSENKEN IM ARBEITSZYKLUS BEI PART/ROUGH
   T12     1.1000  DURCHMESSERFAKTOR BEIM FLACHSENKEN IM ARBEITSZYKLUS BEI PART/ROUGH
   T13     1.2000  DURCHMESSERFAKTOR BEI BEVEL IM ARBEITSZYKLUS
   T14     2.0000  KLEINSTER BOHRUNGSDURCHMESSER BEIM SPITZSENKEN
   T15    80.0000  DURCHMESSERSTUFE DER SPITZSENKER
   T16    50.0000  DURCHMESSERSTUFE DER SPITZSENKER
   T17    31.5000  DURCHMESSERSTUFE DER SPITZSENKER
   T18    20.0000  DURCHMESSERSTUFE DER SPITZSENKER
   T19    12.5000  DURCHMESSERSTUFE DER SPITZSENKER
   T20    25.0000  DURCHMESSERGRENZE FUER ERSTE VORBOHRUNG
   T21    50.0000  DURCHMESSERGRENZE FUER ZWEITE VORBOHRUNG
   T22      .5000  FAKTOR FUER VORBOHRDURCHM. BEI DURCHM. GROESSER T21
   T23      .2000  FAKTOR FUER VORBOHRDURCHM. BEI DURCHM. GROESSER T20
   T24      .7000  FAKTOR FUER MIN. DURCHM. BEIM VORBOHREN
   T25     1.3000  FAKTOR FUER MAX. DURCHM. BEIM VORBOHREN
   T26     2.0000  MIN. SPIRALBOHRERDURCHMESSER
   T27   100.0000  MAX. SPIRALBOHRERDURCHMESSER
   T28      .1000  OBERES ABMASS VOM ERMITTELTEN DURCHM. BEI KERNLOCHBOHRERN
   T29      .1000  UNTERES ABMASS VOM ERMITTELTEN DURCHM. BEI KERNLOCHBOHRERN
   T30     1.2000  FAKTOR FUER MIN. DURCHM. BEIM VORBOHREN FUER SPIRALSENKEN
   T31     1.0000  FAKTOR FUER MAX. DURCHM. BEIM VORBOHREN FUER SPIRALSENKEN
   T32    40.0000  VORBOHRDURCHM. UEBER T32 WERDEN AUF 5MM AUFGERUNDET
   T33    20.0000  VORBOHRDURCHM. UEBER T31 WERDEN AUF 2MM AUFGERUNDET
   T34      .7500  FAKTOR FUER VORBOHRDURCHM. BEIM SENKEN IM ARBEITSZYKLUS
   T35      .7000  FAKTOR FUER TIEFENVERRINGERUNG BEIM FLACHSENKEN (SINK) UND SACKLOCHBOHREN MIT EBENEM BODEN (BLIND2)
   T36     2.0000  KLEINSTER GEWINDEBOHRERDURCHMESSER
   T37    68.0000  GROESSTER GEWINDEBOHRERDURCHMESSER
   T38     1.0000  FAKTOR FUER KERNLOCHDURCHM. BEI WHITWORTH- GEWINDEN
   T39     1.1000  FAKTOR FUER KERNLOCHDURCHM. BEI METRISCHEN GEWINDEN
   T40      .2600  UEBERTIEFENFAKTOR BEIM GEWINDESCHNEIDEN MIT SACKLOCH
   T41     3.0000  UEBERTIEFENSUMMAND BEIM GEWINDESCHNEIDEN MIT SACKLOCH
   T42   100.0000  GROESSTER DURCHMESSER BEIM REIBEN
   T43     2.0000  KLEINSTER DURCHMESSER BEIM REIBEN
   T44    10.0000  DURCHM.GRENZEN FUER DIE UNTERMASSE DER VORSENKUNGEN T47 BIS T50
   T45    30.0000  DURCHM.GRENZEN FUER DIE UNTERMASSE DER VORSENKUNGEN T47 BIS T50
   T46    50.0000  DURCHM.GRENZEN FUER DIE UNTERMASSE DER VORSENKUNGEN T47 BIS T50
   T47      .2000  UNTERMASSE DER VORSENKUNGEN BEIM REIBEN
   T48      .2500  UNTERMASSE DER VORSENKUNGEN BEIM REIBEN
   T49      .4000  UNTERMASSE DER VORSENKUNGEN BEIM REIBEN
   T50      .5000  UNTERMASSE DER VORSENKUNGEN BEIM REIBEN
   T51      .1300  UEBERTIEFENFAKTOR BEIM REIBEN MIT SACKLOCH
   T52     3.0000  UEBERTIEFENSUMMAND BEIM REIBEN MIT SACKLOCH
   T53     4.8000  KLEINSTER SPIRALSENKERDURCHMESSER
   T54    30.0000  GROESSTER SPIRALSENKERDURCHMESSER
   T55     2.0000  KLEINSTER FLACHSENKERDURCHMESSER
   T56    63.0000  GROESSTER FLACHSENKERDURCHMESSER
   T57     2.3000  FAKTOR ZUR BERECHNUNG DER ZENTRIERTIEFE
   T58     1.0300  FAKTOR FUER UNTERE GRENZE DES SPITZENWINKELS
   T59      .9700  FAKTOR FUER OBERE GRENZE DES SPITZENWINKELS
   T60    10.0000  NORMALWERT FUER DEN SICHERHEITSABSTAND
   T61    15.0000  ERSTE DURCHM.GRENZE DER VORSCHUBKURVE
   T62    30.0000  ZWEITE DURCHM.GRENZE DER VORSCHUBKURVE
   T63     2.3000  FAKTUR ZUR BERECHNUNG DER REVERSIERUNGEN
   T64    14.0000  FAKTOR ZUR BERECHNUNG DER REVERSIERUNGEN
   T65     6.0000  FAKTOR ZUR BERECHNUNG DER REVERSIERUNGEN
   T66      .0500  FAKTOR ZUR BERECHNUNG DER REVERSIERUNGEN
   T67     9.0000  FAKTOR ZUR BERECHNUNG DER REVERSIERUNGEN
   T68      .0500  FAKTOR ZUR BERECHNUNG DER REVERSIERUNGEN
   T69      .7000  FAKTOR ZUR BERECHNUNG DER REVERSIERUNGEN
   T70     6.0000  FAKTOR ZUR BERECHNUNG DER REVERSIERUNGEN
   T71      .8000  VORSCHUBSVERMINDERUNGSFAKTOR
   T72      .5000  VORSCHUBSVERMINDERUNGSFAKTOR
   T73      .5000  VERZOEGERUNGSZEIT FUER DELAY IM SPIRET .1 UND .3
   T74     2.0000  SPIRET FUER GEWINDESCHNEIDEN
   T75             ZUR ZEIT NICHT VERWENDET
   T76             ZUR ZEIT NICHT VERWENDET
   T77             ZUR ZEIT NICHT VERWENDET
   T78             ZUR ZEIT NICHT VERWENDET
   T79             ZUR ZEIT NICHT VERWENDET
   T80             ZUR ZEIT NICHT VERWENDET
   I1       4      ZAHL DER SPANNMITTEL
   I2       1      SPANNMITTELNUMMER
   I3       2      SPANNMITTELNUMMER
   I4       3      SPANNMITTELNUMMER
   I5       4      SPANNMITTELNUMMER
 *****************************************************************************************************
```

Figure 15.9   Technology data for post processor PP1 in EXAPT 1

By processing the instruction MACHIN/PP1, the technological data for machine tool PP1 are produced. These include the technological data (Fig. 15.9), the material card indices for material groups 1 and 2 (Fig. 15.10) and the tool card index (Fig. 15.11). (Note: as this program was developed in Germany, much of its printout is currently in German.) The tool card index contains a listing of the data required for the technological calculations in EXAPT 1, especially the free length L1, diameter D1, tip length A, and tip angle PHI.

WERKSTOFFKARTEI - EXAPT1

WERKSTOFFGRUPPENNUMMER 1

BEARBEITUNGS-ART	KUEHL MITT.	SCHN. STOFF	WZ. TYP	SPITZ. WINKEL	V (M/MIN)	A1	B1	A2	B2	A3	B3
ZENTRIEREN	1	1	0	0	26.0	.0137	.1	0 0000	0.0	0 0000	0.0
SPIRALBOHREN	1	1	0	118	26.0	.0137	.1	.0107	.1	.0022	.3
REIBEN	0	1	0	0	11.0	.0207	.2	.0147	.3	.0100	.4
SPIRALSENKEN	1	1	0	0	22.0	.0113	.2	.0080	.2	.0050	.3
STIRNSENKEN	1	1	0	0	14.0	.0076	.1	.0049	.1	0 0000	0.0
SPITZSENKEN	1	1	0	0	18.0	.0097	.1	.0043	.2	.0021	.3
GEWSCHNEIDEN	3	1	0	0	11.0	0 0000	0.0	0 0000	0.0	0 0000	0.0
BOHRSTANGE	1	0	0	0	0.0	0 0000	0.0	0 0000	0.0	0 0000	0.0
FRAESEN	1	0	0	0	0.0	0 0000	0.0	0 0000	0.0	0 0000	0.0
EINSTECHEN	1	0	0	0	0.0	0 0000	0.0	0 0000	0.0	0 0000	0.0
SONST,BEARB.	1	0	0	0	0.0	0 0000	0.0	0 0000	0.0	0 0000	0.0

WERKSTOFFGRUPPENNUMMER 2

BEARBEITUNGS-ART	KUEHL MITT.	SCHN. STOFF	WZ. TYP	SPITZ. WINKEL	V (M/MIN)	A1	B1	A2	B2	A3	B3
ZENTRIEREN	1	1	0	0	22.0	.0093	.0	0 0000	0.0	0 0000	0.0
SPIRALBOHREN	1	1	0	118	22.0	.0093	.0	.0073	.1	.0017	.3
REIBEN	0	1	0	0	7.2	.0193	.1	.0127	.3	.0082	.4
SPIRALSENKEN	1	1	0	0	20.0	.0110	.1	.0067	.2	.0028	.3
STIRNSENKEN	1	1	0	0	12.6	.0076	.1	.0054	.1	0 0000	0.0
SPITZSENKEN	1	1	0	0	18.0	.0097	.1	.0040	.2	.0019	.3
GEWSCHNEIDEN	3	1	0	0	8.0	0 0000	0.0	0 0000	0.0	0 0000	0.0
BOHRSTANGE	1	0	0	0	0.0	0 0000	0.0	0 0000	0.0	0 0000	0.0
FRAESEN	1	0	0	0	0.0	0 0000	0.0	0 0000	0.0	0 0000	0.0
EINSTECHEN	1	0	0	0	0.0	0 0000	0.0	0 0000	0.0	0 0000	0.0
SONST,BEARB.	1	0	0	0	0.0	0 0000	0.0	0 0000	0.0	0 0000	0.0

Figure 15.10 Two material cards for EXAPT 1.

After reading in instruction A = REAM/. . . , the work sequence is determined in reverse order.

(a) *Reaming*

Tool: Chucking reamer
Diameter: DR = 20 mm
Free length: L1R > 20 + A_{TOOL} mm
(A_{TOOL} = Tip length of tool)

(b) *Core drilling*

Tool: Core drill
Diameter: DS = DR − T_{48} (since T_{44} < DR < T_{45})
DS = 19·25 mm
Free length: L1S > 20 + A_{TOOL} mm
(See Fig. 15.9 for T values)

WERKZEUGKARTEI - EXAPT1

BOHRWERKZEUG	SYSTEMNUMMER FTG VER	WZ BEZ	SCH WST	WZ TYP	SPANN MOEGL	MASSANGABEN C (MM)	L1 (MM)	D1 (MM)	A (MM)	PHI GRD	TOL	STEIG	K L/K	L2 (MM)	D2 (MM)	EINSATZBED FAKTOR S	V	DREH RTG	IDENTNR
SPIRALBOHRER	1	24	1	N	2	360,00	260,00	80.00	26.70	118	-0	0,00	-0	0.00	0.00	.9	.9	0	101
SPIRALBOHRER	1	24	1	N	2	300,00	230.00	60.00	20.00	118	-0	0,00	-0	0.00	0.00	.8	.8	0	102
SPIRALBOHRER	1	24	1	N	2	300,00	230.00	25.00	8.40	118	-0	0,00	-0	0.00	0.00	1.0	1.0	0	104
SPIRALBOHRER	1	24	1	N	2	200,00	160,00	24.90	8.40	118	-0	0,00	-0	0.00	0.00	1.0	1.0	0	103
SPIRALBOHRER	1	24	1	N	2	105,20	80,00	10.00	3.40	118	-0	0,00	-0	0.00	0.00	1.0	1.0	0	105
SPIRALBOHRER	1	24	1	N	2	105,20	80,00	9.50	3.40	118	-0	0,00	-0	0.00	0.00	1.0	1.0	0	2105
SPIRALBOHRER	1	24	1	N	2	105,20	80,00	12.00	3.40	118	-0	0,00	-0	0.00	0.00	1.0	1.0	0	1105
SPIRALBOHRER	1	24	1	N	2	150,00	135.50	20.00	6.70	118	-0	0,00	-0	0.00	0.00	1.2	1.2	0	106
SPIRALBOHRER	1	24	1	N	2	163,00	110.00	15.00	5.00	118	-0	0,00	-0	0.00	0.00	1.0	1.0	0	107
SPIRALBOHRER	1	24	1	N	2	80,00	57,00	5.60	1.90	118	-0	0,00	-0	0.00	0.00	1.0	1.0	0	108
SPIRALBOHRER	1	24	1	N	2	80,00	57.00	4.80	1.90	118	-0	0,00	-0	0.00	0.00	1.0	1.0	0	1108
ZENTRIERBOHRER	1	70	1	N	2	30,00	15,00	6.30	0.00	60	-0	0,00	-0	0.00	0.00	1.0	1.0	0	223
SPIRALBOHRER	1	24	1	N	2	210,00	155,00	23.10	8.00	118	-0	0,00	-0	0.00	0.00	1.0	1.0	0	109
SPIRALBOHRER	1	24	1	N	2	210,00	155.00	23.30	8.00	118	-0	0,00	-0	0.00	0.00	1.0	1.0	0	150
SPIRALBOHRER	1	24	1	N	2	210.00	155.00	23.90	8.00	118	-0	0,00	-0	0.00	0.00	1.0	1.0	0	110
SPIRALBOHRER	1	24	1	N	2	100,00	75.00	8.40	3.10	118	-0	0,00	-0	0.00	0.00	1.0	1.0	0	111
SPIRALBOHRER	1	24	1	N	2	160,00	121,00	10.10	3.00	118	-0	0,00	-0	0.00	0.00	1.0	1.0	0	212
SPIRALBOHRER	1	24	1	N	2	210,00	180,00	35.00	10.50	118	-0	0,00	-0	0.00	0.00	1.0	1.0	0	213
SPIRALBOHRER	1	24	1	N	2	170,00	150.00	23.00	6.90	118	-0	0,00	-0	0.00	0.00	1.0	1.0	0	214
SPIRALBOHRER	1	24	1	N	2	150,00	120.00	18.00	5.40	118	-0	0,00	-0	0.00	0.00	1.0	1.0	0	215
SPIRALBOHRER	1	24	1	N	2	140,00	120.00	16.00	4.80	118	-0	0,00	-0	0.00	0.00	1.0	1.0	0	216
SPIRALBOHRER	1	24	1	N	2	250,00	180,00	40.00	12.[illegible]0	118	-0	0,00	-0	0.00	0.00	1.0	1.0	0	217
SPIRALBOHRER	1	24	1	N	2	150,00	100.00	14.00	4.20	118	-0	0,00	-0	0.00	0.00	1.0	1.0	0	218
SPIRALBOHRER	1	24	1	N	2	100,00	75,00	8.00	2.40	118	-0	0,00	-0	0.00	0.00	1.0	1.0	0	219
SPIRALBOHRER	1	24	1	N	2	90,00	70,00	7.00	2.10	118	-0	0,00	-0	0.00	0.00	1.0	1.0	0	220
SPIRALBOHRER	1	24	1	N	2	180,00	130,00	11.00	3.30	118	-0	0,00	-0	0.00	0.00	1.0	1.0	0	221
MASCHINENREIBAHLE	1	41	1	N	2	100.00	50.00	20.00	3.[illegible]	[illegible]	-0	0,00	-0	0.00	0.00	1.0	1.0	0	[illegible]12
MASCHINENREIBAHLE	1	41	1	N	2	120,50	70.00	30.00	4.2[illegible]	[illegible]	-0	0,00	-0	0.00	0.00	.9	.9	0	[illegible]13
SPIRALSENKER	1	44	1	[illegible]	2	190,00	145.00	19.75	3.[illegible]	[illegible]	-0	0,00	-0	0.00	0.00	1.0	1.0	0	[illegible]14
SPIRALSENKER	1	44	1	[illegible]	2	240,00	190.00	29.75	4.[illegible]	[illegible]	-0	0,00	-0	0.00	0.00	1.0	1.0	0	[illegible]15
SPIRALSENKER	1	44	1	[illegible]	2	240.00	190.00	30.00	4.[illegible]	[illegible]	-0	0,00	-0	0.00	0.00	1.0	1.0	1	[illegible]16
ZENTRIERBOHRER	1	70	1	[illegible]	2	20,00	6.[illegible]0	2.50	0.[illegible]	6[illegible]	-0	0,00	-0	0.00	0.00	1.0	1.0	0	[illegible]17
GEWINDEBOHRER MET.	1	7	1	[illegible]	2	50,00	26.[illegible]0	10.00	5.[illegible]	[illegible]	-0	1,50	-0	6.00	6.00	1.0	1.0	0	[illegible]22
GEWINDEBOHRER MET.	1	7	[illegible]	[illegible]	2	50,00	26.[illegible]0	6.00	5.[illegible]	[illegible]	-0	1,00	-0	6.00	6.00	1.0	1.0	0	1[illegible]22

Figure 15.11 _EXAPT 1 tool cards.

(c) *Drilling*

A predrilling operation is required, since the core drill cannot work in the unmachined material.

Tool: Twist drill

Diameter: $DB_{th} = T_{34}.DS$

$DB_{th} = 15$ mm

The diameter of the drill used for predrilling need not be exactly 15 mm. It must lie between

$$DB_{min} = T_{31}.DB_{th} = 15 \text{ mm}$$

and $\quad DB_{max} = T_{30}.DB_{th} = 18$ mm

thus $\quad DB_{min} < DB < DB_{max}$

Free length: $L1B > 20 + A_{TOOL}$ mm

Further predrilling is not necessary, since $DB_{th} < T_{20}$

Tip angle: $PH1 = 118°$ (from material card index No. 1—Fig. 15.10).

(d) *Centre drilling*

Since the modifier TOLPO is given, a centre drilling operation must be determined.

Tool: Centre drill

Diameter: $DZ_{th} = T_4$, since $DB_{th} > T_6$ ($T_6 = 10$ mm)

$DZ_{th} = 4$ mm

$DZ_{min} = DB_{th}.T_8 = 0.163$ mm

$DZ_{max} = DB_{th}/T_9 = 6.00$ mm

therefore $\quad DZ_{min} < DZ < DZ_{max}$

Summary of tool sequence

Tool	D_{min} (mm)	D_{max} (mm)
Centre drill	0·163	6·000
Drill	15·000	18·000
Core drill	19·750	19·750
Chucking reamer	20·000	20·000

After determination of the work sequence, tools are chosen from the tool card index (Fig. 15.11). The tool material is obtained from the material card index No. 1 (code number 1) (Fig. 15.10, column 3: cutter material).

Tool	Identity number	Diameter D1 (mm)
Centre drill	117	2·50
Drill	107	15·00
Core drill	114	19·75
Cutting reamer	112	20·00

The determination of feeds is carried out according to the formula $s = A_n.D1 + B_n$. Since the centre drill diameter is less than T_{61}, the index n is made equal to 1. All the remaining tools have diameters larger than T_{61}, so that $n = 2$. The cutting fluid number and cutting speed are derived from the material card index (Fig. 15.10).

The following machining operations are therefore carried out:

(a) *Centre drilling*

      Tool: Identity number 117

      Depth: $T = T_{57}.D1$

                    $T = 2·3 \times 2·5$

                    $T = 5·7$ mm

      Cutting fluid: No. 1 (drilling oil)

      Cutting speed: $v = 26$ m/min

      Feed: $s = A_1.D1 + B_1$

                    $s = 0·0137 \times 2·5 + 0·1$

                    $s = 0·135$ mm/rev

(b) *Drilling*

      Tool: Identity number 107

      Depth: $T = 20·0 + A = 25$ mm

      Cutting fluid: No. 1

      Cutting speed: $v = 26$ m/min

      Feed: $s = A_2.D1 + B_2$

                    $s = 0·0107 \times 15 + 0·1$

                    $s = 0·261$ mm/rev

      The depth at first reversal would be

                $L1 = T_{63}.D1 + T_{64}$

                $L1 = 2·3 \times 15 + 14$

                $L1 = 48·5$ mm

      and at first feed reduction

                $L1V = T_{65}.D1 + T_{60}.D1^2$

                $L1V = 6·15 + 0·05 \times 15^2$

                $L1V = 101·25$ mm

Since depth T is less than $L1$ and $L1V$, it is drilled with 100 per cent of feed and without reversal.

(c) *Core drilling*

      Tool: Identity number 114

      Depth: $T = 20 + A = 23$ mm

      Cutting fluid: No. 1

      Cutting speed: $v = 22$ m/min

      Feed: $s = A_2 D1 + B_2$

                    $s = 0·008 \times 19·75 + 0·2$

                    $s = 0·354$ mm/rev

(d) *Reaming*

Tool: Identity number 112

Depth: $T = 20 + A = 23$ mm

Cutting fluid: No. 0, i.e., without cutting fluid

Cutting speed: $v = 11$ m/min

Feed: $s = A_2D1 + B_2$

               $s = 0.0147 \times 20 + 0.3$

               $s = 0.594$ mm/rev

These operations are carried out in sequence at the points given by the GOTO instructions (in the example GOTO/100, 100, 0). If several holes are to be machined, the following work sequence is used:

centre drill all holes;
drill all holes;
core drill all holes.

If the modifier PH is given in the WORK instruction (WORK/PH, A), a different machining sequence is used:

centre drill, drill, core drill, and ream the first hole;
same process on second hole;
etc.

The determination of the technological data is carried out for other work cycles in a similar manner to that used in this example.

An EXAPT 2 example

The cutting-value determination in the EXAPT 2 processor will be explained by means of an example. In Fig. 15.12, which is the part program for the turned part shown in Fig. 15.13, the individual statements are numbered serially on the left-hand side. These numbers are referred to in the following explanation.

As in EXAPT 1, the machining data (i.e., material, tool, and machine cards) are made available by the MACHIN/... statement (line 5). The MACHDT statement (line 6) provides a part-programming means of altering individual machining values.

Lines 9 to 14 describe the raw material contour, while the contour of the finished part is dealt with in lines 17 to 31. Individual intersection points on the finished part contour have markers allocated (e.g., M1 in line 18) by means of which particular cuts can be defined later relative to the finished part.

The statement PART/MATERL, 203 (in line 33) causes the turning and drilling machining values to be taken from the material cards appropriate to the machine tool for material group 203 (steel C45). These values (see Fig. 15.14) are then available in the computer core store for the use of the processor. A similar process occurs for the characteristics of the tools whose identification numbers occur in the turning and drilling definitions in lines 42, 45, 49, 52, 61, 64, 67, and 72. These tool identity numbers must be programmed at the present stage of development of the processor; the tools may be selected by the processor in the future, after an investigation.

```
              **** E X A P T 2 - P R O C E S S O R ****

              I B M  SYSTEM/36C,SECTION 1

1                  PARTNO/ROLLF,W3 452                          10
2      MACHIN/PLOTTC
3      PPFUN/PLCTT,SCALE,,6
4      PPFUN/PLCTT,BLANC,CLTT
5      MACHIN/EX2PP
6                  MACHOT/3C,250,C,1,3,5,3CCC,C,8,2C0            EXAPT2
7                  CLPRNT
8                  REMARK/PCHTEILBLSCHREIBUNG                    30
9                  CONTLR/BLANCO                                 40
10                 BEGIN /-2.5,C,YLARGE,PLAN,-2.5               50
11                 RGT    /DIA,150                              60
12                 RGT    /PLAN,112.5                           70
13                 RGT    /DIA,C                                80
14                 TERMCO                                       90
15                 REMARK/IERTICTEILDESCHREIBUNG              1G0
16                 SURFIN/FIN                                 110
17                 CONTLR/PARTCC                              120
18     M1,         BEGIN /1C,IPC,6/21,YLARGE,PLAN,10          130
19                 LFT    /DIA,11C,BEVEL,C,5
2C     M2,         RGT    /PLAN,C
21                        C1=CIRCLE/40,0,7C                   160
22     M3,         RGT    /C1,ON,PLAN,40
23     M7,FWD/C1,3EVEL,C,3
24     M4,         RGT    /C1A,(18     ),BEVEL,C,3            180
25     M5,         RGT    /PLAN,54
26                 LFT    /DIA,5C,BEVEL,C,3
27                 RGT    /PLAN,6C
28                 LFT    /DIA,62,BEVEL,C,3
29     M6,         RGT    /PLAN,E4                             230
30                 LFT    /DIA,8C,BEVEL,C,3
31                 TERMCO                                      250
32                 REMARK/TECHNCLOGISCHE AUSSAGE               260
33                 PART/MATERL,2C3
34                 PART/CORREC,8C,1CC
35                 CLDIST/1.5
36                 FROM/1CCC,-5C
37                 CSRAT /EC                                   300
38                 CHUCK /4CC,183.5,200,2C,12C,-2C
39                 CLAMP /112.5,INVERS                         320
4C                 CUTLOC/BEFORE                               280
41                 PPRINT/FLTTER,KSH,40C,DURCHM,,150,20TIEF,HARTE,BACKEN,NCRMALSP.   330
42     A1=         CDRILL/SC,TCOL,22075C,1,SPEED,10            340
43                 WORK  /A1                                   350
44                 CUT   /CENTER,C                             360
45     A2=         DRILL /SC,DIAMET,45,DEPTH,86,1CCL,226450,2,FEED,0.25,SPEED,25   370
46                 COOLNT/CN                                   380
47                 WORK  /A2                                   390
48                 CUT   /CENTER,C                             400
49     A3=         TURN  /SC,CPCSS,TCOL,506C2C,3,SETANG,90,ROUGH   410
5C                 WORK  /A3                                   420
51                 CUT   /M3,RE,M2                             43C
52     A4= CONT/SC,TOCL,5CCC45,12,SETANG,90,RCLGH
53                 WORK/A4
54     CUT/M3,TC,M7
55                 STOP    11  MESSEN UND KORRIGIEREN           590
56                 CHUCK /4CC,155.5,200,-10,62,20              *63C
57                 CLAMP /112.5                                640
58                 PPRINT/KSPANNEN,WEICHE BACKEN,DURCHM.80X3CLANC,ABGESETZT AUF   65C
59                 WORK  /A1                                   67C
6C                 CLT   /CENTER,112.5                         680
61     A7=         DRILL /SC,DIAMET,17.8,DEPTH,25,TCCL,226178,7   650
62                 WORK  /A7                                   700
63                 CUT   /CENTER,112.5                         710
64     A8=         REAM  /SC,DIAMET,18,DEPTH,25,TCCL,225180,8
65                 WORK  /A8                                   650
66                 CUT   /CENTER,112.5                         750
67     A5= CONT/SD,TCCL,5CCC45,13,SETANG,90,RCLGH              760
68                 WORK/A5
69     CLT/M4,RE,M7
70                 CHUCK /4CC,1E3.5,200,2C,14C,C               *770
71                 CLAMP /112.5,INVERS
72     T3=         TURN/SO,LCNG,RCLGH,TCCL,563 1C4,5,     SETANG,180
73     WORK/T3
74     CLTLOC/BEFIND
75     CLT/M3,RE,M5
76                 FINI                                        800
```

Figure 15.12   An EXAPT 2 part program.

The PART/CORREC, 80, 100 statement (line 34) will cause all the calculated feedrates to be reduced by 20 per cent while the cutting speed will remain unaltered at 100 per cent. The statement CSRAT/80 prevents the cutting speed from falling below 80 per cent of the value calculated from cutting and load criteria; it would otherwise do so during a facing cut. In such cases, the processor calculates intermediate points on the tool path and writes new **SPINDL/** records at these points.

After specifying a simplified inside and outside shape of the chuck in the **CHUCK/...** statement (line 38), the chuck dimensions can be used to check for a collision-free tool path. The part to be turned is to be clamped in the chuck in the position specified by the **CLAMP/112, 5, INVERS** statement (line 39) and is then ready to cut.

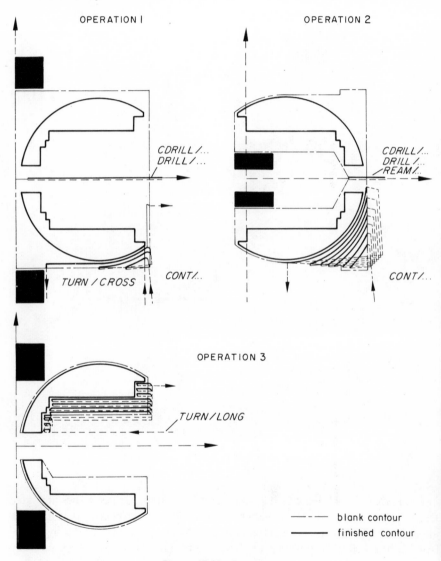

Figure 15.13   Turned part.

WERKSTOFFLISTE

CODE-NUMMER	WERKSTOFF	KS11	EXPON. 1-f	FINK	CROSSX	GROOVK	RKAPK	SPITZENWINKFL
203	C 45	180.	0.80	0.50	0.50	0.30	0.70	118.00

DREHWERKSTOFFE

SCHNSTOFF	CODE	KUEHLUNG	B/H	B/H.MIN	V.MAX	V.MIN	A.MINF	MITTELKUF.	MITTELG.	MITTELH.MITTEL	TL	IFEC	FH.EMPF	FH.EMP
SS	1	1	4.0	0.1	40.	10.	0.20	-0.2389	-0.2807	-0.1485	0.4000	120.0	5.00	0.36
P 10	5	0	14.0	0.1	250.	109.	0.10	-0.2367	-3.1878	-3.2094	0.1753	107.0	2.00	0.20
P 20	6	0	12.0	0.1	200.	70.	0.10	-0.2773	-0.1572	-0.2732	0.3387	230.0	2.13	0.16
P 20	7	0	10.0	0.1	180.	40.	0.20	-0.1981	-3.1432	-0.1836	0.2498	228.0	4.62	0.24
P 30	9	0	10.0	0.1	150.	50.	0.25	-0.1497	-0.1844	-0.1528	0.4134	201.0	5.66	0.37
P 40	9	0	10.0	0.1	120.	40.	0.10	-0.1807	-0.1000	-0.1260	0.3929	63.0	5.66	0.56
P 50	10	0	10.0	0.1	110.	30.	0.30	-0.1732	-0.1000	-0.1260	0.3929	61.0	5.66	0.37
10	11	0	17.0	0.1	160.	70.	0.10	-0.1806	-0.0985	-0.7143	0.7320	120.0	2.00	0.20

BOHRWERKSTOFFE

BEARP.ART	CODE	SCHNSTOFF	CODE	KUEHLUNG	V	A1	B1	A2	B2	A3	B3
CR ILL	1	SS	1	1	22.0	0.93	0.0095	0.73	0.0008	0.17	0.0
CR ILL	2	SS	1	0	7.2	1.93	0.0015	1.27	0.0025	0.82	0.0
REAM	3	SS	1	1	23.0	1.10	0.0013	3.67	0.0020	0.28	0.0
SISINK	4	SS	1	1	12.6	0.76	0.0008	0.54	0.0012	0.0	0.0
SINK	5	SS	1	1	18.0	0.97	0.0010	0.40	0.0010	0.19	0.0
COSINK	6	SS	1	1	18.0	0.90	0.0010	0.40	0.0010	0.15	0.0
TAP	7	0	0	0	0.0	0.0	0.0	0.0	0.0	0.0	0.0

Figure 15.14   Workpiece material list (No. 203 Material).

The cutting value determination in EXAPT 2 is now illustrated by the CLTAPE 2 output of the technological processor, for the facing cut of lines 49 to 51 of the part program.

The work call-up, WORK/A3 in line 50, causes the output of the TOOLNO/ and STAN/ records with the statement numbers 84 and 85 in Fig. 15.15. The preceding CARDNO/ record gives the line number in the part program (50 in this case) which generated these statements 84 and 85. By the RAPID/ move in the following statement 87, derived from the CUT/ of line 51 in Fig. 15.12, the tool is moved to the starting position for the following cut.

78	3C	BLANCO			75.CC		0.0	-C.0
79	3C	ELANCO	2	115.00	22.5C	0.0	0.0	1.00
8C	30	ELANCO	2	26.50	22.5C	0.0	-0.87	C.50
81	3C	ELANCO	2	13.52	-C.C	0.0	0.0	1.00
82	3C	BLANCO	2	C.0	-0.C	-C.0	1.00	-C.0
83	1	CARDNO	50	4	20			
84	2	TOOLNO	1025	3.00	3.CO			
85	2	STAN	1C8C	5C.00				
86	1	CARDNO	51		30			
87	2	RAPID						
88	5	GOTO	5	INTERN	C	121.90	-171.50	
89	5	GOTO	5	INTERN	C	121.90	-167.18	
90	5	GOTO	5	INTERN	C	118.19	-168.17	
91	2	FEDRAT	10C9	C.19				
92	2	SPINDL	1031	262.51	6C			
93	5	GOTO	5	INTERN	C	117.79	-165.80	
94	2	SPINDL	1C31	27C.73	6C			
95	5	GOTO	5	INTERN	C	117.79	-150.64	
96	2	SPINDL	1031	338.41	6C			
97	5	GOTO	5	INTERN	C	117.79	-148.C5	
98	2	RAPID	5					
99	5	GOTO	5	INTERN	C	119.29	-149.20	
1CC	3C	ELANC1	1	112.79	75.CC			
1C1	3C	BLANCO	2	C.0	75.CC	-0.C	0.0	-1.00
1C2	3C	BL		112.79	75.CC	C.0	-1.00	-0.0
1C3	3C			479	22.5C			

Figure 15.15   Part of CLPRINT for Fig. 15.12.

It should be noted that the co-ordinates, given in the GOTO records of Fig. 15.15 for the tool movements, refer to the zero point of the tool description co-ordinate system (i.e., the preset tool length will be corrected for the setting angle). In order to relate the co-ordinates to the tool point in the case of tool

15

506020 (see Fig. 5.16) for the setting angle of $+90°$, a DX $= -4\cdot2$ mm and a DY $= 90$ mm must be added to the co-ordinate values in the GOTO/ records. In the remainder of this example, all values relate to the tool point.

The permissible width of cut for the tool 506020 is 7 mm on the tool card (not illustrated). As no cut depth was programmed in the machining definition (line 49, Fig. 15.12), the division of cuts will be started with 7 mm. The excess dimensions of the raw material, relative to the material left for the finishing cut, permits only a $2\cdot29$ mm broad cut in this case.

```
                           W E R K Z E L G L I S T E
                           ••••••••••••••••••••••••••
      CREHWERKZEUGE
      •••••••••••••
•••••••••••••••••••••••••••••••••••••••••••••••••••••••••••••••••••••••••••••••••••••••
•    SCHNEICH   •  SCHAFT   •  SCHNSTOFF  •  EINSFAHNG  •  IDENTNR  •  UNIT  •  EINSTELL X  •  EINSTELL Y  •
•      31       •    4      •     6       •     0       •  563004   •   M    •   100.00     •    10.00     •
•      24       •    5      •     5       •     C       •  500045   •   M    •    89.00     •     4.00     •
•      14       •    5      •     5       •     0       •  506020   •   M    •    90.00     •     4.20     •
•••••••••••••••••••••••••••••••••••••••••••••••••••••••••••••••••••••••••••••••••••••••••

      BCHRWERKZEUGE
      ••••••••••••••
•••••••••••••••••••••••••••••••••••••••••••••••••••••••••••••••••••••••••••••••••••••••••
•   BEZEICHNG   •  WKZTYP   •  SCHNSTOFF  •  DREHRICHTG  •  IDENTNR  •  UNIT  •  EINSTELL X  •  EINSTELL Y  •
•      70       •    0      •     1       •     1       •  220750   •   M    •   120.00     •     0.0      •
•      41       •    0      •     1       •     1       •  225180   •   M    •   215.3C     •     0.C      •
•      24       •    0      •     1       •     1       •  226178   •   M    •   208.30     •     0.0      •
•      24       •    0      •     1       •     1       •  226450   •   M    •   291.30     •     C.C      •
•••••••••••••••••••••••••••••••••••••••••••••••••••••••••••••••••••••••••••••••••••••••••
```

Figure 15.16   Tool list.

A thickness of cut is determined next in the cutting value model with regard to cutting theory and to the cutting force criteria. In this case it must be determined from the following values:

(a) Preferred thickness from shape-of-cut criteria

$$H1 = 0\cdot49 \text{ mm/rev}$$

(b) Maximum thickness for which the machine loading is not exceeded and which does not call for less than the minimum cutting speed for this combination of cutting tool and workpiece

$$H2 = 6\cdot46 \text{ mm/rev}$$

(c) Maximum thickness, so that the permitted cutting force on the tool is not exceeded

$$H3 = 5\cdot02 \text{ mm/rev}$$

(d) The maximum and minimum thickness specified on the appropriate tool card

$$HZUL = 0\cdot6 \text{ mm/rev}$$
$$HMIN = 0\cdot1 \text{ mm/rev}$$

The smallest cut thickness, which lies within the range HZUL and HMIN suitable for the tool, is therefore in this case the thickness H1 determined from the shape-of-cut criteria, and this value serves as a basis for further calculation. Finally, the machine torque is checked to see whether it is sufficient to deal with the chosen cross-section of cut at the turning diameter. This is so in this example.

In this case, the work is facing and the feed has to be reduced to 80 per cent of the calculated value (as a result of the PART/CORREC, 80, 100 statement of line 34 in Fig. 15.12). After the feed value has been calculated from the cut thickness and the setting angle of the tool, it has to be multiplied by the correction factor for facing cuts and by the feed-reduction factor derived from the last CORREC value. The result of this calculation is given (in statement 91 on the CLTAPE 2) as a feedrate (FEDRAT) of 0·19 mm/rev.

The cutting speed is calculated in another part of the program dependent on the following criteria:

(a) Tool life and width of tool wear mark, considering the effect of depth of cut and feed on wear,

$$V1 = 128\cdot9 \text{ m/min}$$

(b) Machine tool load; limits cutting speed to

$$V2 = 792\cdot6 \text{ m/min}$$

(c) Limits set in the card files for the combination of workpiece and cutting-tool materials,

$$VMAX = 250\cdot0 \text{ m/min}$$
$$VMIN = 100\cdot0 \text{ m/min}$$

The lowest value for cutting speed for this example is thus $V1 = 128\cdot9$ m/min and this is within the limits set by (c). This value is then converted to spindle speed for the current diameter of turning, and written on the CLTAPE 2 (Fig. 15.15) as records such as statements 92, 94, and 96.

Because of the CSRAT/80 statement (in line 37 of Fig. 15.12), the cutting speed is not allowed to fall below 80 per cent of the value at the beginning of a cut. Thus, in statement 95 (of Fig. 15.15), an intermediate point is output on the tool path, allowing a new spindle speed to be output. Considering the movement of the tool point, the ratio of the radii for the initial and intermediate points becomes:

$$\text{Initial Y} = -165\cdot80 + 90 = -75\cdot80$$

From line 93, Fig. 15.15, $Y = -165\cdot8$ and (see above) the correction to obtain tool point $DY = +90$

$$\text{Intermediate Y} = -150\cdot64 + 90 = -60\cdot64.$$
$$\text{Ratio of radii} = -60\cdot64 / -75\cdot80$$
$$= 0\cdot8$$

The determination of the cutting values for the remainder of the turning work is carried out in a similar fashion. The cutting values for the internal work are determined as in EXAPT 1.

Summary

The introduction of NC machine tools removes the optimizing effect of the co-operation between the planner and the machine operator. Since the part programmer has no means of affecting the machining process which he has programmed, he will (because of his natural caution) tend to program less stringent cutting conditions than are possible. This can have an appreciable effect on the economy of the capital-intensive NC machines. In the case of computerized programming, electronic data-processing installations are used which can now carry out the determination of the optimum machining sequence.

The EXAPT programming system contains both appropriate part programming scope and also corresponding processor routines which determine toolpath cutting conditions, choice of tools to be used, and work sequence.

References

1. Opitz, H., Budde, W., Engelskirchen, W. -H., Hirsch, B., and Reckziegel, D. 'The EXAPT Programming System', *TZ für praktische Metallbearbeitung*, **61** (1967) Heft 8.
2. Opitz, H. and Simon, W. *EXAPT 1—Language Description* (Aachen 1967).
3. Hirsch, B. 'Determination of Optimum Cutting Conditions for the Automatic Programming of NC Lathes with EXAPT 2', *Industrie-Anzeiger*, **24** (1968).
4. Opitz, H., Simon, W., Spur, G., and Stute, G. *EXAPT 2—Language Description* (Aachen 1967).
5. Opitz, H. *Production Methods as Elasticity Factors* (Dynamische Betriebsfürung, Berlin 1959).

16. Trend-setting developments

D. T. N. Williamson

D. T. N. Williamson, FRS, is Director of Research and Development, Molins Machine Company Ltd, London. He studied engineering at Edinburgh University and subsequently spent 3 years in the Development and Applications Laboratories of the Marconi-Osram Valve Company, London. He joined the Research Department of Ferranti in Edinburgh in 1947 and was concerned with the first applications of wartime technology to industrial requirements. In 1951, he started the development of the world's first commercial computer-controlled machine tool system. He became manager of the Ferranti Machine Tool Control Division in 1959, leaving to take up his present appointment. For the past 7 years, he has been concerned with the application of advanced technology and systems engineering to the problem of improving the equipment supplied to printing and tobacco processing plants, and also with the design of advanced systems of manufacture for the components used in such machinery. He was elected Fellow of the Royal Society in March, 1968.

The NC scene

The development of NC promised a revolution in batch manufacture; yet ten years' workshop experience of NC tools has shown little fulfilment of this promise. Sections of the aircraft industry have applied NC quite widely, and they can show very significant improvements over their old way of working, so much so that it is safe to say that no aircraft manufacturer who has adopted NC methods would ever turn back. Yet even in this most favourable area of manufacture, the nettle has not been grasped. Most aircraft manufacturers subcontract the majority of their components and make only the most difficult themselves. Numerical control has come to their aid for these difficult parts, but neither they nor their subcontractors make wide use of it for their simpler and smaller parts.

Turning to the more mundane area of the batch-engineering industry, the total number of NC machine tools has not yet reached even 1 per cent of the number of conventional tools installed and, as one might expect with such low numbers, the NC tools are just dumped down in a corner of a conventional workshop and treated as new and rather clever toys which will do the odd difficult job. They impress the visitors, but make no real impact on the business. Life would go on in much the same way if they were uprooted tomorrow.

What are the reasons for this apparent neglect of such potentially powerful techniques? The reasons are manifold, but they will arise basically from the consideration of NC tools just as new types of tool, supplementing conventional machines of all types, ranging from the simple lathe to the exotic EBM, ECM, and EDM equipment. The programming and support organization which NC tools require is thought of as a necessary adjunct to their use, rather like the toolmaking, testing, and materials control facility which surrounds an injection moulding machine, or template-making in conjunction with copying machines, but not as a means whereby the information flow in a factory can be radically changed. There is rarely any understanding that it is only by completely changing the environment and building a new organization around the concept of NC that its real potential can be exploited. The short lead time and machining time is frequently cited as giving a reduction of work in progress and inventory, but what practical use can be made of this if the majority of the components which go to make up a finished machine have a 100-day manufacturing cycle, and only the small minority is made on one or two overworked NC machines? Clearly, the NC components will sit on a shelf in the stores for an average of 50 days. No great breakthrough here!

Numerically controlled machine tools should be treated merely as components in a greater whole, a system of manufacture. Take, for example, an average engineering company engaged in the manufacture of sophisticated mechanical engineering products by conventional methods, each containing several thousand components. The company exists to make these products, and it is only by marketing them successfully that it keeps in business. It is likely that the overhead in such a company will be about 300 per cent of the direct manufacturing cost of the product. This overhead covers the whole running of the company, including research and development (not usually a very large proportion!), purchasing, storekeeping, marketing, administration, and all the paraphernalia of a modern business. Thus, with the inclusion of some profit, we are down to a direct manufacturing cost of about 20 per cent of the selling price. The local workshop overheads on direct manufacturing are likely to be around 100 per cent, which means that the work done in shaping the components represents about 10 per cent of the selling price. Depending on the types of machine tools used, therefore, the cost of component manufacture lies between 10 and 20 per cent of the selling price of the finished product. Whatever improvement technology can make to the direct metal-cutting or shaping process, if it leaves the overhead structure unchanged it will do no more than lower the price of the product by this amount, even if it creates the parts by magic.

This reasoning might (and sometimes does) lead enthusiastic accountants to deduce that workshop technology is not the important place to look at all, but rather in the areas of cost control and management, which look as if they offer bigger pickings. Nothing, however, could be further from the truth. What they do not realize is that the overheads are structured as they are because of basically inefficient methods of manufacture. Everything stems from this.

The production department is usually the largest part of an engineering company, and the majority of the overhead costs of the company is due to the way in which this part is run, and how it impinges on other areas of the company. The capability of the manufacturing organization sets limits to the quality of the designs which can be developed, determines the delivery dates which Marketing can offer and the type of organization and administration which is necessary, and so on. Its efficiency largely determines the success or failure of the company. The overhead costs are very largely determined by the technology of manufacture, as a particular technology tends to grow a corresponding sets of overheads necessary to control and manage it, and to interface it with Design and Marketing, Purchasing, and the other areas of the company. If one examines the growth of companies from small beginnings to large-scale operations, it can be seen how the necessity of controlling methods of manufacture which were not designed to grow large causes the control and support organization to multiply at a faster rate than the direct manufacture, until the number of people 'controlling' and interfacing exceeds (sometimes greatly) the number of people 'doing', and so a top-heavy structure is created. Counting heads, of course, does not tell the whole story, because it is usual for the people in the control structure to be paid more than the machine operators.

Attempts to change this by superimposing computer production scheduling and control techniques are only a palliative, and to date have not achieved signal success, since they leave the basic problem unchanged. Only a radical change in manufacturing technology and its organization can correct this situation[1], and it is here that NC has its greatest part to play. On this part of the drama, the curtain is only starting to rise.

It is necessary to examine the whole structure of manufacture, and with it the whole structure of a company, and to reshape it to take advantage of all the possibilities which a new technology of manufacture can give, if the maximum advantage is to be gained. This having been done and what is eventually possible having been realized, it is then practicable to tackle individual areas one by one, making sure that, as the redesign proceeds, the interfaces with the other areas allow these in their turn to be redesigned and to be linked up into a total system. Thus the prospect need not be too frightening, because the problem can be tackled section by section, *but only after the overall pattern has been determined*. The remainder of this chapter will show how the most difficult area of component manufacture in a medium-size engineering company, which makes 70 per cent of the world's cigarette-making machinery, was tackled in this manner, and shows how, in subsequent development stages, this will react back into the Development and Design Department, and into other areas of the organization.

(1) See references on p. 457.

SYSTEM 24

The system of manufacture which was evolved (called SYSTEM 24[2, 3] because it enabled round-the-clock manufacture to take place substantially with day-shift staffing) is based on three principles:

(a) The first proposition is that the use of light alloy as a constructional material wherever possible, removes a fundamental barrier to manufacturing progress—the low metal-removal rates currently achievable with ferrous materials. This enables the performance of computer-controlled machine tools to be raised by an order of magnitude. Clearly, the use of light alloy is not mandatory in a manufacturing system, but the advantages are so great that it should in future be used as a constructional material wherever possible.

(b) The second is that a big improvement in efficiency and flexibility can be obtained by discarding the current conception of single-machine working in NC, and replacing it by a multi-tool concept, which is, in effect, a flexible transfer line.

(c) The third proposition is conditioned by the other two. Having, by their use, achieved very short machining times, it is no longer possible to tolerate the dead time associated with manual decision and manual handling of materials. This leads to a totally computer-controlled process, in which the computer has the executive power to carry out its decisions.

SYSTEM 24 stems logically from these three fundamental concepts, and it remained to design a range of machine tools and auxiliary equipment, including 'electronic hardware' and 'software', to carry out the overall concept. In this, we tried to make the best use of the available technology to design equipment that has a degree of reliability and expectation of life adequate for continuous, round-the-clock working.

Background
The aim of the development programme which led up to SYSTEM 24 was to simplify and reduce the cost of manufacture of components for a wide range of cigarette-making and packing machinery. This consisted of perhaps 30 major machines, each containing between 1,000 and 3,000 components, which at that time were mainly machined from solid or from castings, principally ferrous.

The components were currently being made by conventional methods using mainly hand-operated tools. In order to gain experience with our own components, one or two NC machine tools were obtained and put to work on a selection of the most suitable components. This number has now risen to over 20, and considerable experience has been gained. This experience can be summarized by saying that, although NC manufacture as generally practised today has numerous advantages over conventional manufacture, a really substantial reduction in cost is not one of them. It is possible to show good cost reductions by selecting the most suitable components but, when taken across the board to cover a substantial percentage of the total parts, costs are comparable to conventional manufacture.

The main barrier to increasing productivity with NC tools is the metal-removal rate of ferrous materials. Ferrous alloys make up the bulk of manufacture, for the good reasons that they are available in a wide range of strengths and properties and that they are cheap, and possibly for the bad reason that 'machines have always been made of steel and cast iron'. Cutter materials have improved over the years, with the introduction of sintered carbides and, more recently, ceramic materials, but there is no sign that this improvement will continue, and attention has been diverted to chemical, electrical, and electrochemical methods of metal removal, with some success in specialized areas but little sign of general applications, certainly not for precision work. The production engineers' old axiom that 'you should put metal where you want it, rather than remove metal from where you don't want it' has been called into play, with the revival of precision-casting processes and even suggestions for NC forging; but it is doubtful whether these will ever have the wide application which precision cutting tools can provide.

The reason for the impasse is, of course, that the attention of the majority of engineering manufacturers has always been focused on ferrous materials. One industry, aircraft manufacture, has not encountered these problems because the use of aluminium as a structural material has been mandatory for the past 40 years, from simple strength-weight considerations, and is only recently being challenged because of the elevated temperatures associated with gas turbines and supersonic speeds, which do not apply to more down-to-earth equipment. The aircraft industry has never been troubled by metal-removal limitations, at least not since it enlisted the help of the makers of very high-speed woodworking machinery to apply their techniques to the cutting of light alloy. The metal-removal rate for light alloys, particularly those free-cutting alloys containing traces of bismuth and lead, appears to be unlimited, and provided good chip-formation conditions are observed, 'the faster you go, the better' appears to be the order of the day.

It seems obvious that the main barrier to progress in NC machining would disappear if the articles being machined were made of light alloy. At Molins, therefore, we asked ourselves 'Why should our machines not be made of light alloy?' It was quite clear why they had hitherto been made mainly in cast iron and steel: it was simply that these were the traditional materials, and that with current production methods there seemed to be no advantage in deviating from them, apart from reduction in weight. If two identical components are made by conventional methods, one in steel and the other in light alloy, it is likely that the light alloy component will, if anything, be slightly more expensive because of increased material cost. In fact, in Molins the weight consideration alone has led to considerable use of light alloy, because tobacco machinery is frequently

15*

crowded together on floors that have become overloaded, so weight can be important. Upon investigation, the answer to the question turned out to be that there was no good reason whatsoever why the machines should not be made mainly of light alloy, if there was any manufacturing advantage in doing so. We set out to obtain just such an advantage.

Design in light alloy

In the capital goods industry, unlike the consumer goods industries and heavy industries such as shipbuilding, raw material cost is not usually a significant factor; it is the added value which really counts. In Molins, for instance, we buy raw materials at perhaps £200 to £400 per ton and sell the finished product at £5,000 to £10,000 per ton, which is not untypical. In fact, changing over from ferrous material to light alloy, material cost does not rise significantly because components are designed largely by volume, and the ratio of density of 2·5 goes far to equalize the cost per unit volume.

A number of industries, especially those where refugee designers from the aircraft industry are to be found, have already appreciated the advantages of light alloy, and in the instrument and optical industries it is the preferred metal. This trend is also noticeable in the computer peripheral and business machines industries, and it can clearly spread to other industries, such as textile, packaging and printing machinery and many others. The advent of SYSTEM 24 should accelerate this trend by providing a powerful economic incentive.

We carried out surveys at the start of the development programme in order to determine the size range of components and to try to decide how universal light-alloy construction could be. We found that 70 per cent of our machined components were within a size of 300 × 300 × 150 mm (12 × 12 × 6 in.), and that the vast majority of metal components, between 80 and 90 per cent, could be made in light alloy, many with functional advantages in addition to those to be obtained by faster manufacture. Experience has confirmed that this survey was correct, and that these proportions are to be found in many other industries. The metal components on our new range of machines are almost entirely of light alloy, where advantage can be taken of high machining speed.

The group of complementary machine tools

Most engineering parts are geometrically designed, and the majority of the machining on these parts is therefore relatively simple, usually within a three-axis capability. There is, however, frequently a small amount of machining on each part which is more complex, needing a five-axis capability if it is to be accomplished without changing the set-up. It is axiomatic, of course, that re-setting the part must be avoided if at all possible, as otherwise most of the benefits of NC will be lost. The five-axis machining may only consist of a hole at an awkward angle, or a tiny face, or a pipe-connection flange at an angle; but, unless this can be done, the set-up has to be broken down and carefully reset, and the additional tiny piece of machining may cost as much as the rest of the part, even though it only represents 5 per cent of the total metal-removal. A check at Molins showed that about 50 per cent of potential NC parts require a five-axis capability, but usually for less than 10 per cent of the total machining on the part.

It is obvious, therefore, that a 'machining centre' consisting of a single machine should have a five-axis capability or otherwise be restricted to only half the potential work. It is also true that a five-axis machine has necessarily many compromises in its construction and, apart from its high cost, is not likely to be as efficient a metal-removal tool in the three-axis mode as a simpler and more robust three-axis machine. Also, while it is easy to make a three-axis machine with multiple spindles, to improve the metal-removal capability, it is much more difficult to do this with a five-axis machine.

The solution is to design a group of complementary machine tools, each one of which carries out its own category of manufacturing operations as swiftly and efficiently as possible, and to pass the workpieces through any or all of these machines, as its complexity demands. For example, a part containing 50 per cent three-axis milling, 40 per cent hole manufacturing (including boring, reaming, drilling, and tapping) and 10 per cent five-axis milling or hole manufacturing (perhaps a small face containing two holes at an awkward angle), would spend time at each of three specialized machines. This does not mean necessarily that time in these proportions would be spent on each machine tool, because each of the three machines will operate at different speeds. In the SYSTEM 24 machine tools, for example, the effective speed of the three-axis milling machine is about 4 times that of the six-axis milling machine when high metal-removal rates are required, in spite of the fact that the latter has been designed using exactly the same technology. Because of its complexity and flexibility, many compromises have to be made in the design of the six-axis machine which reduce its metal-removal capability by comparison with a simpler machine tool.

This illustrates the great advantage to be obtained from this concept of transferring the part through a group of complementary machines. Each machine in the group can be designed to give the best possible performance in its own manufacturing area, whether it is to be the removal of metal at high speed, the boring of accurate holes, or the sculpturing of complex shapes. It will be clear to anyone who has ever designed anything, that this idea gives a far greater scope to achieve efficiency without the performance being whittled away by extensive compromise. Another advantage is that the system is open-ended, and a new machine to carry out a particular process better, or to include a new process in the group capability, can be added when required.

A key to the concept is to be found in the design of a pallet and a method of locating this pallet to an accuracy considerably higher than that demanded of the machining process, so that the pallet and the worktable on which it locates become the link between the complementary machines in the group. In SYSTEM 24, this accuracy is achieved by an electronic servo-location system, because it did not appear practicable to design a mechanical location system which would repeat to an accuracy of ± 0.0001 in. over long periods in the presence of machining swarf and other dirt.

An integrated batch-manufacturing system

It is one thing to conceive and design hardware for NC manufacture. It is quite another to ensure that this is operated efficiently. One has only to visit an NC workshop and to count the percentage of machines cutting metal at any one time to understand what is meant. As the machining time is reduced, with work-fixing removed away from the machine on to pallets, efficiency and accuracy of transfer become of crucial importance to the overall efficiency. A six-machine SYSTEM 24 installation is capable of producing between 2,000 and 20,000 components per day. The likelihood of manually handling this amount of material and information without chaos is remote.

The only solution is to bring the whole process under the control of an on-line digital computer, to which the handling of such information is child's play. In order for this to be effective, the computer must not only instruct, it must be physically capable of executing the instructions to ensure that they are carried out expeditiously. This requires the provision of computer-controlled automatic handling equipment designed to enable complete flexibility to be obtained under all conditions. It demands, in fact, electromechanical systems engineering, with the whole process designed so that the computer can cope with any likely eventuality, calling for manual assistance only on rare occasions. Every process, from the initial demand, through materials preparation, work-fixing, the design of tooling systems for work-fixing, the automatic computer-controlled transfer of materials and pallets from place to place, the design of the machine tools to work fully automatically, looking after cutter change, swarf disposal, program change, the resetting of workpieces, and the unloading of finished material and its subsequent handling to rejoin the main manufacture and assembly, must be thought out in the greatest detail, with a procedure to cover every possible eventuality.

The opportunity has been taken to design as advanced a system as seems possible with present technology, in order that its obsolescence should be slow. This is highly desirable because the components of the system have been designed for an indefinite life, using oil hydrostatic and air bearings wherever possible. An integrated batch-manufacturing system represents a large step from present-day production technology, and, to cushion the impact of this, the equipment has been designed in units, so that manufacturing systems varying from a single machine up to the full complex can be assembled to suit prevailing conditions.

Design concepts

The basic concepts which led to the development of SYSTEM 24 have been outlined in the preceding sections. Remembering that the original requirement was for Molins's own use, in order to save time and cost it would have been desirable to adapt existing equipment as far as possible. This possibility was examined and abandoned. Experience derived from using one of the biggest installations of NC machine tools in Europe showed quite conclusively that such machines were not suitable for use as building bricks in a fully automatic system. Where human intervention is to be kept to a minimum, a different order of machine accuracy, consistency, reliability, and life is essential, quite apart from the need for automatic work-transfer devices, which are not to be found on present-generation machines. Although the majority of cutting on engineering work is not done to close tolerances, many parts have features, such as holes for gear centres, which must be to jig-borer accuracy. Existing designs of machine tools, whether NC or not, require regular attention by their operators to produce accurate work consistently. No one today uses such machines without skilled operators in attendance, even though many, including Molins, have tried. Deficiences which pass almost unnoticed, because they are overcome by the operator, can bring unattended machines to a stop. If tolerances of ± 0.05 mm (0·002 in.) are required, constant attention to detail is necessary. Only if the tolerance is greater than ± 0.125 mm (0·005 in.) can suitable machines be left without skilled attention. Machines required to run unsupervised and to produce work of high accuracy must be designed to higher standards.

From considerations of accuracy alone, the build-up of errors from all sources (cutter diameter and eccentricity, spindle runout errors, cutter positioning slideway variations, lag of servomechanisms, etc.) means that if components are to be produced within an acceptable tolerance band on fully-automatic machines, all individual sources of machining error must be reduced to between 10 and 30 per cent of those values currently achieved on commercially available NC machine tools. Accuracy of this order can only be attained by starting afresh, incorporating the latest technology, and paying extremely close attention to the detailed design of all aspects of the machining installation such as pallets, loading and unloading mechanisms, location procedures, slideway geometry and support, cutter-changing methods, machine spindles and drives, swarf clearance, control systems, interlocks, and a host of equally important features.

In short we are seeking machine tools with the following characteristics:

(a) A rather small capacity (300 × 300 × 150 mm).

(b) An ultra-high reliability for round-the-clock operation, which demands hydrostatic bearings everywhere.

(c) At least two simultaneous cutting spindles to increase productivity.

(d) A vertical or upside-down machining surface to give adequate chip clearance.

(e) Extremely high acceleration rates, to avoid cutter rubbing when accelerating to machining speed and to reduce wasted time during air movements.

(f) Extremely high spindle horsepower to give fast metal removal. Power in the region of 15 to 20 bhp at speeds up to 30,000 rev/min. This requirement cannot be met by electrically driven spindles because of the size and the limitations on accuracy of the thermal cycling involved.

(g) A five- or six-axis machine of similar size range, in addition to three-axis machines.

Clearly, it was not possible to purchase machine tools which met this specification, so we decided to engineer the whole system from the ground up, without regard to current conceptions of what a machine tool looked like, but using the best present-day technology to ensure the highest attainable level of productivity.

This has involved a great deal of research and development into metal cutting and the optimum material-cutter relationship, advanced hydrostatics and fluid dynamics, and modular structural engineering. The programme has been a rewarding one, producing some new techniques which can be applied to fields wider than the present system. Considerable advances in control engineering have been made in collaboration with Ferranti, Edinburgh; and the productivity of the system owes much to the capabilities of the NC system.

In order to obtain outstanding performance with total freedom from wear and minimum maintenance under arduous, around-the-clock conditions, the design of the family of machine tools is unconventional. An important factor which influenced their design was the desire to make very accurate tools, but to do this without the usual highly skilled metal-scraping and figuring operations which are characteristic of today's precision tools such as jig borers. The jig-borer companies who practise this art maintain a tradition of skill which it is impracticable to imitate when starting from scratch. We, therefore, tried to design the machine tools on the basis of using slideway members which could be machined to very high accuracy, making provision for the individual slide-ways to be adjusted in precise relationship to each other.

The construction of the machine tools is based on a new form of combined hydrostatic slide and hydraulic actuator, and this forms the basic method of location and control of all slides. The form of construction can be seen from Fig. 16.1. A parallel rod or tube is formed from alloy steel by turning and grinding, with a larger-diameter section in the centre to form a piston. The difference in cross-sectional area between the central piston and the main section of the rod forms the effective area of the actuator. The central piston is fitted with hydraulic seals. The whole surface of the rod is hard chrome-plated and ground very accurately cylindrical and straight.

This rod operates in a cylinder formed by boring and sleeving the slide casting. Two end caps containing hydrostatic bearings and their drains and seals complete the assembly, which together forms a cylinder-and-piston unit with a greatly enlarged piston rod. The piston rod and the hydrostatic bearings form together a cylindrical hydrostatic slideway, which constrains the casting or slide to move in a straight line. To prevent rotation, this slideway is used in conjunction with a flat caliper hydrostatic slide, and together the cylinder and flat combination are perfectly kinematic, and can be designed to give very high rigidity and accuracy. This construction is used in one form or another on every slide of the machine-tool family, and provision is made to be able to adjust the

Figure 16.1 Exploded view of combined hydrostatic actuator and slideway system: the end caps include complete hydrostatic bearing systems. (*Courtesy Molins Machine Co. Ltd.*)

slides to be perfectly parallel or at right-angles as required. Light alloy has been used for the moving parts wherever possible in order to reduce their mass and thereby improve servo performance.

A Pelton turbine (Fig. 16.2) using oil at 2,000 lbf/in.² has been designed to give a high-powered, high-speed spindle, to maximize metal-removal. The machine tools are fully automatic and do not require an operator. Work is loaded and unloaded automatically on pallets, and cutters are changed and set to length automatically, thereby avoiding the need for preset tooling. Chip disposal is fully automatic, and chips are conveyed to a central point and baled for collection. Machine tools are controlled by hydraulic servo-mechanisms, and the measuring feedback is by means of linear and circular gratings.

In addition to machine tools which were fully automatic and had a high-accuracy pallet-location system as a common feature, a mass of auxiliary equipment to handle pallets and material under the command of a computer, work-fixing stations and tooling systems to simplify and speed up a necessary process, workpiece-resetting machines to enable the difficult process of resetting to be carried out as routine by girls, and many other devices, have been developed.

Figure 16.2   Pelton turbine wheel providing a 20 bhp spindle drive at speeds up to 24,000 rev/min. (*Courtesy Molins Machine Co. Ltd.*)

Their operation is described briefly in the following sections. This is by no means a complete and detailed account, but it will serve to illustrate the possibilities which a thorough treatment of a particular area of manufacture can provide.

Besides the hardware, SYSTEM 24 involves a considerable amount of computer software development. The issue and scheduling of parts and material into the system has been made automatic, and the use of the IBM 1130 as an on-line control computer has required the development of input/output equipment and software to cope with the complex situations which can arise in an automatic system. Special software to ease NC parts-programming difficulties is vital to speed the huge flow of components through the system.

In the first Molins SYSTEM 24 installation in Deptford, we have been fortunate enough to need a new building, so the opportunity has been taken to design this to suit the system. The hydraulic power station and the swarf disposal system are installed in a cellar beneath the main machine room, the work-piece preparation area is at the end of the machine room, and above this a mezzanine floor houses the data-processing and control equipment. One would hazard a guess that the pleasant working conditions have never before been approached in a metal-cutting machine shop.

Description of a typical SYSTEM 24 installation

The following is a description of a six-machine unit designed for continuous 24-hour production and 8-hour day-shift loading, which is being installed in a new building at the Company's works in Deptford. This layout has the advantage that it is modular, and can be extended to any number of machines in the group. By simply altering the pitch between machines, storage and work-fixing space is altered to suit the average cycle time.

Equipment

The layout is shown in Fig. 16.3. The machine tools are arranged in a single row at 3,285 mm (10 ft 9·3 in.) pitch, and the swarf disposal and power-supply arrangements, including the hydraulic pumping station, are installed in a cellar beneath the machines. By the side of each machine stands a *loading unit*.

Alongside the loading units stands a large pigeon-hole *pallet rack*, on either side of which runs a servo-controlled *pallet conveyor*, known as a MOLAC (Molins On-Line Automatic Conveyor). The rack stores 72 pairs of pallets per 3,285 mm run, sufficient for 18 hours continuous operation at a consumption rate of 4 pairs of pallets per hour per machine, an average cycle time of 15 minutes.

On the other side of this rack/conveyor complex runs a line of *work-setting stations* in groups of three, interposed with *resetting/inspection* machines. The work-setting stations are designed to give every facility to make it easy for girl operators to attach workpieces to pallets. This is a semi-skilled operation, but it must be done without error if the system is to run at high efficiency. Consider-able attention has therefore been paid to ergonomic design and foolproof com-munication. The work-setting station has a display panel in front of the operator with the following facilities:

(a) Function display, indicating *set* or *reset* or *unload*.
(b) A two-digit numerical indicator, including plus/minus signs.

(c) An intercommunication loudspeaker-microphone, connecting with the *control room*. This also relays music during working hours, if desired by. the operator.

Figure 16.3   Model of the first SYSTEM 24 installation at Deptford. (*Courtesy Molins Machine Co. Ltd.*)

The work-setting stations are in units three pallet pitches wide—1,095 mm (43 in.)—and 1,460 mm (58 in.) deep, the 730 mm wide rear section adjacent to the MOLAC being designed as a pallet delivery, collection, and storage rack capable of holding two pairs of pallets on the delivery and one on the collection side. This forms a buffer store between the MOLAC and the work-setting station.

Figure 16.4   Model of workpiece preparation centre. (*Courtesy Molins Machine Co. Ltd.*)

Material is delivered from, and returned to, the adjacent *workpiece-prepara-tion centre* (Fig. 16.4) by means of two linear conveyors: the *routing conveyor*, which enables bins full of material to be delivered automatically from the *bin rack* in the workpiece-preparation centre to each work-setting station; and the *return conveyor*, which enables bins, finished material, tooling, etc. to be sent back. Each work-setting station is equipped with electronic data-acquisition systems which read pallet numbers and bin numbers and transmit these to the computer.

Parts scheduling and data processing

The issue of parts into SYSTEM 24 manufacture is made simultaneously with the issue of parts to the remainder of the factory. Part numbers for SYSTEM

SYSTEM 24

DAILY MASTER FILE EXTRACT

PART NO	QTY	BILLET CODE	BILLET SIZE	PARTS/PAIR OF PALLETS	WORKSET NORM	FIRST OPERATION				SECOND OPERATION				THIRD OPERATION			CLEAN
						M/C FIRST TOOLBAR	PRGM NO	RUN TIME MIN	RESET NORM	M/C FIRST TCOLBAR	PRGM NO	RUN TIME MIN	RESET NORM	M/C FIRST TOOLBAR	PRGM NO	RUN TIME MIN	
424760	24	006	240225	6	5	2	4764	5.2	0	1	4765	7.5	0				0
317298	12	002	190100	8	8	2	291	11.1	0	1	292	4.2	11	2	293	9.8	0
398712	64	011	270275	8	4	3	1090	4.9	0	2	1091	6.1	0	1	1092	7.8	0
471332	16	003	250150	8	6	2	2302	10.5	12	1	2303	6.7	0				0
388644	12	018	210200	12	6	4	1771	5.6	0	3	1772	13.8	0	1	1773	4.9	0
563233	80	007	185125	4	5	2	934	9.3	0	2	935	3.8	0				0
410869	28	006	290225	2	3	1	3827	5.4	0	1	3828	11.3	0				0
515423	96	012	200175	10	7	3	878	9.4	0	2	879	6.3	7	3	880	5.2	0
342453	16	008	280250	4	8	1	2791	7.6	0	1	2792	4.9	0				0
465772	22	011	265275	12	4	2	4621	6.2	10	3	4622	8.7	0				0
385491	64	012	290175	8	11	3	785	11.2	0	2	786	3.6	0	2	787	7.6	0
550762	6	007	250125	6	10	1	1616	4.6	0	1	1617	9.8	0	5	1618	3.9	0
419769	20	010	220250	10	9	4	2765	9.3	0	1	2766	5.5	16	2	2767	12.8	0
541198	36	004	195150	6	5	3	701	9.1	11	2	702	3.9	0				0
324878	48	018	260200	12	12	2	4030	3.6	0	2	4031	6.6	0	1	4032	5.9	0

Figure 16.5 Daily master file extract. (*Courtesy Molins Machine Co. Ltd.*)

24 have a suffix or other means of recognition which enables them to be sorted from the main issue of parts into manufacture. The issue, in terms of *part number* and *quantity* for a period of say ten days or more, is then run with the SYSTEM 24 *master file* tape, and an extract from this made for those part numbers and quantities. A sample extract from this master file tape is shown in Fig. 16.5. The SYSTEM 24 master file tape contains all the manufacturing information necessary for all part numbers which have been made on SYSTEM 24, and is updated continually as new parts are programmed. The quantity, of course, is not included on the master file, but is added from the issue when the extract is made.

The next step is to schedule the parts in groups which provide daily batches of suitable size and which give reasonably uniform machine-tool utilization for the particular arrangement of SYSTEM 24 machines concerned. This is done by extracting machine numbers and tape-running time for each operation by part number, and running these with a scheduling program to group the avail-

SYSTEM 24

DAILY MATERIAL SCHEDULE

PART NO.	QUANTITY	BILLET CODE	BILLET LENGTH	BILLET WIDTH	BILLET QUANTITY
424760	24	006	240	225	8
317298	12	002	190	100	4
398712	64	011	270	275	16
471332	16	003	250	150	4
388644	12	018	210	200	2
563233	80	007	185	125	40
410869	28	006	290	225	28
515423	96	012	200	175	20
342453	16	008	280	250	8
465772	22	011	265	275	4
385491	64	012	290	175	16
550762	6	007	250	125	2
419765	20	010	220	250	4
541198	36	004	195	150	12
324878	48	018	260	200	8

Figure 16.6   Daily material schedule. (*Courtesy Molins Machine Co. Ltd.*)

SYSTEM 24

DAILY MACHINE PROGRAM LIST

MACHINE NO.1	MACHINE NO.2	MACHINE NO.3	MACHINE NO.4	MACHINE NO.5	MACHINE NO.6
4765	4764	292	1091	293	1092
291	2302	1090	1771	2303	1773
935	1772	3828	2792	880	2767
3827	934	878	2765	1618	
2791	879	785	702		
4622	4621	1617			
786	787	701			
1616	2766	· 4032			
4031	4030				

Figure 16.7   Daily machine program list. (*Courtesy Molins Machine Co. Ltd.*)

able parts into optimum daily batches to give a specified degree of system loading.

The output of this process, called the *daily master file extract*, is run either simultaneously or sequentially with two programs which produce two printouts:

(a) The *daily material schedule*, an example of which is shown in Fig. 16.16.
(b) The *daily machine program list*, listing the tape programs by machine number, as shown in Fig. 16.7.

The daily master file extract is converted into punched cards or other suitable form for transfer to the SYSTEM 24 computer.

Material and workpiece preparation

The material for SYSTEM 24 manufacture can be prepared in a single stage on the day before it is to be used, which implies that the material store is combined with workpiece preparation, or in two stages, in which case the material store is separate from final workpiece preparation. Company organization plays a large part in determining which method is selected. In Molins we have selected the second alternative, and this will be described.

The Material Stores is supplied, preferably several days in advance, with copies of the daily material schedule for a suitable period. The most generally used material for SYSTEM 24 manufacture is extruded light alloy billet, followed by rolled sheet, and occasionally castings. Extruded material is stocked in standard cross-sections selected to cover the required range. Each cross-section, in combination with the material, carries a code number, and sheet is similarly coded by material and thickness. The basic workpiece is a length cut from an extruded bar of a given cross-section, or alternatively an area cut from sheet, and the size and quantities are specified for each job in the daily material schedule. The advantage of two-stage preparation is that batches with the same billet code can be grouped over a longer period than the daily batch appropriate to single-stage workpiece preparation, with some economies.

Billets are cut to length on a high-speed circular saw, which cuts the material like wood, and one surface is faced on a specially adapted milling machine. The complete cutting to size and surfacing takes less than two minutes per piece, and probably less than one minute if the quantities are grouped. To assist this process, a *weekly material schedule* can be prepared by the computer, grouping together all parts from the same billet cross-section or sheet thickness.

The cut blocks of material are stored on trolleys for transfer to the SYSTEM 24 workpiece-preparation centre on the appropriate day.

The workpiece-preparation centre, which is adjacent to the main SYSTEM 24 workshop, is concerned with batching the metal blanks, work-fixing devices, and information appropriate to each part number in numbered bins on a cafeteria system, which are then stored in the bin rack to be sent to the appropriate work-setting station on the following day. The workpiece-preparation centre receives the daily material schedule and the daily machine program list one day in advance of actual machining. The whole system is controlled by a small on-line digital computer (IBM 1130 or similar), which must already have the daily master file extract assembled in its store, and the *modus operandi* is as follows.

The appropriate plastic job envelope corresponding to a part number is taken from a file. This contains all the necessary information for workpiece preparation and work-fixing, which has previously been prepared and checked, consisting of:

drawing of the workpiece blank;
drilling template or punched card;
work-setting drawing or photograph;
work-setting aperture card transparency;
list of necessary work-setting devices.

If the workpiece requires to be reset, the envelope will also contain:

resetting aperture card transparency;
list of necessary resetting devices.

A bin rack accessed by a MOLAC runs along one wall of the workpiece-preparation centre. Bins placed on a conveyor leading to one position in this rack are automatically stacked in their correct position as numbered. The bin-assembler takes an empty bin at random. He places the completed metal blanks in the bin along with the drawing of the workpiece blank and the drilling template. The bin is passed to the drill-fixing section, where the metal blanks are drilled and tapped—or prepared in one of the other preferred ways—for work-fixing, using the drawing and template provided. Drilling and tapping are intended eventually to be carried out by means of a multi-spindle drill, controlled by punched card, drilling all holes simultaneously to the grid pattern corresponding to the pallet fixing holes. At the next stage, the template and drawing are removed, and a toolbox containing the appropriate work-fixing devices and the aperture card is placed in the bin. The final stage is a checking station, where the bin is positioned over an electronic number reader. When the checker is satisfied that the contents are correct, she places the aperture card into a reader, and the simultaneous appearance of the card number (which is the same as the part number) and the bin number causes the bin number to be entered against the part number in the computer memory. If a batch requires more than one bin, each number is coupled up with the aperture card as before, and the card must then be placed in the first bin to be accessed. The completed bin is now placed on the conveyor, to be stored in the correct rack position by the MOLAC. In this way the rack is filled up with numbered bins containing workpieces, tooling, and information for the following day, when they will be conveyed automatically on demand by the computer to the appropriate worksetters.

Worksetting

When the time arrives, on the following day, for a particular job to be machined, the computer allocates a group of pallets, sufficient for the quantity of parts required, in a section of the pallet rack near to the machine tools to be used, and selects a work-setter in the same region. This is to minimize the movement of the MOLACS. The work-setter is warned by the illumination of her 'set' display. These decisions having been taken, the computer issues instructions to the conveyor system linking the workpiece-preparation centre to the work-setting stations to deliver the appropriate bin to her. Other bins will be despatched at appropriate intervals, determined by the work-setter's progress.

The number of the bin is read automatically upon arrival at the work-setting station, and the recognition of this number causes the computer to instruct the work-setting MOLAC to deliver the first pair of pallets from the group which it has allocated for the job. The arrival of the first pair of pallets initiates the timing of the work-setting process. The *work-set norm* (see Fig. 16.5) appears on a digital display in front of the girl, and starts counting down in real time. She places the aperture card into the slide projector, which projects the image onto the pallet, and studies the information, which is in a simple pre-arranged form corresponding to a limited number of standard methods of work-fixing. Each tooling member has a different, clearly recognizable shape, and she places these to correspond with the projected image on the pallet. She then takes a torque-limiting pneumatic screwdriver and spins the tooling screws into the tapped grid holes on the pallet. The workpiece is offered up to the tooling as indicated by the transparency. Spring-loaded screws in the tooling are spun up into the tapped holes in the workpiece, retaining it securely but allowing freedom to machine the whole top surface. The completed pallet is then pushed into the 'out' rack. The 'out' rack is equipped with two electronic pallet number readers, and immediately the second pallet of the pair is pushed into this rack, the count-down stops and the following information is automatically signalled to the computer:

a pallet pair, with identical (check) numbers, is waiting for collection;
the bin number (as a check);
the aperture card number (as a check);
the work-setting station number (as a check).

On receipt of this information, the computer instructs the work-setting MOLAC to collect the pallets and carry them to their position in the storage rack. The opportunity would be taken, if appropriate, to deliver another pair of pallets on the same trip. Both pallet MOLACS can carry two pairs of pallets simultaneously, because most operations require simultaneous delivery and collection, and the provision of capacity for two pairs of pallets roughly halves the amount of travelling which they have to do. The work-setter continues until all the blanks for a particular job have been mounted on pallets. Empty pallets are delivered as required from the storage rack in sympathy with the appearance of completed pairs of pallets in the 'out' section, until all the pallets required for the particular job number and quantity have been delivered.

In this manner, sufficient work is fed into the storage racks during an 8-hour day-shift to keep the machine operating during this shift plus the 16-hour un-attended night-shift, including the overlap for the beginning of the following day-shift. The number of work-setters required to do this is inversely pro-portional to the cycle time, and in this example three positions have been pro-vided for each machine installed in the group.

The work input rate of the system is controlled by the pace of the work-setters. In order that a check can be kept that this pace is reasonable, the com-puter keeps a total of the residual time for each setting operation, and displays the cumulative total to the worksetter on the same display in between the work-setting count-downs. This cumulative total for each work-setter is also displayed continuously in the control room, to enable the System Controller to see how the schedule is progressing and to take appropriate action if required.

Machining

The machining process taking place on the other side of the storage rack is also under the control of the computer, as follows. The MOLAC supplying the machine tools receives an instruction from the computer, determined by the work schedule, to transfer a pair of pallets to a particular machine tool. It goes to the appropriate rack position, withdraws the pair of pallets from the rack, reads the numbers and, if they are correct, delivers the pallets to the appropriate machine tool-loading unit. This information will, in general, have been given by the computer at a time when there is a pair of pallets to be collected from that machine tool, in which case the MOLAC picks up the completed pair of pallets from the loading unit before it delivers the next pair into the same position on this unit.

Nothing further happens to this pair of pallets until the machine tool has completed the job at which it is presently engaged. Immediately this occurs, the pallets which have just been machined are ejected on to the top vertical member of a rotatable cruciform section on the loading unit. Simultaneously, the cutter magazine or toolbar is replaced in the section of the turret from which it came. The pallets which have just been delivered by the MOLAC are sitting on the rear horizontal section of the cruciform. The cruciform is now rotated through 270° in 90° steps, which takes the newly machined workpieces through the cleaning section, where they are cleaned by compressed air blasts, and into the position previously occupied by the most recently delivered work. This automatically brings the first pair of pallets in the buffer store queue into the machine loading position. In this position the pallet numbers are again read, and the computer interrogated, to find the number of the toolbar and the number of the machining program. Receipt of the toolbar number causes the five-position magazine turret to index until a reader senses this number magnetically on the end of the toolbar, and the pallets and toolbar are loaded simultaneously, and the pallet location process commences.

The machining programs are recorded on half-inch magnetic tape contained in metal cassettes, each one of which has a unique electronically read program number on its upper edge. Each machine tool has its own control unit fitted with a random-access cassette magnetic-tape reader, all this equipment being housed in the control room. The random-access units have been specially designed for SYSTEM 24, and contain up to 60 cassettes, each one of which can be accessed and changed within 15 seconds, as demanded by the computer. It is a girl's job to ensure that each machine tool's random-access reader contains the tapes listed on the daily machine program list. These can be in any order, but keeping them in the same order as listed reduces the search time considerably.

Immediately the *correct pallet location* and *correct cassette location* signals are received simultaneously, the tape starts, and from that time the tool is under the control of the magnetic tape until one of two signals is received by the computer, either the *end of tape* signal, which means that the tape has successfully completed, or the *machine fault* signal, which indicates that something untoward has happened. Receipt of the 'end of tape' signal causes the pallets and toolbar to be ejected, and the cycle continues. Receipt of the 'machine fault' signal is recorded by the computer typewriter, along with the time, and signals the maintenance staff to attend the machine. A display at the machine tool, repeated at the Controller's desk, indicates all kinds of likely faults, and the maintenance engineer takes the appropriate action. He can, if necessary, by throwing a switch, remove the machine from the control of the computer, if he wishes to carry out any testing. If, in the opinion of the maintenance engineer, the fault will exceed thirty minutes, he presses a button which signals the computer that the machine will be out of action for at least that length of time, which causes the computer to modify its schedule, if possible, to make the best use of remaining capacity, and to. try to prevent the starvation of other machines in the line. When the machine comes back 'on the air', the computer reverts to its original program as far as possible.

When the machine tool has completed its operations on the workpieces fixed on the pair of pallets, as indicated by the 'end of tape' signal, these are then ejected as described before. The loading unit, as in the previously described sequence, now has these pallets in the collection/delivery position. The reading of the numbers of the next pair of pallets signals the completion of the cycle to the computer. This calls up the MOLAC, which arrives as soon as its commitments will allow, to remove the pallets and replace them with another pair. Meanwhile, operations have begun on the next pair of pallets in the queue. When the MOLAC has collected the pallets from the loading unit, the numbers are again read, and they are replaced in the rack ready for the second operation. Upon the instruction of the computer, the MOLAC then delivers the pallets to the second machine, and the second part of the sequence proceeds similarly to the first—and so on until all machining which the workpiece set-up allows has been completed. Because of the cutter-access flexibility of the system, this usually means (80 per cent of parts) that the component has, in fact, been completed. The pallets are then returned to the rack to be unloaded or reset, as the case may be.

Toolbars can be selected from the turret at will during the running of the magnetic tape program, controlled by signals from the tape. By this means, each machine has access to 70 cutters per spindle and, since a part will go through two or three machines on average, between 140 and 210 cutters are available. Should others be required, the toolbars in the turrets can be interchanged automatically by the MOLAC from a *toolbar store* at one end of the pallet rack. This store can be of unlimited size, and, in addition to giving access to special cutters, complete cutter replacement can be programmed at any time, if required.

The group of machine tools may contain a tape-controlled automatic inspection machine. This machine is structurally identical to the twin-spindle milling machine, except that the spindles are replaced by a pair of measuring probes which, when moved by the control system to an appropriate position, indicate the error of the machining surface at that position. A magnetic tape checking program is made by a computer subroutine from the cutting program, to check the accuracy of the important features of the part. When the first pair of pallets of a batch of parts has completed machining, it is routed by the computer to the inspection machine, where the accuracy of the job is checked. If the accuracy is satisfactory, machining can proceed on the rest of the job. If not, the computer typewriter will indicate at what point and by how much the part deviates from the correct value, so that appropriate action may be taken, and the remainder of the batch is inhibited until this is cleared.

The decisions for unloading and resetting are under the control of the computer, which will select the most favourable sequence of events by examining the state of the pallet store. Because work-setting involves people, and there will clearly be human problems arising, it is desirable in exceptional circumstances to enable the sequence of events for this section to be overridden by the System Controller, in charge of the workshop. In order that he should be *au fait* with the state of the art at any given time, the System Controller is kept in touch with any deviation from the normal pattern by pre-arranged messages from the computer typewriter, supplementing other visual displays. These messages are in plain language, and cover almost every eventuality. They also provide a log of the day's performance.

Resetting

Because of the cutter-access flexibility of the system, which incorporates a six-axis machine, the only resetting process which is necessary is to turn the part through 180° to bring the underface, which was previously sitting on the pallet, uppermost. A process, including a special piece of equipment, has been devised to enable this to be done swiftly and with very high accuracy. Workpieces which require resetting have two small conical identations made in the faced surface of the underside of the billet during initial setting by an additional work-fixing device, which is a pillar, fitting into a single hole in the pallet grid, carrying a conical diamond, the tip of which sits 0·1 mm (0·004 in.) above the level of the tooling. Two of these are used in appropriate grid positions for each part to be reset, and are assembled with the rest of the tooling as indicated by the transparency. When machining has taken place to the point of resetting, the pallet pair is delivered to the workpiece-resetting machine, which sits in between groups of work-setting stations and has built into its table a standard pallet-locating worktable. Above this worktable are two long-working-distance microscopes, fitted with Zeiss double-image oculars placed at a convenient viewing angle. These are carried on arms which can be positioned anywhere over the area of the worktable. The movements of the arms are measured by means of Ferranti diffraction grating measuring systems to an accuracy of 0·0025 mm (0·0001 in.). Great care has been taken in the design of the structure, and of the microscope carriages, to enable this accuracy to be realized on the axes of the microscopes. The positions of the microscopes from the datum point are displayed continuously by means of illuminated numerical indicators.

In use, the pallet is located accurately on the worktable and the workpieces are removed from the tooling. The microscopes, which have previously been set to datum, are now moved so that each is located accurately over the diamond points like a camera range-finder. The numerical displays will now be reading grid-location positions, but probably with a slight error, since the diamonds are only nominally on the grid. The readings are noted and the microscopes moved to the new positions indicated by the resetting instructions. They are now corrected as a mirror image by the amount of deviation first measured, giving the precise resetting positions. If necessary, the tooling on the pallet is changed, or, alternatively, a new pallet located which already contains the reset tooling. A part is offered up to the tooling and its position adjusted by tapping, with the clamps lightly in place, until the indentations are positioned correctly in the microscopes. The clamping screws are then tightened. The mirror-image correcting procedure is only used if the specified accuracy demands it.

The workpiece-resetting machine may also be used for the inspection of a component, in which case only one microscope is used, focused in turn on features which require to be inspected and the co-ordinates noted.

The work-setting machine may be staffed continuously, or, alternatively, used as required by work-setters who leave their normal stations for the purpose, when instructed to do so by the appearance of the 'reset' light at their station.

Unloading

The pallets are handed from the storage rack to the work-setting station by the MOLAC, and, to avoid congestion at the start of the day-shift, it fills the two pallet-pair positions on the work-setting stations towards the end of the night-shift. The work-setter's display signals 'unload', and an empty bin is delivered by the computer. This bin is filled with finished parts and tooling, and returned to the workpiece-preparation centre on the return conveyor. As this bin arrives at the end of the return conveyor, it pauses for a moment, its number is read, and a printer operated by the computer drops a slip into the bin indicating the further routing of this batch of parts, either to another process, e.g., deburring or anodising, or to the Finished Parts Stores, in which case the document is a stores entry note. The bins queue up in a rack at the rear of a bench where the parts are removed from each bin, and placed, along with the note, in an unnumbered bin, which is then loaded onto an appropriate trolley, depending on its routing. The work-fixing devices are cleaned and returned to the appropriate containers. Components which require a simple manual operation, such as inserting a bush, to finish them, are routed to a small finishing unit within the environs of the computer-controlled workshop.

Software

As mentioned before, the computer software includes a measure of optimization so that, if a machine failure occurs, this is automatically signalled to the computer, which then rearranges its workload to make the best use of the remaining facilities, until the maintenance squad restores the machine into service. If a part had priority, it would be routed to a similar machine in the group, if one existed, and the workload rearranged to accommodate this. If no alternative facility exists, then the pallets will be returned temporarily to the rack until the machine is restored into service, when they will automatically be dealt with to catch up with the original schedule. The computer here does nothing more than an intelligent man would do under similar circumstances, and the rules to follow are easily laid down. This, of course, is true of the whole system operation, but the advantage which the computer-controlled system would have over a similar manually controlled system is that it is capable of handling a great deal more data at lower cost without error or fatigue and without night-shift work, and of ensuring that no time is wasted.

Priority has little meaning in a two-day cycle. Basic priority is established by the sequence in which the production control computer lists the parts. In order to give maximum flexibility, the software is arranged so that parts can be injected into the system at any time, merely by preparing a bin and reading the cards appropriate to the daily master file extract for that part into the computer. Parts so injected are assumed to require priority, and they go immediately to the head of the queue.

It is, of course, mandatory that only checked jobs be fed into SYSTEM 24, because it is clearly impossible for it to cope with large numbers of human errors. All jobs must previously have been checked out and corrected where necessary, either on an auxiliary system, which would normally be manually controlled, or on the automatic system, but at a time set aside for manual program checkouts, e.g., weekends. Any machine tool in the computer-controlled group can be switched to manual operation for program testing, in which case the computer ignores it, but this is only practicable where machines are duplicated. If this procedure is followed, the only failures which are likely to occur are system failures caused by mechanical or electrical breakdown. Experience with our present NC workshop suggests that this will not produce a time loss greater than 5 per cent.

Variations on the procedure described here may be made simply by changing the computer software. The hardware will remain unchanged.

Performance

The bar chart (Fig. 16.8) shows a comparison between conventional manufacture and a SYSTEM 24 unit of the same capacity. The SYSTEM 24 unit has six active machine tools: the conventional installation has 300.

A vivid illustration of the productivity is given by Figs. 16.9 and 16.10, each of which shows the output of one twin-spindle milling machine for one hour.

The advantage of the complementary family over single-machine working is illustrated in Fig. 16.11, in which four aircraft components, selected as needing a five-axis milling machine for their economical manufacture, are compared on a single-machine versus system basis. It should be noted that the disparity will be even greater for simpler components requiring a smaller proportion of five-axis machining.

The gains shown here apply to direct manufacturing only, and do not take account of savings due to reduction of inventory and overheads. These could, with total reorganization, exceed those shown by several times.

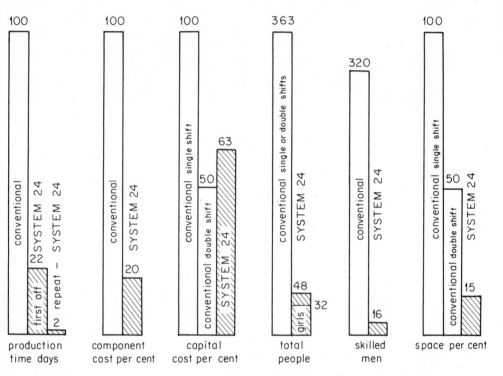

Figure 16.8 Comparison between conventional and SYSTEM 24 manufacture. (*Courtesy Molins Machine Co. Ltd.*)

Summary

Sufficient has been said to show the radically different approach which is necessary to fulfil the real requirements of batch production in a modern age. No longer are individual tools to be dumped down in a building and put to work, but on the contrary the tools are now only components in a complete system in which it is desirable that even the building is specialized in order to extract the maximum benefit from each facet of the system. This philosophy is not different from that successfully applied in the oil and chemical industries, or in the manufacture of cans; it is merely that the problem is more complex because of the randomly changing nature of the process and therefore requires a more

Figure 16.9 One hour's output from a twin-spindle milling machine. (*Courtesy Molins Machine Co. Ltd.*)

Figure 16.10 One hour's output from a twin-spindle milling machine. (*Courtesy Molins Machine Co. Ltd.*)

1. Sud-Aviation Pt. No. D251532		
Unit 1	9½ min	£0·165
Unit 6	8 min	£0·40
TOTAL:	17½ min	£0·565
UNIT 6 ONLY:	48 min	£2·40

3. Sud-Aviation Pt. No. D253421		
Unit 1	19 min	£0·34
Unit 6	10 min	£0·50
TOTAL:	29 min	£0·84
UNIT 6 ONLY:	46 min	£2·30

2. Sud-Aviation Pt. No. D251793		
Unit 1	15 min	£0·265
Unit 6	65 min	£3·25
TOTAL:	80 min	£3·515
UNIT 6 ONLY:	180 min	£9·00

4. Sud-Aviation		
Unit 1	62½ min	£1·11
Unit 6	20 min	£1·00
TOTAL:	82½ min	£2·11
UNIT 6 ONLY:	240 min	£12·00

Figure 16.11 Economic comparison between single-machine and system manufacture of four aircraft components.

flexible self-optimizing solution. The type of solution described is capable of application, with suitable modification, to many industrial processes.

For the future, much of the work-setting procedure is so standardized that it is entirely feasible to envisage a programmed work-setting machine which could handle all but the most complex parts which the work-setter has to cope with. This device would be a tape-controlled manipulator, which would be programmed at the same time as the machining program for the part is done. It is also possible to envisage automatic machines which would prepare workpieces from billets under tape control. The development of such devices will make the day-to-day working of this system substantially independent of human labour.

Continuing integration

The preceding description of SYSTEM 24 shows how it interfaces with computerized production issue systems, when the components have already been processed into the system, and all the data verified. This corresponds to the normal issue arrangements for conventional manufacture, and it assumes that the part programming and other data preparation will already have been done by the usual methods.

The next stage is to eliminate the repetitive part-programming process for each new component as far as possible by a computer-aided process, and it is by the use of this technique, which can virtually eliminate the present complicated chain of command between design and manufacture in which so much of the overhead cost lies, that the next big improvement will come.

The development of computer-aided design technology will eventually make this possible over quite a wide range of engineering design, but at Molins we have spotted a number of frequently used component types for which we can write self-contained design programs which will enable components of these

16

general types to be made, but with dimensional changes to suit the mechanism being designed. It should be possible to do this in any given organization, but clearly the programs will vary in complexity from company to company. Criteria other than size changes can, of course, be incorporated into such a program at will; for example, strength, weight, or stiffness criteria could be added and called for by the designer as required.

It is practicable to write programs for these selected component ranges such that they determine the complete manufacturing information required by SYSTEM 24 at the design stage. The following is an outline example of one such program which we have written for the design of a simple, but much used, bracket.

Figure 16.12   Fixing bracket.(*Courtesy Molins Machine Co. Ltd.*)

The bracket (Fig. 16.12) is conventionally made from light alloy sheet or strip by marking out the out line, cutting to size, rounding the corners, bending, marking out the hole positions, which are usually ±0·005 in. tolerance and occasionally closer, and drilling the holes. The one shown has slotted holes, which entails a double setting operation on a milling machine, as the holes have to be measured from the surface on which the bracket sits.

This is a process which takes about 206 minutes for a quantity of 12 brackets. The first step to improving this was to develop a bending technique that would produce predictable and repeatable metal flow. The material selected was NS5, which is available as sheet with very good cold-flow properties, enabling a bend radius of 1.5t to be achieved. By developing a hydraulically controlled bender which eliminated friction in the bending process, it was found that bends could be reproduced to an accuracy which would allow holes made in the flat condition to be placed within ± 0.002 in. of their correct position in the limbs after bending. This clearly enables a number of similar brackets to be cut, drilled, and slotted from a single sheet of material, and subsequently bent into shape.

By providing a simple computer program incorporating this information, the designer merely has to list the length, breadth, thickness, corner radius, bend radius, and the position and types of hole—in all about 10 numbers, which can be further reduced by standardization. No drawing need be made. This results in an entirely automatic process of making a montage of the maximum number of brackets that will fit into a sheet of material of pallet size, and the automatic correction of the developed position of the holes to allow for the extension in the bending process, depending on the thickness of sheet and bend radius chosen. The output of the computer consists of the magnetic tape to make the bracket, a sketch of the bracket, and print-outs for the material, and the time and precise cost of manufacture. A comparison of design, planning, and manufacturing times for conventional, conventional NC, and SYSTEM 24/CAD methods is shown in Fig. 16.13.

Function	Time (minutes) for 12 components		
	Conventional	Conventional NC	SYSTEM 24/CAD
Design, drawing, and specification	120	120	20
Methods planning	30	30	—
Part programming	—	480	—
Tape preparation (inc. punched tape)	—	60	5
Worksetting	⎫	50	10
Machining	⎪	96	12
Unloading	206	20	4
Bending	⎭	6	6
Total time for first batch	356	862	57
Total time for subsequent batches	206	172	32
Total machine time for subsequent batches	206	96	12

Figure 16.13 Comparison between conventional, conventional NC, and SYSTEM 24/CAD design and manufacture for the bracket shown in Fig. 16.12.

While this is a simple part, it nevertheless shows the powerful nature of the technique, because its manufacture is expensive by any normal methods in small quantities. The point to be emphasized is that SYSTEM 24 allows this new type of interface to be established precisely and without confusion, because under certain conditions it is possible to avoid the human element at the interface stage.

Purchasing

One of the most unpredictable and frequently expensive items of services in an engineering plant is the purchasing of materials. Since SYSTEM 24 manufacture relies on a limited range of materials and sizes, and since parts are made mainly from billet stock and therefore castings are only rarely necessary, the normal operating procedure is to make an arrangement with suppliers to keep maximum and minimum controlled stocks for SYSTEM 24 in use. The material supplier can be given computer print-outs at appropriate intervals which show the exact material usage, and from these he can deliver replacement stock as required. By comparison with the normal practice of purchasing castings from outside suppliers, which necessitates the production of pattern drawings, patterns, test castings, modifications to patterns, scrap, porosity difficulties, and all the agony and cost that chasing this cycle represents, life becomes very peaceful indeed.

Modular preform castings and swarf recirculation

Machining from billet begins to be wasteful when more than half of the billet has to be cut away. Accordingly a method of casting has been developed in which the 'preform' shape, which is a rough outline of the finished shape of the component, is formed from a pattern assembled from simple modules arranged together in a sequence specified by a number, which can be telephoned to the foundry, or transmitted by any other method. The metal is poured into a plaster or sand mould duplicated from the pattern. Because the shape is invariably simple, no casting problems arise with reasonable control. The 'preform' is machined all over like a billet, but now much less metal-removal is necessary.

It has been found possible, when only one alloy is being used on SYSTEM 24, to get very high-quality castings using melted swarf machined from billet. Swarf not used for 'preform' casting can be cast into ingots and returned to the extruder for reprocessing into billet. These are highly desirable ways of utilizing swarf, being vastly superior economically to normal scrap-disposal methods.

They lead to a high-efficiency recirculating process, whereby, to a first approximation, the only metal consumed is represented by the finished weight of machined components.

The effect on an engineering business

Thinking about ways of improving engineering manufacture in general has convinced the writer that the fundamental crux should be to turn the capital over more frequently, i.e., to make better use of plant or arrangements of plant which enable the manufacturing cycle time to be reduced. The production of x machines every month instead of xy machines every y months changes the whole economic picture. The consequent reduction of inventory and work-in-progress represents money in the bank, freeing capital to pay for the plant improvements which have brought it about.

SYSTEM 24 and similar methods applied to the range of components made by a given firm will enable parts-manufacturing cycle times to be reduced from the norm of about 100 days to 10 days, or even, in the extreme, two days. Clearly, to make full use of this, the assembly process has to be closely synchronized with parts manufacture. This highlights the need for close control of the delivery of bought-out parts, if again the full benefits are to be obtained. Nevertheless, it is still possible to keep a stores stock of the necessary size as a buffer. It merely costs money to do this, but not more than before.

It is taken for granted that the issue and scheduling of parts through the system will be done by a computer from pre-arranged information, and close tie-up between this department and the sales department will lead to much swifter issue, parts manufacture and assembly, giving a vastly improved delivery to the customer. The effect on the production program of executing a potential order can be forecast accurately by the computer, thus enabling the sales department to give promises which are realistic and can be met. This alone would be a great and welcome change from present practice. The foregoing applies, of course, only to established machines which are 'in the system'. New machines cannot be handled quite so swiftly and predictably, but the potential is there to do so, and freeing the production department from their presently overwhelming day-to-day worries would enable them to concentrate on new machines and get them as swiftly as possible into the flow line.

The manufacturing cycle in an average engineering works is rarely less than nine months from receipt of order to delivery, and is frequently more than twelve months. It should be possible by these new methods to get the cycle down below three months, resulting in an enormous improvement in the effectiveness and profitability of the business. The technology is now available to do this, certainly in the light machinery, instrument, optical, business machine, computer peripherals and other similar industries. The factor which is lacking, perhaps, is not technology, but far-seeing managements who understand that there is a problem, and have the will and drive to solve it.

References

1. Williamson, D. T. N. 'The Pattern of Batch Manufacture and its Influence on Machine Tool Design', *Proceedings of the Institution of Mechanical Engineers,* **182**, Part 1 (1967–68).
2. Williamson, D. T. N. 'SYSTEM 24—A New Concept of Manufacture', *8th Int. Mach. Tool Des. Res. Conf.* (University of Manchester) Sept. 1967; Conf. Proc. published by Pergamon Press 1968.
3. Williamson, D. T. N. '"New Wave" in Manufacturing', *American Machinist,* **3** (1967, No. 19, Sept.), 143–154; Spec. Rep. No. 607.

Glossary

Terms in italics are defined in this glossary.

Absolute co-ordinates   The distance (or dimensions) of the current position from the origin (or zero point) of a *co-ordinate system* (or measuring system) measured parallel to each axis of the system.

Accuracy   A statement of the maximum *error* which exists in specified circumstances. As high accuracy corresponds to small error, and the same number is used to express both, the use of the term accuracy should be avoided in favour of the use of error.

Adapt   A *2C,L* type continuous path or *contouring program* using a selection from the *APT* vocabulary. Available in various versions from various computer manufacturers.

Address   A name (or label, or number) identifying a storage area in a *control system* or computer memory. On *control tapes*, the address is usually a letter which precedes the numerical information. Thus X07352 may indicate an X dimension of 7·352 in. to be entered into the X dimension store and F618 may indicate a feedrate of 180 in/min to be entered into the feedrate store.

Alphanumeric characters   Although strictly this should be the set of characters which include all the letters of the alphabet and all *decimal digits*, it is also loosely used in the more general sense defined under *characters*.

Analogue   The use of a physical quantity (like voltage) whose amplitude represents that of another physical quantity (like distance). Although analogue quantities appear to be capable of continuous variation, in fact they usually alter in small discrete steps which can be ignored.

APT   Automatically Programmed Tools: A *5C* continuous path or *contouring program* maintained by the Illinois Institute of Technology Research Institute in Chicago, USA. See chapter 10.

ASCII   American Standard *Code* for Information Interchange.

Assembler program   See *Program*.

AUTOMAP   A *2C,L* type continuous path or *contouring program* using *APT* language type statements. An IBM program.

AUTOPOL   A *2C* type *program* available from IBM for lathes.

AUTOPROPS   A *2P,L* type *program* for drilling available from IBM.

458

AUTOSPOT A *2P,L* type *program* available from IBM and written initially for the Kearney & Trekker Milwaukee-Matic.

Auxiliary function Another name for *Miscellaneous function*. The part *programming* word AUXFUN is used for this purpose in *APT*.

Axis When associated with machine tools, an axis is a direction in which a machine tool table or head can move. Axes are either linear (a straight line) or rotary (rotation about a straight line). When associated with describing the shape of a part, the axes are straight lines at the intersection of the co-ordinate system planes or on the axis of a cylinder.

Backlash The maximum movement at one end of a mechanical system (such as a gear train) which does not cause the other end to move.

Base number A base number (or radix) is an implied number used when expressing a value numerically. In the normal *decimal* system, the base is 10 and 72 is a short way of representing

$$7 \times 10^1 + 2 \times 10^0 = 70 + 2 = 72 \text{ in decimals}$$

Similarly in *binary* notation 1011 represents

$$1 \times 2^3 + 0 \times 2^2 + 1 \times 2^1 + 1 \times 2^0 = 8 + 2 + 1 = 11 \text{ in decimals.}$$

Binary Involving 2. A binary numbering system has a *base number* of 2 and represents any numerical value in terms of a string of binary digits.

Binary coded decimal number (BCD number) The representation of a number by groups of four *binary digits* for each *decimal digit* in a number.

$$\begin{array}{cc} \text{BCD} & \text{Decimal} \\ 0111 \quad 1001 = & 79 \end{array}$$

Binary digits The characters 0 and 1 used in the binary system to express any value (see *Base number*).

A binary 1 can be represented, in a pre-allocated place, by a hole, or a pulse, or a steady voltage.

A binary 0 could then be represented by the absence of the hole, pulse, or voltage.

Bit A *binary digit* or its representation.

Block A collection of *words* in some agreed form. Usually on a *control tape*, and separated from succeeding blocks by an *End-of-block code*. One block on a control tape often provides enough information for a complete cutting operation.

Block format See *Format*.

BSI British Standards Institution. The official body in Britain who devise and maintain Specifications with the aid of committees representing government and industry. Address: 2 Park Street, London, W1Y 4AA.

Buffer A temporary storage area where information is held until it is moved into an operating area.

C See *Contour* (C) and also *2C, 2C,L*, etc.

Canned cycle See *Cycle*.

Card See *Punched card*.

CCW See *Counterclockwise*.

Change points Points at the junction of two elements (lines, circles, curves) of a contour, either on the part or on the tool path.

Channel   See *Track*.

Characters   The set of letters, *decimal digits*, signs (such as $+$ $-$: % etc.) which have been agreed for use in preparing *control tapes* or *part programs*.

Circular interpolation   A *contour* control system with circular *interpolation* cuts an arc of a circle from one *block* on a *control tape*. The control circuit varies the relative velocities of the *axes* to generate the circular motion.

CLDATA   The output from the *APT*-like processor. Information about the Cutter Location relative to the *part programming co-ordinate system*.

CLFILE   A complete set of CLDATA for a *part*.

Clockwise (CW)   A term for the negative *direction of rotation* (reserved for *spindles*). Clockwise rotation advances a right-hand screw away from a cutting spindle into a stationary workpiece. It permits a normal right-handed drill to be used with lathes and drilling machines. Clockwise should *not* be applied to the positioning of rotary tables on 3A to 5A machine tools (see *Direction of rotation*).

Closed-loop control system   See *Feedback control*.

CLTAPE   A *magnetic tape* record of a *CLFILE*. Also loosely used instead of *CLDATA*.

Code   An agreed method of representing characters by means of a pattern of bits on tape (see chapter 6). The *EIA* Code has been in use in USA since about 1960 and in Europe since 1965. The *ISO* (International Standards Organization) Code was proposed for International use in 1962 and ratified internationally in 1967. It should render the EIA code obsolescent.

Coding   Sometimes used in place of *program* in statements such as computer coding, machine tool coding, etc.

Compatible   An ideal relationship. Compatible *control systems* should accept identical *control tapes* and perform the same task. Compatible computer *programs* should accept identical *part programs* and issue the same *CLDATA*. In practice the difference between compatible systems has been minimized rather than eliminated.

Compiler program   See *Program*.

Computer program   See *Program*.

Continuous path   See *Contour*.

Contour (C)   A contour *control system* continuously monitors the positions attained by a machine tool moving about its axes, and ensures that a desired tool path involving 2 or more *axes* is obtained. 3C is used, for example, to indicate that 3 axes can be used simultaneously to form a contour.

Control system   An arrangement of electrical and/or mechanical components which causes a desired effect to occur. NC systems accept a numerical description of the task. See *Contour(C)*, *Manual(M)*, *Positioning(P)*, and *Straight-line(L)*.

Control tape   A control tape is a *magnetic* or *punched* paper tape containing the *coded* representation of a *machine tool program*. The program contains all the instructions for the tool in the order required to have the part machined.

Controller   See *Director*.

Co-ordinates   The distances or angles which specify the position of a point with respect to a *co-ordinate system*.

Co-ordinate system   A series of intersecting planes, or planes and cylinders (usually three), which form a reference system on which the position of a point, line, circle, or other geometric feature can be specified.

CPU   Central Processing Unit. The controlling unit in a digital computer.

Counterclockwise (CCW)   The rotation of a spindle in the opposite direction to *clockwise*.

Cutter diameter compensation   Provision in the *control system* to modify the *cutter offset* by entering a numerical correction to the cutting tool diameter. This can be used to accommodate a tool slightly different from that programmed, or to take roughing and finishing cuts with double use of one sequence of *blocks* on the *control tape*. See also *Tool length compensation*.

Cutter offset   Position of reference point on the tool, relative to the point of tangency between the cutter and the part surface.

Cutting speed   The velocity of the cutting edge of the tool relative to the workpiece. For circular tools, the cutting speed is related to the tool at its maximum cutting diameter. Usually the effect of *feedrate* on cutting speed is ignored.

Cut vector   A single straight-line cutter-path motion in a sequence of instructions for a *contouring* control system.

CW   See *Clockwise*.

Cycle   A sequence of operations which are frequently repeated. Usually a complete sequence of machine tool movements from one *block* on the *control tape* to drill, tap, or bore, etc.

Damping   The operation of causing oscillatory conditions to be modified to steady conditions.

Data   Information.

Debug   To remove faults in a *program* or in a *control system*.

Decimal   Involving 10.
   A decimal numbering system has a *base number* of 10 and represents any value by a string of *decimal digits*.

Decimal digits   The ten characters 0, 1, 2, 3, 4, 5, 6, 7, 8, 9 used in the decimal system to express any value. See *Base*, *Number*.

Depth of cut   The amount of metal (in mm or in.) removed perpendicularly to the direction of *feed* in one pass of the cutter over the workpiece.

Diagnostic   In the use of a computer *program*, diagnostic information is printed out to indicate how the input *part program* has broken one or more of the computer *program* rules.

Digit   A character used in a numbering system. See *Binary digit*, *decimal digit*.

Digital   The use of a series of discrete states of a physical quantity (such as voltage) to represent each *digit* of a number. The number is the desired value to which another physical quantity (such as distance) is to be set. In a *binary digital* system, the presence of a voltage on each of a set of wires may represent a ' 1 ' and the absence a ' 0 '. Given enough wires (or circuits) such a system may specify a value with any desired *precision*, but there is always a discrete step between adjacent values.

Digital/Analogue converter   A device for providing an *analogue* quantity corresponding to a *digital* value.

Digitizer   A device for providing a *digital* representation of an *analogue* quantity.

Direction of rotation   The terms positive and negative indicate direction of rotation (but for *spindles*, the terms *clockwise* and *counterclockwise* are also used). A positive rotation about the Z axis is in the direction which moves from positive X to positive Y direction, with a 90° rotation. Similarly, positive rotation about X and Y axes move from $+Y$ to $+Z$ and from $+Z$ to $+X$ respectively (see chapter 6 and Fig. 6.11).

Director That part of the *control system* which accepts the change points in the tool path for a *contour* and calculates the *interpolated* closely spaced information required to control the position. Usually consists of special-purpose digital control circuits followed by *digital-to-analogue converters*.

Double precision The use of two computer *words* to represent a value with greater *precision* than can be attained by the use of one word.

Downtime Time during which equipment is out of action because of faults.

Dwell A pause of programmed duration, usually to ensure that a cutting action has time to be completed.

Dynamic response The output of a *control system* versus time following a step in the input signal.

EIA The Electronic Industries Association in USA. A trade association representing manufacturers of electronic equipment which issues standards documents. Address: 2001 Eye Street NW, Washington DC.

End-of-block code An agreed *code* which indicates the completion of a *block* of input information on *punched tape*. This code usually also causes a line feed operation on a tape-operated typewriter. Thus one block of information is printed on one line by the typewriter.

Error The preferred term to be used when discussing *accuracy*. Error is the deviation of an attained value from a desired value.

EXAPT 1 a *2P,L* type of *computer program* for drilling, available from the EXAPT Verein in Aachen. See chapter 7.

EXAPT 2 A *2C* type of *computer program* for turning, available from the EXAPT Verein in Aachen. See chapter 8.

Feedback control A *control system* in which a portion of the controlled value (such as position) opposes the desired input value. When the two are not equal the control system causes a change of output to reduce the *error*. Also known as 'closed-loop control'.

Feed The movement of a cutting tool into a workpiece.

Feedrate The rate, in mm/min or in./min, at which the cutting tool is advanced into the workpiece. For milling and drilling, the feedrate applies to the reference point on the end of the axis of the tool. For turning, it applies to an agreed reference point on the tool.

Feedrate override An optional extra on many *control systems*. The operator can dial in and obtain a desired percentage of the programmed *feedrate*.

Fixed block format The position of each type of word in a *block* of data on a *control tape* is fixed in this system. Thus, even where no change of data is involved for a particular *word*, the word must be repeated to keep the remaining words in an identifiable position. For a given machine tool and control system, each block has exactly the same number of words and characters.

Fixed cycle See *Cycle*.

Fixed point number A number with specified numbers of digits before and after the point. Those before the point represent an *integer number* and those after the point represent a *fractional* part (total value less than one integer). Normally such numbers are represented in computers by *floating point numbers*.

Floating point number A number consisting of *integer* and/or *fractional* parts can be stored in the computer in floating point form as a purely fractional part and the positive or negative exponent of the *base number*, e.g.,

$$24 = 0.24 \times 10^2 = 0.24 \text{ E2 in the } decimal \text{ system}$$
$$\text{or} \quad 11000 = 0.11 \times 2^5 = 0.11 \text{ E5 in the } binary \text{ system.}$$

Floating zero Often an optional facility in a *control system* which enables the controlled position which corresponds to an input value of zero, to be set to any desired place within the controlled range (see *Zero shift*).

FMILL/APTLOFT A *program* developed by Boeing and available as an addition to *APT*. See chapter 10.

Format An agreed order in which the various types of *words* will appear within a *block*, and thus also along the length of a *magnetic* or *punched tape*. See *Fixed block, Variable block*. See also chapter 6.

Format classification A method of indicating which of a wide variety of options have been exercised in specifying a particular use of a 'standard' *format*.

FORTRAN See *Program, Source language*.

FPM Feet per minute (ft/min)—same as *SFPM*.

Fractional number A number with no *integer* part (total value less than 1). For example,

$$0.1101101, \text{ or } \tfrac{3}{4}, \text{ or } 0.9998, \text{ etc.}$$

Frequency response The amplitude and phase plot of the output of a *control* system, related to the frequency of a sinusoidal input signal.

Gain A measure of the amplification of a *control system*. For a *controlled axis* on a machine tool, it is often stated as the thrust developed by the control system divided by the error at its output. It could then be stated as, say, 200 lbf per 0·001 in. or 200,000 lbf/in.

Hardware Equipment—e.g., Machine tool, or Control, or Computer (see *Software*).

Hold A programmed delay of indefinite duration, terminated by operate or interlock action.

Hunting An unsuccessful attempt by a *control system* to reach the desired output value. Hunting is typified by a sequence of movements towards the desired value, in each case resulting in an *overshoot* past the value, followed by another attempt.

Hysteresis Except for a perfect control system, the static response to a given input is different, depending on whether the previous input was larger or smaller than the current input. This difference in response is called hysteresis.

Inch (or Jog) control A manual control button which permits the position of a machine element to be altered by a very small step at a time. Used only when setting up a machine.

Incremental co-ordinates The distance of the current position from the preceding position, measured in terms of axial movements in the *co-ordinate system*.

Input equipment A device which accepts *data* into a *control system*. This can be a *magnetic* or *punched-tape* reader, a *card* reader, or manually set decade switches.

Input medium The method by which information is presented to the input equipment. This can be *magnetic tape, punched tape, punched cards*, or an operator.

Integer A number with no *fractional* part (no part less than 1 in value). For example

$$1101101, 3, 9998, \text{ etc.}$$

Interpolation   The process of supplying the positions of a set of more closely spaced points between more widely spaced points such as change points. Linear interpolation occurs when the interpolated points lie on the straight line joining each pair of supplied points. *Circular interpolation* implies that the interpolated points lie on specified circles between pairs of points.

IPM   Inches per minute (in./min).

IPR   Inches per revolution (in./rev) of a cutter or workpiece.

ISO   International Organization for Standardization. An organization through which national standards organizations work together to devise internationally acceptable standards.

L   See *Line motion (L)*.

Language   The sum of the agreed forms of statements which are accepted by a system for all legitimate commands.

Leader   The first part of a *control tape*, a piece of *tape* containing no information meaningful for control purposes. A leader may contain no punching (other than sprocket holes), or it may have null codes, or space codes, or information meaningful to the operator (punched in visible letters or handwritten) or meaningful to the computer as a check. The leader is used to thread the tape on to the tape-reading equipment, and to be spooled without missing any control information.

Leading zero suppression   See *Suppressed zero*.

Line mill   See *Line motion (L)*.

Line motion (L)   A line motion *control system* allows the *feedrate*, and next desired position of each *axis* thus controlled, to be numerically specified on the *control tape*. The resulting movements of the machine tool in the different axes of motion are not co-ordinated, and take place parallel to the linear or circular machine tool ways. 3L indicates that line motion control applies to three axes on a tool.

Linear interpolation   See *Interpolation*.

Machine language or **Machine instructions**   The instructions and data for computers are based on patterns of bits. A 24-bit word might appear as:

$$001111010011110111111001.$$

Instructions and data in this basic form are in machine language, and can be executed by the computer without further processing.

Machine program   See *Program*.

Macro   A *subroutine* composed of *part-programming statements* instead of *computer programming* statements.

Magic three code   A method of expressing feeds and speeds as codes. See chapter 6.

Magnetic tape   Usually a thin plastic *tape*, coated on one side with a thin layer of magnetic material. Information is usually stored on this by magnetizing a *row* of 7 or 9 spots across the tape to represent a *character*. Sequences of characters are stored along the tape in *blocks*, and the information corresponding to one spot in a row, when repeated along the tape, forms a *track*.

Manual (M)   In the context of NC, an *axis* is manually controlled if the position the machine takes up on that axis is set manually. The instruction to take up the position may be on the *control tape*, for example, but the position may have been determined by manually setting limit switches, cams, etc.

Manual data input   A means, on the control panel, of inserting numerical information into the *control system*. Usually decade switches.

Manual feedrate override   See *Feedrate override*.

Manuscript   Handwritten *program*.

Measuring system   Equipment which gives a *digital* or *analogue* signal representing a physical quantity such as position.

Memory or Store   Equipment which can retain data for use later.

Miscellaneous function   A *control tape* term for codes such as M03 used to control machine tool functions such as 'rotate spindle clockwise'. See chapter 6. The *APT part programming* word AUXFUN, for auxiliary function, is used for this.

Mnemonic codes   Instruction *words* which are written in an easily remembered form.

NC   Numerical control.

NELNC processor   A *computer program* designed for preparing *control tapes* for NC milling machines, lathes, and drilling machines, and also for machining centres which combine these operations. See chapter 9.

Number system   A systematic method for expressing values in terms of *digits*. See *Base number*.

Numerical control system   See *Control system*.

Numerical data   Information which is principally expressed in combinations of numbers.

Numerical display   Usually a *decimal digital* display of some of the data from the *control tape*. For example, the sequence number of the current *block* and the desired co-ordinates (in an absolute co-ordinate form) might be displayed. Very occasionally, the actual position might be displayed by a numerical readout.

Object program   See *Program*.

Off-line operation   An operation indirectly carried out by a device. Thus, in computing, a *control tape* would be punched off-line if the tape punch were not directly coupled to the computer. The computer might prepare a temporary record on *magnetic tape* or *punched cards*. The tape or cards would then be read on a completely separate converter which would cause the tape to be punched.

On-line operation   An operation directly carried out on a device. Thus, in computing, a *control tape* is punched on-line if the tape punch is directly coupled to the computer, which commands each punching operation.

Open-loop system   A *control system* in which the final output value is not directly measured and checked against the desired value. The final value may, however, be implicitly known providing there has been no malfunction. For example, if the final motion is obtained by a pulsed stepping motor driving a lead-screw, and the correct number of pulses have been sent, the desired motion has usually occurred.

Overshoot   If the controlled value passes the desired value before coming back towards it, overshoot occurs.

P   See *Positioning (P)*.

Paper tape   See *Punched tape*   The term paper tape also embraces plastic tape of the same dimensions, intended to be used where paper tape is not sufficiently durable.

Parallel   A number of operations handled at the same time. Thus, in the case of parallel input, a set (or *block*) of *data* is entered simultaneously.

Parity   In order to provide a check on the accuracy of writing (or punching) and reading equipment, parity checking can be used. With *paper* and *magnetic tape*, one *track* is often allocated for a parity checking *bit*. With an even number of tracks, it is best practice to write or punch a bit in the parity track in order to make the total number of bits in the row even. The all-holes-punched code can then be used to erase information. Thus, every correct code has an even number of bits and can be checked. If an odd number of bits is read, then it is known that an error has occurred. With an odd number of tracks, odd parity is usually employed.

Part A term used for a mechanical component which has to be produced by an NC machine.

Part program See *Program*.

Part programmer A person who prepares *part programs*.

Pendulum machining Either machining back and forwards over a workpiece to minimize idle movements, or arranging two workpieces at opposite ends of a machine tool table so that one can be set up (or removed) while the other is being machined.

PMT2 An ICL *program* which enables a surface to be defined on a mesh of points and to be machined in a *2C* mode.

Point-to-point Used to describe a *control system* capable of *positioning*. See *Positioning*.

Positioning (P), also called **Point-to-point** A type of *control system* which moves the machine to the next position without controlling the path. Often, motion is along one axis at a time and the speed is at an uncontrolled maximum speed. 2P is used to indicate that two axes of the machine can be positioned numerically.

Post processor A *computer program* which adapts the output *CLDATA* from an NC *computer program* to suit a particular machine tool *control tape*. See chapter 11.

Precision The degree of discrimination with which a quantity can be stated. Thus a decimal number of four digits can discriminate between 10,000 different values and thus has a precision of one part in ten thousand of maximum value.

Preparatory function word A *word* near the beginning of a *control tape block* which calls for a change in mode. In a common format, the word takes a form like G01 which may call for linear interpolation. The *APT part programming* word PREFUN is used for this purpose.

Preset tools The setting of tools in special holders away from the machine tool. The tools are carefully positioned in the holders so that when the holders are manually or automatically clamped in the machine tool their effective cutting positions will correspond to those assumed when *part programming*.

Preventive maintenance A maintenance system which, by the use of regular tests, enables incipient failure to be detected before it occurs, so that it may be avoided.

Processor Another name for a computer *program*.

Program A systematic arrangement of instructions or information to suit a piece of equipment.

    Assembler program A *computer program* which converts a *mnemonic* assembler *source program* into an *object program* in *machine language*. Usually one assembler source language statement generates one machine language statement.

    Compiler program A *computer program* which converts a *source program* into an *object program* in *machine language*. The *source language* is usually a higher one such as *FORTRAN*, *APT*, etc., and normally one source language statement generates several machine language statements.

    Computer program A sequence of instructions which cause a computer to perform desired types of operation.

FORTRAN program A computer program which accepts FORTRAN *source language* statements and converts them to *machine language* statements. Most usually to suit the *USASI* FORTRAN 4 *source language* with perhaps a few facilities added and/or subtracted.

Machine program Another name for *object program*.

Machine tool program A sequence of instructions in the form required for a machine tool and its *control system*. This is normally *coded* to form a *control tape*.

Object program The sequence of *machine language* instructions which result from the processing of a *source program* (which has been written in a *source language*) using the appropriate computer and *computer program*.

Part program A sequence of instructions which describe the work which has to be done on a *part*, in the form required by a computer under the control of an NC *computer program* (e.g., in a *source language* such as *APT*).

Source program A *computer program* written in a *source language*, which has to be converted into an *object program* in *machine language*, before execution.

Programming Preparing programs.

Punched card An agreed size of card in which data is represented by punched holes (commonly in 80 columns of 12 positions). Punched cards were primarily developed for business tabulators and for computers, but have been used for input to numerical control systems.

Punched tape A *tape* (usually 1 in. wide) and usually of opaque paper material. Information can be represented on the tape in agreed arrangments of punched holes. See *Code, Format, Parity*.

Radix Same as *base number*.

Random access Access to information stored (for example, on magnetic disc) so that there is roughly equal access time to each storage location (in contrast to storage on magnetic tape where access time is proportional to the length of tape searched before the information is located).

Rapid traverse In a machine whose *feedrate* can be controlled, provision is usually made for a rapid movement at the maximum speed available in order to position the machine before and after cutting.

Read To transfer data from one storage medium to another. Commonly the term 'Read' is used for the transfer of data from a *control tape* into the *control system* (to READ IN) or for the transfer of data from a *measuring system* to a *numerical display* (to READ OUT).

Repeatability The maximum difference within a series of results when the same demand is repeatedly applied under identical test conditions during a short span of time.

Reproducibility The *repeatability* when the demand is either applied to a range of similar pieces of equipment, or to one piece of equipment from time to time.

RPM Revolutions per minute (rev/min), a measure of *spindle* speed.

Resolution The smallest movement which can be specified and realized by a *control system*.

Row A line of information in the form of *bits* across a *paper* or *magnetic tape*. A row of bits usually represents a *decimal digit* or other *character*.

Sense To *read* holes in *punched tape* or *cards*.

Sequence number   The number allocated to a *block* or group of blocks to identify them. Commonly takes a form like N278.

Sequence number readout   A control panel display of the *sequence number* of the *block* currently being obeyed.

Serial   A number of operations handled one after the other. Thus, in the case of serial input, *data* is entered one *character* at a time.

Servomechanism   A *closed-loop positioning-control system*. The higher the *gain*, or sensitivity, the smaller the static error, providing the system does not *hunt*.

SFPM   Surface feet per minute (for *Cutting speed*).

Sign   A + or − used to indicate that a positive or negative valued number follows.

Software   *Programs*, sequences of instructions, etc. Not equipment (see *Hardware*).

Source language   Computers are normally programmed in a source language because it would be tedious and uneconomic to program in *machine language*. A source language uses *mnemonic* or *symbolic* notation which requires to be converted to machine language using an *assembler* or *compiler* type *computer program* and a computer.

Source language—FORTRAN   A *symbolic* computer programming language (developed originally by IBM) called Formula Translating system. Designed primarily for programming mathematical problems, e.g.,

$$A = B^2 + C^2$$

can be written in FORTRAN as

$$A = SQRT(B * 2 + C * 2)$$

A standard FORTRAN has been defined by *USASI* and is known as USASI FORTRAN 4. It is prudent to enquire about the detailed variations to this USASI FORTRAN 4 which apply for any particular *FORTRAN program*.

Source program   See *Program*.

Spindle speed   The rotational speed in *RPM* of the spindle or shaft which supplies the cutting power.

This is the drilling or milling *spindle* or the lathe headstock speed. For *direction of rotation*, see *Clockwise*.

Split   A computer *program* provided for use with the Sunstrand OM3 multi-axis machine tool.

Statement   An agreed arrangement of *words* and/or *data* which is accepted by a system to command one particular function. This usually applies to *computer program* statements (or instructions).

Static response   The final steady response of a *control system* to a steady input condition.

Storage capacity   The amount of *data* which can be stored at any one instant by a *control system* or computer. Usually expressed as a number of *words* or *characters*. For larger systems, particularly computer stores, it may be expressed in thousands of words. Thus 32K = 32,000 words. But beware, this may be an octal number.

Straight-line cut (L)   See *Line motion*.

Subroutine   A sequence of *computer-programming statements* or instructions which perform an operation frequently required. The instructions are written, tested, given a name, and stored on *cards*, *tape*, or magnetic drum. When the operation is required again in a *program* the subroutine is called (or named in the appropriate form of instruction) together with the required data, without having to write out the sequence of instructions involved.

Suppressed zeros   The elimination of the zeros before or after the significant figures in a *word*. If the number 005000 is written as 005 the trailing zeros have been suppressed. If it is written as 5000, the leading zeros have been suppressed.

Symbolic   The representation of a more complicated instruction or *statement* by a few *characters* which can thereafter be more simply used by man. The symbolic name may also be *mnemonic*. Thus, in *part programming*, C17 may be given to a particular circle when it is first defined. Thereafter, in that part program, a reference to C17 avoids the need to redefine the circle when using it again (perhaps as the pitch circle for some holes).

Tab code   An agreed *paper tape code* which is used to precede the place where each type of *word* should appear in a *block* of control data. The word itself need not appear if the function represented by that word is not required. The code is also used to cause a tape-operated typewriter to tabulate. Tab or tabulation action can be used to print each type of control word in its own column when typed.

Tape   See *Magnetic tape, Punched tape.*

Tool word   A *word* on a *control tape*, usually of the form T05, used to select a tool mounted on a particular turret position (position 5 in this case). With tool changers, the tool word would supply the necessary information to enable the tool changer to locate the desired tool. For single spindle machines, the tool word can be displayed numerically on the control panel so that the operator can select the tool.

Tool length compensation   In drilling and turning, the effective length of a tool may not correspond exactly to that assumed when preparing the *control tape*, unless *preset tools* are used. Tool offset permits the difference in effective length between the programmed and actual tools to be set manually on decade dials corresponding to the particular tool. The *control system* subtracts the tool offset from the programmed tool positions on the control tape and positions the machine tool slide to the amended values thus obtained. The actual length of a tool may either be measured away from the machine tool and communicated to the operator, or the operator may determine the effective length by trial cuts. Compare with *cutter offset.*

Track   A path along the length of a *punched tape*, or *magnetic tape*, on which information can be recorded in a sequence of *bits*. Sometimes also known as a channel.

Trailer   The terminating piece of a *control tape* which contains no meaningful information for the control system. The trailer is required to allow some tape to be in the reading equipment and tape spool when the reading head has reached the end of the control information. See also *Leader.*

Trailing zero suppression   See *Suppressed zero.*

Transducer   A component which converts a signal, or energy, from one physical quantity to another. An electric motor converts electrical to mechanical power. A *digitizer* converts a measurement to a *digital* electrical representation.

Transient   A temporary condition or disturbance which, after it is over, leaves the steady state or static condition.

USASI   The United States of America Standards Institute. The American equivalent of *BSI*. Address: 10 East 40th Street, New York, NY.

Variable-block format In this system the order of *words* within a *block* on a *control tape* are agreed, but particular words need appear only when specifying a new value. Thus, the number of words in a block varies. In order to enable the type of word to be identified it must either be *addressed* or *tab* codes must mark where each word can appear. (The number of tab codes in each block is fixed although the number of words can vary.)

Velocity lag An error in position during *contour* controlling caused by the velocity of the various axes of a machine tool.

Verify To check for absence of *errors* in *data*, by preparing it twice in a recorded form such as *punched tape*. The first record is compared with the second, either after both have been completed, or as each *character* of the second is produced. Any errors thus disclosed are manually considered and corrected.

Wind-up Lost motion in the mechanical part of a *control system* due to elastic deflection. Typically due to torsion and longitudinal strain in a long leadscrew.

Word An agreed arrangement of *characters* and *digits* (usually less than 10) which conveys one instruction, piece of information, or idea. Thus

$$X + 37500 \text{ or } 1259 \text{ or } GO1$$

are agreed forms of *control tape* words. CIRCLE and SMALL are common *part program* words.

Word address format See *Format, Address*.

Write To transfer data from one storage medium to another. Commonly the term 'Write' is used for the transfer of data from a computer or a typewriter to a *control tape*.

Zero shift, or **Zero offset** Usually on absolute *control systems* values can be set in manually on decade dials to cause the position of the machine slides to be shifted by the dimensions set on the dials. In effect, the settings on the zero shift dials are added to the corresponding values read from the control tape (also known as *Floating zero*).

2C Two *axes controlled* simultaneously to follow a *contour*.

2C Subset A 2C *program* available as a subset of the *NELNC program*. See chapter 9.

2C,L Two *axes controlled* simultaneously to follow a desired contour (2C) and independent control of the third axis (usually perpendicular to the other two) in a *Straight-line cut* mode (L).

2C,L subset A 2C,L *program* available as a subset of the *NELNC program*. See chapter 9.

2P,L Two *axes controlled* to *position* (2P) and independent control of the third axis (usually perpendicular to the other two) in a *Straight-line cut* mode (L).

2P,L subset A 2P,L *program* available as a subset of the *NELNC program*. See chapter 9.

2P,M Two *axes controlled* to *position* (2P) and independent control of a third axis (usually perpendicular to the other two) in a *manual* mode (M). The most common type of drilling machine is 2P,M in capability, *positioning* in X and Y and drilling in Z to depths set manually but initiated by control words.

3C Three *axes controlled* simultaneously to follow a desired *contour*.

3P,L   Three *axes controlled* to *position* (3P); usually an X and Y and a rotary table, B. Independent control of another axes in the *Straight-line cut* mode (L). (See Fig. 6.10 for example.)

5C   Simultaneous control of up to three orthogonal and two rotary *axes* to cause the cutting tool to cut the desired *contour*.

Index

*Terms in italics are defined in the Glossary (e.g., *Accuracy, ADAPT*, etc.)
  Terms in capitals are computer program names and words (e.g., ABS, *ADAPT*, etc.)

Terms in italics are defined in the Glossary (e.g., *Accuracy, ADAPT*, etc.)

Terms in capitals are computer program names and words (e.g., ABS, *ADAPT*, etc.)

Terms in italics are defined in the Glossary (e.g., *Accuracy, ADAPT*, etc.)

Terms in capitals are computer program names and words (e.g., ABS, *ADAPT*, etc.)

Terms in italics are defined in the Glossary (e.g., *Accuracy, ADAPT*, etc.)

Terms in capitals are computer program names and words (e.g., ABS, *ADAPT*, etc.)

Terms in italics are defined in the Glossary (e.g., *Accuracy, ADAPT*, etc.)

Terms in capitals are computer program names and words (e.g., ABS, *ADAPT*, etc.)

Terms in italics are defined in the Glossary (e.g., *Accuracy, ADAPT*, etc.)

THIS BOOK HAS BEEN SET IN MONOPHOTO TIMES NEW ROMAN
AND PRINTED AND BOUND IN GREAT BRITAIN BY
WILLIAM CLOWES AND SONS, LIMITED, LONDON AND BECCLES